VOLUME SIXTY TWO

ADVANCES IN MARINE BIOLOGY

Advances in Sponge Science: Physiology, Chemical and Microbial Diversity, Biotechnology

Advances in MARINE BIOLOGY

VOLUME SIXTY TWO

ADVANCES IN MARINE BIOLOGY

Advances in Sponge Science: Physiology, Chemical and Microbial Diversity, Biotechnology

Edited by

MIKEL A. BECERRO*, MARIA J. URIZ, MANUEL MALDONADO AND XAVIER TURON

Center for Advanced Studies (CEAB-CSIC)
Acc Cala S Francesc 14
17300 Blanes, Girona, Spain
**Current address: Natural Products and Agrobiology Institute (IPNA-CSIC)*
Avda Astrofisico Francisco Sanchez 3
38206 La Laguna, Tenerife, Canary Islands, Spain

ELSEVIER

AMSTERDAM • BOSTON • HEIDELBERG • LONDON
NEW YORK • OXFORD • PARIS • SAN DIEGO
SAN FRANCISCO • SINGAPORE • SYDNEY • TOKYO
Academic Press is an imprint of Elsevier

Academic Press is an imprint of Elsevier
32 Jamestown Road, London NW1 7BY, UK
Radarweg 29, PO Box 211, 1000 AE Amsterdam, The Netherlands
Linacre House, Jordan Hill, Oxford OX2 8DP, UK
225 Wyman Street, Waltham, MA 02451, USA
525 B Street, Suite 1900, San Diego, CA 92101-4495, USA

First edition 2012

ISBN: 978-0-12-394283-8
ISSN: 0065-2881

For information on all Academic Press publications
visit our website at elsevierdirect.com

Printed and bound by CPI Group (UK) Ltd, Croydon, CR0 4YY
Transferred to Digital Print 2012

CONTRIBUTORS TO VOLUME 62

Christopher J. Freeman
Department of Biology, University of Alabama at Birmingham, Birmingham, Alabama, USA

Grégory Genta-Jouve
University of Nice-Sophia Antipolis, Institut de Chimie de Nice UMR 7272 CNRS, PCRE, Parc Valrose 06108 Nice, France

April Hill
Department of Biology, University of Richmond, Richmond, Virginia, USA

Sally P. Leys
Department of Biological Sciences, University of Alberta, Edmonton, Alberta, Canada

Werner E.G. Müller
ERC Advanced Investigator Grant Research Group, Institute for Physiological Chemistry, University Medical Center of the Johannes Gutenberg University Mainz, Duesbergweg 6, 55128 Mainz, Germany

Manuel Maldonado
Department of Marine Ecology, Centro de Estudios Avanzados de Blanes (CEAB-CSIC), Blanes, Girona, Spain

Dirk E. Martens
Bioprocess Engineering, Wageningen University, Wageningen, The Netherlands

Ronald Osinga
Aquaculture and Fisheries, Wageningen University, Wageningen, The Netherlands

Shirley A. Pomponi
Harbor Branch Oceanographic Institute/Florida Atlantic University, Fort Pierce, FL, USA

Marta Ribes
Institut de Ciències del Mar (ICM-CSIC), Passeig Marítim de la Barceloneta, Barcelona, Spain

Klaske J. Schippers
Bioprocess Engineering, Wageningen University, Wageningen, The Netherlands

Ute Schloßmacher
ERC Advanced Investigator Grant Research Group, Institute for Physiological Chemistry, University Medical Center of the Johannes Gutenberg University Mainz, Duesbergweg 6, 55128 Mainz, Germany

Heinz C. Schröder
ERC Advanced Investigator Grant Research Group, Institute for Physiological Chemistry, University Medical Center of the Johannes Gutenberg University Mainz, Duesbergweg 6, 55128 Mainz, Germany

Detmer Sipkema
Laboratory of Microbiology, Wageningen University, Wageningen, The Netherlands

Hauke Smidt
Laboratory of Microbiology, Wageningen University, Wageningen, The Netherlands

Robert W. Thacker
Department of Biology, University of Alabama at Birmingham, Birmingham, Alabama, USA

Olivier P. Thomas
University of Nice-Sophia Antipolis, Institut de Chimie de Nice UMR 7272 CNRS, PCRE, Parc Valrose 06108 Nice, France

Fleur C. van Duyl
Department of Biological Oceanography, Royal Netherlands Institute for Sea Research, P.O. Box 59, Texel, The Netherlands

Xiaohong Wang
National Research Center for Geoanalysis, Chinese Academy of Geological Sciences, Beijing 100037, China, and ERC Advanced Investigator Grant Research Group, Institute for Physiological Chemistry, University Medical Center of the Johannes Gutenberg University Mainz, Duesbergweg 6, 55128 Mainz, Germany

Matthias Wiens
ERC Advanced Investigator Grant Research Group, Institute for Physiological Chemistry, University Medical Center of the Johannes Gutenberg University Mainz, Duesbergweg 6, 55128 Mainz, Germany

René H. Wijffels
Bioprocess Engineering, Wageningen University, Wageningen, The Netherlands

Contents

Preface

The idea of this special contribution reviewing the latest advances in sponge science was conceived during the World Sponge Conference held in Girona (Spain) in September 2010. Dr. Michael Lesser, editor of the Advances in Marine Biology book series, first suggested the production of a sponge-dedicated monograph. As organizers of the conference, we realized that the amount of information available had increased exponentially in recent years. As this overwhelming new information is scattered over an enormous volume of scientific papers published in journals of very different disciplines, we agreed that a thorough compilation and comprehensive review would be appropriate and useful. The monograph could convey the latest advances in sponge science to sponge specialists besides providing a comprehensive overview to a wider audience with interest in invertebrate biology, marine ecology, molecular ecology, or phylogeny among others. This contribution is timely because we lack reviews in some topics, while in other aspects, reviews were either too old or have become outdated because significant progress has been achieved in the past years. So we took the bait and you have in your hands the results of our efforts to sum up the most relevant and up-to-date scientific literature on the Phylum Porifera.

Sponges are extraordinary animals. With over 8000 extant described species, these organisms are major players in many scientific disciplines. Sponges have relevant roles in shaping the ecological functioning of many marine benthic communities, hold a strategic position for understanding the evolutionary origin of animals, and produce a great variety of secondary metabolites and skeletal structures that have made them preferred targets in biotechnological research. This contribution, split in two thematic volumes, comprises a representative selection of the most active fields of sponge research. Even if not exhaustive, this multiauthor blend of visions offers a wide portrait of the state of the art in sponge science. We have intended the volumes to highlight recent developments in multiple scientific fields, while identifying current limitations and knowledge gaps and delineating challenges and foreseeable future directions.

More specifically, the contributions include an overview of the titanic research work performed on taxonomy and ecology of Caribbean sponges over the past decades. The amazing array of ecological interactions in which sponges engage, with special emphasis on the diversity and functionality of their associated microbiomes, are dealt with in other chapters. The revolution that new molecular tools have represented in ecological studies is also covered in a dedicated chapter. The role of sponges in biogeochemical nutrient cycling is reviewed for the first time. The cell and molecular biology of sponges is a rocketing field, which gets its most recent advances and insights discussed from a modern perspective. Some chapters deal with sponge systematics and phylogeny, which are being hotly debated from several points of view, including a variety of hypotheses to interpret the relationships between sponge groups, other basal invertebrates, and early bilaterian animals. The rich chemical warfare featured by sponges, which has made this group a prolific source of new active natural products, has also been addressed, as well as the sponge machinery for processing and accumulating silica and its implications in tissue engineering. Although some of these chapters provide a good balance between basic and applied research, more biotechnologically oriented issues related to the culture of sponges, sponge cells, or symbionts for the production of chemicals have also found its place in the monograph. The chapters have been organized in two volumes: one covering the topics of phylogeny, systematics, and ecology, and the other dealing with physiology, chemical and microbial diversity, and biotechnology.

We address these volumes to both sponge specialists and nonspecialists, pursuing a twofold goal. We have intended to make the forefront of sponge research easily accessible to the nonspecialist, illustrating the state of the art of the field, and presenting current controversial issues. For the specialist, we wanted this monograph to be a handy, valuable update on the most recent advances in sponge science. We hope we have achieved our goals, at least partially. It goes without saying that the value of the volumes is mostly the merit of the contributing authors and the willing reviewers who altruistically devoted much time to read and make useful suggestions on the manuscripts. Our warmest thanks to all of them as well as to the AMB editorial staff who took care of editing and producing these books. We also thank you, the reader, for your interest in sponges and sponge science. We hope this collection of reviews is entertaining, useful, and inspiring for you all.

Mikel A. Becerro, Maria J. Uriz,
Manuel Maldonado and Xavier Turon

Series Contents for Last Fifteen Years*

Volume 32, 1997.
Vinogradov, M. E. Some problems of vertical distribution of meso- and macroplankton in the ocean. pp. 1–92.
Gebruk, A. K., Galkin, S. V., Vereshchaka, A. J., Moskalev, L. I. and Southward, A. J. Ecology and biogeography of the hydrothermal vent fauna of the Mid-Atlantic Ridge. pp. 93–144.
Parin, N. V., Mironov, A. N. and Nesis, K. N. Biology of the Nazca and Sala y Gomez submarine ridges, an outpost of the Indo-West Pacific fauna in the eastern Pacific Ocean: composition and distribution of the fauna, its communities and history. pp. 145–242.
Nesis, K. N. Goniatid squids in the subarctic North Pacific: ecology, biogeography, niche diversity, and role in the ecosystem. pp. 243–324.
Vinogradova, N. G. Zoogeography of the abyssal and hadal zones. pp. 325–387.
Zezina, O. N. Biogeography of the bathyal zone. pp. 389–426.
Sokolova, M. N. Trophic structure of abyssal macrobenthos. pp. 427–525.
Semina, H. J. An outline of the geographical distribution of oceanic phytoplankton. pp. 527–563.

Volume 33, 1998.
Mauchline, J. The biology of calanoid copepods. pp. 1–660.

Volume 34, 1998.
Davies, M. S. and Hawkins, S. J. Mucus from marine molluscs. pp. 1–71.
Joyeux, J. C. and Ward, A. B. Constraints on coastal lagoon fisheries. pp. 73–199.
Jennings, S. and Kaiser, M. J. The effects of fishing on marine ecosystems. pp. 201–352.
Tunnicliffe, V., McArthur, A. G. and McHugh, D. A biogeographical perspective of the deep-sea hydrothermal vent fauna. pp. 353–442.

Volume 35, 1999.
Creasey, S. S. and Rogers, A. D. Population genetics of bathyal and abyssal organisms. pp. 1–151.

* The full list of contents for volumes 1–37 can be found in volume 38

Brey, T. Growth performance and mortality in aquatic macrobenthic invertebrates. pp. 153–223.

Volume 36, 1999.
Shulman, G. E. and Love, R. M. The biochemical ecology of marine fishes. pp. 1–325.

Volume 37, 1999.
His, E., Beiras, R. and Seaman, M. N. L. The assessment of marine pollution—bioassays with bivalve embryos and larvae. pp. 1–178.
Bailey, K. M., Quinn, T. J., Bentzen, P. and Grant, W. S. Population structure and dynamics of walleye pollock, Theragra chalcogramma. pp. 179–255.

Volume 38, 2000.
Blaxter, J. H. S. The enhancement of marine fish stocks. pp. 1–54.
Bergström, B. I. The biology of Pandalus. pp. 55–245.

Volume 39, 2001.
Peterson, C. H. The "Exxon Valdez" oil spill in Alaska: acute indirect and chronic effects on the ecosystem. pp. 1–103.
Johnson, W. S., Stevens, M. and Watling, L. Reproduction and development of marine peracaridans. pp. 105–260.
Rodhouse, P. G., Elvidge, C. D. and Trathan, P. N. Remote sensing of the global light-fishing fleet: an analysis of interactions with oceanography, other fisheries and predators. pp. 261–303.

Volume 40, 2001.
Hemmingsen, W. and MacKenzie, K. The parasite fauna of the Atlantic cod, Gadus morhua L. pp. 1–80.
Kathiresan, K. and Bingham, B. L. Biology of mangroves and mangrove ecosystems. pp. 81–251.
Zaccone, G., Kapoor, B. G., Fasulo, S. and Ainis, L. Structural, histochemical and functional aspects of the epidermis of fishes. pp. 253–348.

Volume 41, 2001.
Whitfield, M. Interactions between phytoplankton and trace metals in the ocean. pp. 1–128.
Hamel, J.-F., Conand, C., Pawson, D. L. and Mercier, A. The sea cucumber Holothuria scabra (Holothuroidea: Echinodermata): its biology and exploitation as beche-de-Mer. pp. 129–223.

Volume 42, 2002.
Zardus, J. D. Protobranch bivalves. pp. 1–65.
Mikkelsen, P. M. Shelled opisthobranchs. pp. 67–136.
Reynolds, P. D. The Scaphopoda. pp. 137–236.
Harasewych, M. G. Pleurotomarioidean gastropods. pp. 237–294.

Volume 43, 2002.
Rohde, K. Ecology and biogeography of marine parasites. pp. 1–86.
Ramirez Llodra, E. Fecundity and life-history strategies in marine invertebrates. pp. 87–170.
Brierley, A. S. and Thomas, D. N. Ecology of southern ocean pack ice. pp. 171–276.
Hedley, J. D. and Mumby, P. J. Biological and remote sensing perspectives of pigmentation in coral reef organisms. pp. 277–317.

Volume 44, 2003.
Hirst, A. G., Roff, J. C. and Lampitt, R. S. A synthesis of growth rates in epipelagic invertebrate zooplankton. pp. 3–142.
Boletzky, S. von. Biology of early life stages in cephalopod molluscs. pp. 143–203.
Pittman, S. J. and McAlpine, C. A. Movements of marine fish and decapod crustaceans: process, theory and application. pp. 205–294.
Cutts, C. J. Culture of harpacticoid copepods: potential as live feed for rearing marine fish. pp. 295–315.

Volume 45, 2003.
Cumulative Taxonomic and Subject Index.

Volume 46, 2003.
Gooday, A. J. Benthic foraminifera (Protista) as tools in deep-water palaeoceanography: environmental influences on faunal characteristics. pp. 1–90.
Subramoniam, T. and Gunamalai, V. Breeding biology of the intertidal sand crab, Emerita (Decapoda: Anomura). pp. 91–182.
Coles, S. L. and Brown, B. E. Coral bleaching—capacity for acclimatization and adaptation. pp. 183–223.
Dalsgaard J., St. John M., Kattner G., Müller-Navarra D. and Hagen W. Fatty acid trophic markers in the pelagic marine environment. pp. 225–340.

Volume 47, 2004.
Southward, A. J., Langmead, O., Hardman-Mountford, N. J., Aiken, J., Boalch, G. T., Dando, P. R., Genner, M. J., Joint, I., Kendall, M. A.,

Halliday, N. C., Harris, R. P., Leaper, R., Mieszkowska, N., Pingree, R. D., Richardson, A. J., Sims, D.W., Smith, T., Walne, A. W. and Hawkins, S. J. Long-term oceanographic and ecological research in the western English Channel. pp. 1–105.

Queiroga, H. and Blanton, J. Interactions between behaviour and physical forcing in the control of horizontal transport of decapod crustacean larvae. pp. 107–214.

Braithwaite, R. A. and McEvoy, L. A. Marine biofouling on fish farms and its remediation. pp. 215–252.

Frangoulis, C., Christou, E. D. and Hecq, J. H. Comparison of marine copepod outfluxes: nature, rate, fate and role in the carbon and nitrogen cycles. pp. 253–309.

Volume 48, 2005.

Canfield, D. E., Kristensen, E. and Thamdrup, B. Aquatic Geomicrobiology. pp. 1–599.

Volume 49, 2005.

Bell, J. D., Rothlisberg, P. C., Munro, J. L., Loneragan, N. R., Nash, W. J., Ward, R. D. and Andrew, N. L. Restocking and stock enhancement of marine invertebrate fisheries. pp. 1–358.

Volume 50, 2006.

Lewis, J. B. Biology and ecology of the hydrocoral Millepora on coral reefs. pp. 1–55.

Harborne, A. R., Mumby, P. J., Micheli, F., Perry, C. T., Dahlgren, C. P., Holmes, K. E., and Brumbaugh, D. R. The functional value of Caribbean coral reef, seagrass and mangrove habitats to ecosystem processes. pp. 57–189.

Collins, M. A. and Rodhouse, P. G. K. Southern ocean cephalopods. pp. 191–265.

Tarasov, V. G. Effects of shallow-water hydrothermal venting on biological communities of coastal marine ecosystems of the western Pacific. pp. 267–410.

Volume 51, 2006.

Elena Guijarro Garcia. The fishery for Iceland scallop (Chlamys islandica) in the Northeast Atlantic. pp. 1–55.

Jeffrey, M. Leis. Are larvae of demersal fishes plankton or nekton? pp. 57–141.

John C. Montgomery, Andrew Jeffs, Stephen D. Simpson, Mark Meekan and Chris Tindle. Sound as an orientation cue for the pelagic larvae of reef fishes and decapod crustaceans. pp. 143–196.
Carolin E. Arndt and Kerrie M. Swadling. Crustacea in Arctic and Antarctic sea ice: Distribution, diet and life history strategies. pp. 197–315.

Volume 52, 2007.
Leys, S. P., Mackie, G. O. and Reiswig, H. M. The Biology of Glass Sponges. pp. 1–145.
Garcia E. G. The Northern Shrimp (Pandalus borealis) Offshore Fishery in the Northeast Atlantic. pp. 147–266.
Fraser K. P. P. and Rogers A. D. Protein Metabolism in Marine Animals: The Underlying Mechanism of Growth. pp. 267–362.

Volume 53, 2008.
Dustin J. Marshall and Michael J. Keough. The Evolutionary Ecology of Offspring Size in Marine Invertebrates. pp. 1–60.
Kerry A. Naish, Joseph E. Taylor III, Phillip S. Levin, Thomas P. Quinn, James R. Winton, Daniel Huppert, and Ray Hilborn. An Evaluation of the Effects of Conservation and Fishery Enhancement Hatcheries on Wild Populations of Salmon. pp. 61–194.
Shannon Gowans, Bernd Würsig, and Leszek Karczmarski. The Social Structure and Strategies of Delphinids: Predictions Based on an Ecological Framework. pp. 195–294.

Volume 54, 2008.
Bridget S. Green. Maternal Effects in Fish Populations. pp. 1–105.
Victoria J. Wearmouth and David W. Sims. Sexual Segregation in Marine Fish, Reptiles, Birds and Mammals: Behaviour Patterns, Mechanisms and Conservation Implications. pp. 107–170.
David W. Sims. Sieving a Living: A Review of the Biology, Ecology and Conservation Status of the Plankton-Feeding Basking Shark *Cetorhinus Maximus*. pp. 171–220.
Charles H. Peterson, Kenneth W. Able, Christin Frieswyk DeJong, Michael F. Piehler, Charles A. Simenstad, and Joy B. Zedler. Practical Proxies for Tidal Marsh Ecosystem Services: Application to Injury and Restoration. pp. 221–266.

Volume 55, 2008.
Annie Mercier and Jean-Francois Hamel. Introduction. pp. 1–6.
Annie Mercier and Jean-Francois Hamel. Gametogenesis. pp. 7–72.

Annie Mercier and Jean-Francois Hamel. Spawning. pp. 73–168.
Annie Mercier and Jean-Francois Hamel. Discussion. pp. 169–194.

Volume 56, 2009.
Philip C. Reid, Astrid C. Fischer, Emily Lewis-Brown, Michael P. Meredith, Mike Sparrow, Andreas J. Andersson, Avan Antia, Nicholas R. Bates, Ulrich Bathmann, Gregory Beaugrand, Holger Brix, Stephen Dye, Martin Edwards, Tore Furevik, Reidun Gangst, Hjalmar Hatun, Russell R. Hopcroft, Mike Kendall, Sabine Kasten, Ralph Keeling, Corinne Le Quere, Fred T. Mackenzie, Gill Malin, Cecilie Mauritzen, Jon Olafsson, Charlie Paull, Eric Rignot, Koji Shimada, Meike Vogt, Craig Wallace, Zhaomin Wang and Richard Washington. Impacts of the Oceans on Climate Change. pp. 1–150.
Elvira S. Poloczanska, Colin J. Limpus and Graeme C. Hays. Vulnerability of Marine Turtles to Climate Change. pp. 151–212.
Nova Mieszkowska, Martin J. Genner, Stephen J. Hawkins and David W. Sims. Effects of Climate Change and Commercial Fishing on Atlantic Cod *Gadus morhua*. pp. 213–274.
Iain C. Field, Mark G. Meekan, Rik C. Buckworth and Corey J. A. Bradshaw. Susceptibility of Sharks, Rays and Chimaeras to Global Extinction. pp. 275–364.
Milagros Penela-Arenaz, Juan Bellas and Elsa Vazquez. Effects of the *Prestige* Oil Spill on the Biota of NW Spain: 5 Years of Learning. pp. 365–396.

Volume 57, 2010.
Geraint A. Tarling, Natalie S. Ensor, Torsten Fregin, William P. Good-all-Copestake and Peter Fretwell. An Introduction to the Biology of Northern Krill (Meganyctiphanes norvegica Sars). pp. 1–40.
Tomaso Patarnello, Chiara Papetti and Lorenzo Zane. Genetics of Northern Krill (Meganyctiphanes norvegica Sars). pp. 41–58.
Geraint A. Tarling. Population Dynamics of Northern Krill (Meganyctiphanes norvegica Sars). pp. 59–90.
John I. Spicer and Reinhard Saborowski. Physiology and Metabolism of Northern Krill (Meganyctiphanes norvegica Sars). pp. 91–126.
Katrin Schmidt. Food and Feeding in Northern Krill (Meganyctiphanes norvegica Sars). pp. 127–172.
Friedrich Buchholz and Cornelia Buchholz. Growth and Moulting in Northern Krill (Meganyctiphanes norvegica Sars). pp. 173–198.
Janine Cuzin-Roudy. Reproduction in Northern Krill. pp. 199–230.

Edward Gaten, Konrad Wiese and Magnus L. Johnson. Laboratory-Based Observations of Behaviour in Northern Krill (Meganyctiphanes norvegica Sars). pp. 231–254.

Stein Kaartvedt. Diel Vertical Migration Behaviour of the Northern Krill (Meganyctiphanes norvegica Sars). pp. 255–276.

Yvan Simard and Michel Harvey. Predation on Northern Krill (Meganyctiphanes norvegica Sars). pp. 277–306.

Volume 58, 2010.

A. G. Glover, A. J. Gooday, D. M. Bailey, D. S. M. Billett, P. Chevaldonné, A. Colaço, J. Copley, D. Cuvelier, D. Desbruyères, V. Kalogeropoulou, M. Klages, N. Lampadariou, C. Lejeusne, N. C. Mestre, G. L. J. Paterson, T. Perez, H. Ruhl, J. Sarrazin, T. Soltwedel, E. H. Soto, S. Thatje, A. Tselepides, S. Van Gaever, and A. Vanreusel. Temporal Change in Deep-Sea Benthic Ecosystems: A Review of the Evidence From Recent Time-Series Studies. pp. 1–96.

Hilario Murua. The Biology and Fisheries of European Hake, *Merluccius merluccius*, in the North-East Atlantic. pp. 97–154.

Jacopo Aguzzi and Joan B. Company. Chronobiology of Deep-Water Decapod Crustaceans on Continental Margins. pp. 155–226.

Martin A. Collins, Paul Brickle, Judith Brown, and Mark Belchier. The Patagonian Toothfish: Biology, Ecology and Fishery. pp. 227–300.

Volume 59, 2011.

Charles W. Walker, Rebecca J. Van Beneden, Annette F. Muttray, S. Anne Böttger, Melissa L. Kelley, Abraham E. Tucker, and W. Kelley Thomas. p53 Superfamily Proteins in Marine Bivalve Cancer and Stress Biology. pp 1–36.

Martin Wahl, Veijo Jormalainen, Britas Klemens Eriksson, James A. Coyer, Markus Molis, Hendrik Schubert, Megan Dethier, Anneli Ehlers, Rolf Karez, Inken Kruse, Mark Lenz, Gareth Pearson, Sven Rohde, Sofia A. Wikström, and Jeanine L. Olsen. Stress Ecology in *Fucus*: Abiotic, Biotic and Genetic Interactions. pp. 37–106.

Steven R. Dudgeon and Janet E. Kübler. Hydrozoans and the Shape of Things to Come. pp. 107–144.

Miles Lamare, David Burritt, and Kathryn Lister. Ultraviolet Radiation and Echinoderms: Past, Present and Future Perspectives. pp. 145–187.

Volume 60, 2011.
Tatiana A. Rynearson and Brian Palenik. Learning to Read the Oceans: Genomics of Marine Phytoplankton. pp. 1–40.
Les Watling, Scott C. France, Eric Pante and Anne Simpson. Biology of Deep-Water Octocorals. pp. 41–122.
Cristián J. Monaco and Brian Helmuth. Tipping Points, Thresholds and the Keystone Role of Physiology in Marine Climate Change Research. pp. 123–160.
David A. Ritz, Alistair J. Hobday, John C. Montgomery and Ashley J.W. Ward. Social Aggregation in the Pelagic Zone with Special Reference to Fish and Invertebrates. pp. 161–228.

Volume 61, 2012.
Gert Wörheide, Martin Dohrmann, Dirk Erpenbeck, Claire Larroux, Manuel Maldonado, Oliver Voigt, Carole Borchiellini and Denis Lavrov. Deep Phylogeny and Evolution of Sponges (Phylum Porifera). pp. 1–78.
Paco Cárdenas, Thierry Pérez and Nicole Boury-Esnault. Sponge Systematics Facing New Challenges. pp. 79–210.
Klaus Rützler. The Role of Sponges in the Mesoamerican Barrier-Reef Ecosystem, Belize. pp. 211–272.
Janie Wulff. Ecological Interactions and the Distribution, Abundance, and Diversity of Sponges. pp. 273–344.
Maria J. Uriz and Xavier Turon. Sponge Ecology in the Molecular Era. pp. 345–410.

CHAPTER ONE

The Physiology and Molecular Biology of Sponge Tissues

Sally P. Leys*,[1] and April Hill†

Contents

* Department of Biological Sciences, University of Alberta, Edmonton, Alberta, Canada
† Department of Biology, University of Richmond, Richmond, Virginia, USA
[1]Corresponding author: Email: sleys@ualberta.ca

Advances in Marine Biology, Volume 62
ISSN 0065-2881, DOI: 10.1016/B978-0-12-394283-8.00001-1

Abstract

Sponges have become the focus of studies on molecular evolution and the evolution of animal body plans due to their ancient branching point in the metazoan lineage. Whereas our former understanding of sponge function was largely based on a morphological perspective, the recent availability of the first full genome of a sponge (*Amphimedon queenslandica*), and of the transcriptomes of other sponges, provides a new way of understanding sponges by their molecular components. This wealth of genetic information not only confirms some long-held ideas about sponge form and function but also poses new puzzles. For example, the *Amphimedon* sponge genome tells us that sponges possess a repertoire of genes involved in control of cell proliferation and in regulation of development. *In vitro* expression studies with genes involved in stem cell maintenance confirm that archaeocytes are the main stem cell population and are able to differentiate into many cell types in the sponge including pinacocytes and choanocytes. Therefore, the diverse roles of archaeocytes imply differential gene expression within a single cell ontogenetically, and gene expression is likely also different in different species; but what triggers cells to enter one pathway and not another and how each archaeocyte cell type can be identified based on this gene knowledge are new challenges. Whereas molecular data provide a powerful new tool for interpreting sponge form and function, because sponges are suspension feeders, their body plan and physiology are very much dependent on their physical environment, and in particular on flow. Therefore, in order to integrate new knowledge of molecular data into a better understanding the sponge body plan, it is important to use an organismal approach. In this chapter, we give an account of sponge body organization as it relates to the physiology of the sponge in light of new molecular data. We focus, in particular, on the structure of sponge tissues and review descriptive as well as experimental work on choanocyte morphology and function. Special attention is given to pinacocyte epithelia, cell junctions, and the molecules present in sponge epithelia. Studies describing the role of the pinacoderm in sensing, coordination, and secretion are reviewed. A wealth of recent work describes gene presence and expression patterns in sponge tissues during development, and we review this in the context of the previous descriptions of sponge morphology and physiology. A final section addresses recent findings of genes involved in the immune response. This review is far from exhaustive but intends rather to revisit for non-specialists key aspects of sponge morphology and physiology in light of new molecular data as a means to better understand and interpret sponge form and function today.

Key Words: Functional morphology; sponge physiology; signalling molecules; cell biology; developmental regulatory molecules

1. Introduction

Sponges are unusual animals which, due to their ancient heritage, can shed light on fundamental questions such as the origin of multicellularity, the evolution of tissues, signaling pathways, body polarity, and coordination systems. Sponge body plans are so different from those of other animals that it is difficult to compare even basic features, yet their molecular framework—which was revealed with the first full sponge genome from *Amphimedon queenslandica* (Srivastava *et al.*, 2010), as well as from transcriptomic and other gene data from other sponges (Nichols *et al.*, 2006; Harcet *et al.*, 2010)—shows that they have a very similar complement of genes and gene pathways to those in other animals. Although at the time of this review we have only one full sponge genome on which to base our comparison, it appears that some of the molecules that may underlie important morphological innovations in animals might be missing in sponges. Examples are the absence of sodium channels from the *A. queenslandica* genome and yet their presence in choanoflagellates, the closest unicellular ancestor to animals (Liebeskind *et al.*, 2011), and the presence of most components of the post-synaptic density in the sponge genome, including the ligand binding sites, but the absence of the ligands (Sakaraya *et al.*, 2007; Alie and Manuel, 2010). This, together with the suggestion that ctenophores might be basal to sponges (Dunn *et al.*, 2008), implies that sponges may not be 'witnesses to the pre-history' of animal systems as was suggested earlier by Pavans de Ceccatty (1974a) but may instead have lost complex animal characters, even neurons, however, unusual that might seem. Increasingly, it becomes clear that we need to revisit the morphology and physiology of sponges in order to better understand the relationship between gene and protein, and between cell and tissue function.

Sponges are suspension feeders, with a body plan designed to process as much water as possible, or needed, for feeding and respiration. Thus, the evolution of the sponge body plan can be seen as being guided by the fluid environment: sponges in low food environments find ways to process more water (use passive flow, e.g. Maldonado and Young, 1998; Leys et al., 2011) or eat different food as in the case of carnivorous sponges (e.g. Vacelet and Boury-Esnault, 1995); sponges in high food environments, or which have formed symbioses with microbes, may need to process less water (Weisz *et al.*, 2008).

Since the sponge is essentially a piping system, whose pressure differences are essential to generate the proper flow rates over the filter, at the cellular level modifications to a canal filtration system require constant readjustments, and therefore, cells and tissues should have the flexibility to adjust and modify the canal system in response to changes in flow to maintain the correct pressure differential. This dependence on flow penetrating all regions of the animal means that regional specialization is less evident than in other animals. In carnivorous sponges, where the tissues are freed up from that dependence, regionalization is often overt with distinct spatial separation of food capture regions and reproductive regions, separation of male and female gametes; where they are still present, choanocyte (feeding) chambers are also segregated to a small region of the sponge body. But all sponges have some regionalization of tissues, and in many groups, skeletal types are regionalized, while in others, reproductive structures are separated from choanosomal feeding tissues. Despite the commonality of the aquiferous system, there are many ways of building it, and the millennia during which sponges have been doing this have generated enough variations on the theme that interpretation of regions as tissues that carry out a common function requires a good understanding of the cell type, origin, and function. Today, molecular expression data can help fill in the picture.

Simpson's 'The cell biology of sponges' (1984) is an invaluable comprehensive resource on the structure and function of sponges. It covers a period of research that used ultrastructure to study the fine details of sponge cell function, associations, and lineage, and highlights several areas of uncertain knowledge. Some of these areas we now know more about by the use of new techniques, in particular scanning electron microscopy and immunocytochemistry, but also X-ray microtomography, new physiological approaches, and new molecular data. In order to understand what genes and their products do in sponges, we need to have a good idea of sponge cell biology and physiology. The aim of this review is to re-evaluate aspects of sponge structure and function upon which molecular and physiological data have shed new light. We are selective in our approach and do not intend to try to match Simpson's scope, but rather touch on topics in which new advances have been made with respect to our understanding of sponge tissues, tissue function, differentiation, and patterning. We first provide an overview of sponge body organization and then discuss new interpretations of the cell biology and physiology of sponges, highlighting, where relevant, new knowledge yielded by the study of gene expression.

2. General Organization of Sponges

2.1. Gross morphology

Sponges can be massive and spherical, thin and encrusting, tall and tubular, and many variations on these forms. In many texts, one will see sponges described as having ascon, sycon, and leucon grades of structure, terms which refer to the organization of choanocyte feeding chambers in the body wall, ascon, a simple tube lined by choanocytes; sycon, a tube with fingers lined by choanocytes; and leucon, with canals leading to sperhical or ovoid chambers lined with choanocytes (Ruppert *et al.*, 2004). However, since only Calcarea have all three grades (Fig. 1.1A), and there are many variations of leuconoid form in demosponges, hexactinellids, and homoscleromorph sponges (Fig. 1.1B–D), it is not a very useful way of thinking about the body structure. To understand how sponges actually function, we need to get away from this view, and yet at the same time work toward understanding what the implications are of having the variation of grades of structure in the Calcarea.

Body form has been most useful for taxonomists, but for the sponge, it is clearly a matter of feeding and excretion and, therefore, water flow. In the case of hexactinellids, massive incurrent canals penetrate deep into the choanosome which can give a syconoid appearance (Fig. 1.1B), and in the case of homoscleromorph sponges, incurrent canals appear as depressions in the outer surface and penetrate deep into the wall of the sponge, also giving a syconoid-like, 'sylleibid' appearance in section (Fig. 1.1C). In leuconoid forms, where the body wall is filled with many choanocyte chambers (the choanosome), across the body wall some incurrent canals are short, while the respective excurrent canals are long, and vice-versa for chambers deeper into the body wall (Fig. 1.1D): that is to say choanocyte chambers lie in parallel, never in series. The grades of structure were originally considered grades of evolutionary complexity (reviewed in Manuel *et al.*, 2002) presumably since the leuconoid sponges would have less issue with filtering the same water. That is, in asconoid Calcarea, flow is unidirectional down the length of a tube lined by potential feeding cells, so it appears that the same flow would be used by all cells; in syconoid Calcarea, the water current goes into separate, albeit large, chambers, and therefore, fewer cells would compete for the same water flow; in leuconoid sponges, the flow would be divided up, each portion to be filtered by only very few cells. However, if it were more valuable to be able to capture a larger item, we might view this in reverse. In this case, larger

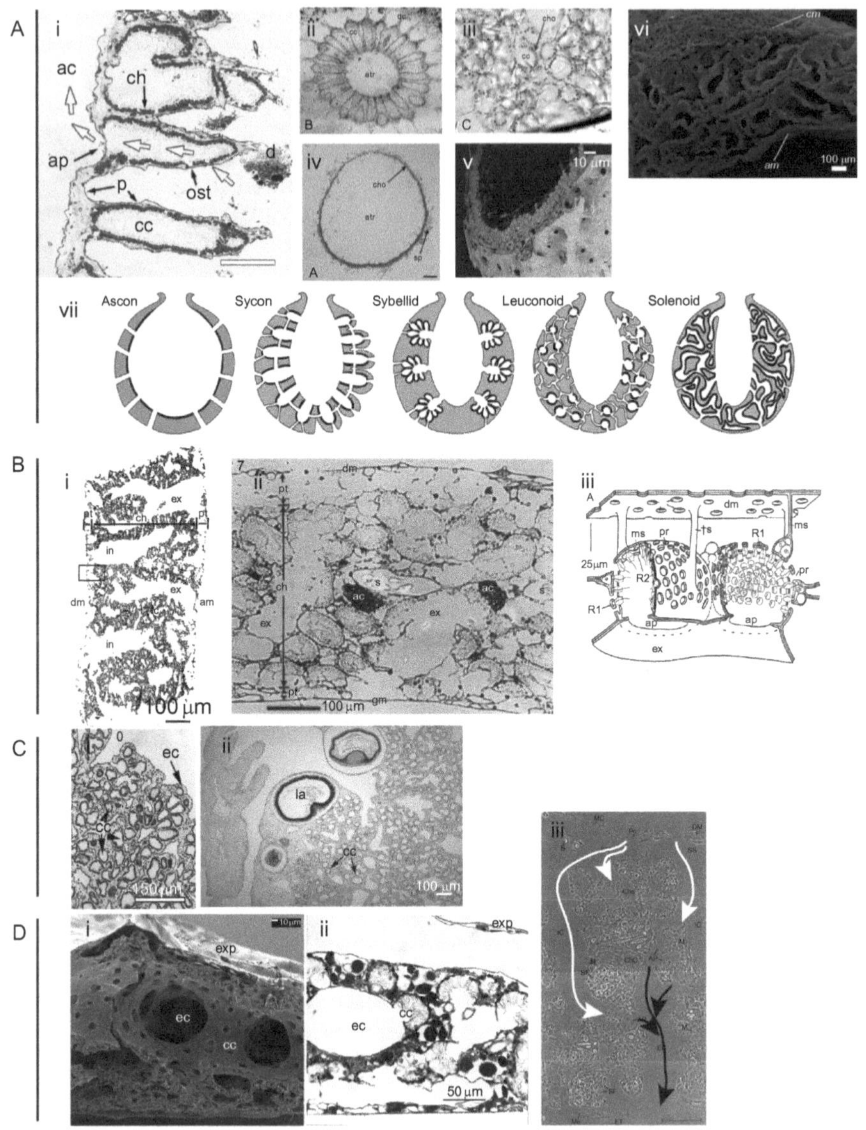

Figure 1.1 Body structure of sponges. A. Calcarea (i) *S. coactum* longitudinal section, (ii) *S. raphanus* cross section, (iii) *Leuconia johnstoni* transverse section, (iv) *Leucosolenia variabilis* cross section, (v) *Leucosolenia eleanor* cross-section (SEM), (vi) *Leucascus roseus* transverse section (SEM), (vii) diagrams illustrating the different body wall structures found in Calcarea. (i, Eerkes-Medrano and Leys, 2006; ii-iv, from Manuel, 2009; v, Leys, unpublished; vi, vii, from Cavalcanti and Klautau, 2011). (B) Hexactinellida. (i) Cross section of the body wall of *Rhabdocalyptus dawsoni* from a composition of images of wax sections, (ii) Cross-section and (iii) diagram of the chamber structure in *Farrea occa* (i, Leys, 1999; ii, iii, Reiswig and Mehl,

chambers would allow several cells to cooperatively capture a protist in syconoid sponges. In asconoid sponges, having a large single tube might even encourage the entry of unwary intruders which could be disabled using toxins, allowing the sponge to feed on something much larger than is normally able to enter via the incurrent flow.

2.2. Body wall overview

Sponges have quite diverse body constructions. We focus here on the haplosclerid-type morphology, many aspects of which can be generalized to other groups. The body wall consists of an outer layer (cortex), a canal system that brings water into chambers where the pumping and feeding cells are (the choanosome), and lacunae (epithelial-lined spaces) or canals that bring water to converge on the excurrent vent or osculum (the atrial tissue) (Fig. 1.2).

The outermost tissue has been called the dermal membrane because early researchers thought the epithelium in all sponges was syncytial (Wilson, 1907). Electron microscopy has confirmed that only hexactinellid sponges have syncytial tissues; however, in some species, boundaries of surface cells are hard to see even by scanning electron microscopy, so it is understandable that they were difficult to see by light microscopy (Fig. 1.2C). The dermal membrane (as used by earlier workers) consists of a thin single layer of cells, pinacocytes (see Section 3.1) that lies on a collagenous middle layer, the mesohyl. It is worth noting that some authors used the term dermal membrane to refer to the entire outer region of the sponge—whether a single exopinacoderm, or an exopinacoderm and endopinacoderm with fine mesohyl inbetween—and others used it to refer to all pinacocytes lining the surface and canals of the sponge. Recent workers no longer use this term and refer instead only to the 'exopinacoderm' (Boury-Esnault and Rützler, 1997).

The nature of the sponge surface layers varies greatly. In some sponges, the exopinacoderm covers a cortex—a distinct rind that consists of particular cell types, spicule types, and occasionally foreign material embedded into the tissue (e.g. Teragawa, 1986). In other sponges, the skin is formed by a thin single layer of exopinacocytes covering a massive collagenous mesohyl, while in others, the skin forms a 3-μm-thin layer over an equally meagre mesohyl, which is sandwiched by another epithelium, the endopinacoderm (or in syconoid and asconoid sponges, the choanoderm). Under the exopinacoderm lies the first of

1991). (C) Homoscleromorpha. Cross-sections of (i) *Oscarella kamchatkensis*, Gazave *et al.*, 2010; (ii) *Corticium candelabrum* (cc, choanocyte chamber, la, larva) (courtesy of A. Riesgo). (D) Demospongiae. (i) Scanning EM and (ii) cross section of *Ephydatia muelleri* (ec, excurrent canal; cc, choanocyte chamber; exp, exopinacoderm) (iii). Cross-section of *Haliclona elegans* showing the different paths water might take through the sponges. (i and ii, Elliott and Leys, unpublished; iii, Langenbruch and Scalera-Liaci, 1986. (For colour version of this figure, the reader is referred to the online version of this chapter.)

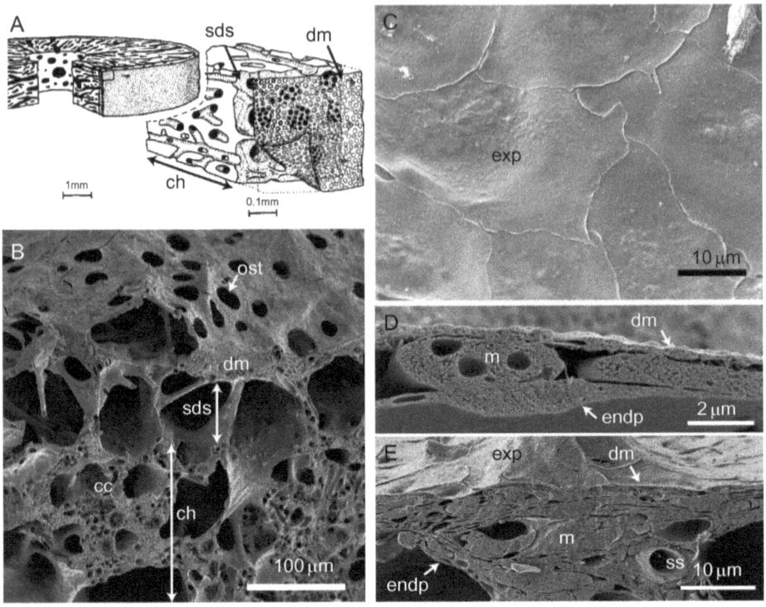

Figure 1.2 Regionalization in sponges. (A) Diagram of a cross-section and cutaway showing the aquiferous system of *Haliclona permollis*. (B) Fractured specimen of *Neopetrosia problematica* showing the dermal tissues with surface dermal membrane (dm) and ostia (ost), subdermal space (sds) and choanosome (ch) with choanocycte chambers (cc). (C–E) Aspects of the dermal membrane. (C) Exopinacoderm (dermal membrane proper) of *Ephydatia muelleri*, SEM. (D) Fracture through the dermal membrane of *E. muelleri*, SEM. (E) Fracture through the dermal membrane and mesohyl of *Haliclona mollis*. (Dm, dermal membrane; sds, subdermal space; ch, choanosome; cc, choanocyte chamber, exp, exopinacoderm; end, endopinacoderm; m, mesohyl cell; ss, spicule space) (A from Reiswig, 1975; B–E, Leys and Adams, unpublished.)

a set of epithelial-lined spaces into which the water first enters. These may be lacunae (small epithelial-lined spaces) or a single large cavity (called the subdermal cavity). Water passes either directly into the subdermal cavity or via short epithelial-lined canals into the lacunae and from there into the choanosome (Fig. 1.2). The choanosome consists of canals and chambers together with spaces that may exist around each chamber. The structure of chambers and the way they are attached to each canal are characteristic of genera and families of sponges (Boury-Esnault *et al.*, 1990).

Like the subdermal tissues, the subatrial tissues may have lacunae in which the excurrent water collects before venting out of the osculum, or they may simply consist of converging excurrent canals, as they increase in diameter before merging at the base of the osculum. The excurrent canals and lacunae may be important for monitoring and adjusting the rate of flow of water through the animal, either by contracting or by providing a reservoir for water before and after the feeding chambers.

2.3. Cells, tissues, and regionalization

The sponge body has functional regions defined by specific tissues and the cell types in them. Tissues are defined after Hyman (1940) as being formed by one or more type of cell and organized so as to form a functioning unit (whole). Most sponge larvae also have well-differentiated regionalized tissues, and in the larva, there is the additional function of motility and direction finding, for which regions are specialized (Maldonado and Bergquist, 2002), but the types of tissues and cells present in the larva for the most part take their names (and we infer their function) from those in the adult. In the adult, sponge tissues include the choanocyte epithelium (choanoderm), including cells supporting the choanocyte chambers; the pinacocyte epithelium (pinacoderm), including dermal tissues, incurrent, and excurrent canals; the skeletal and collagenous support tissues; and the osculum. Here, we focus on the structure and formation of two epithelial tissues only, the choanoderm and pinacoderm. We do not address skeletal tissues or cells of the mesohyl.

3. The Choanoderm Epithelium

3.1. Overview of the aquiferous system

Flow through the sponges is generated by the beating of many thousands of flagella per cubic millimetre of choanosomal tissue. The beat of the flagellum generates a low pressure at its base, drawing water toward and through the collar and from there up along the length of the flagellum away from the cell body (Larsen and Riisgård, 1994). Many choanocytes in a choanocyte chamber, with collars and flagella facing toward the exit of the chamber, generate a current of water through the collar into their bases, and from there into the chamber and out of the chamber exit. The manner in which choanocytes are organized into chambers (how many there are, how many openings there are between them, what the dimensions of the collar slits are) and the manner in which chambers are organized at the end of incurrent canals have been studied in many sponges (e.g., Langenbruch, 1983, 1988; Langenbruch *et al.*, 1985; Langenbruch and Scalera-Liaci, 1986; Langenbruch and Weissenfels, 1987; Langenbruch and Jones, 1990; Boury-Esnault, 2006). The dimensions of the fine incurrent and excurrent canals are features that determine the precise resistance over the collar slit in the choanocyte chamber (Larsen and Riisgård, 1994) (Fig. 1.3). The pressure drop must slow the flow sufficiently for choanocytes (or pinacocytes at the entrance to chambers) to phagocytose food which is primarily bacteria ⅒ to ½ the size of the choanocyte.

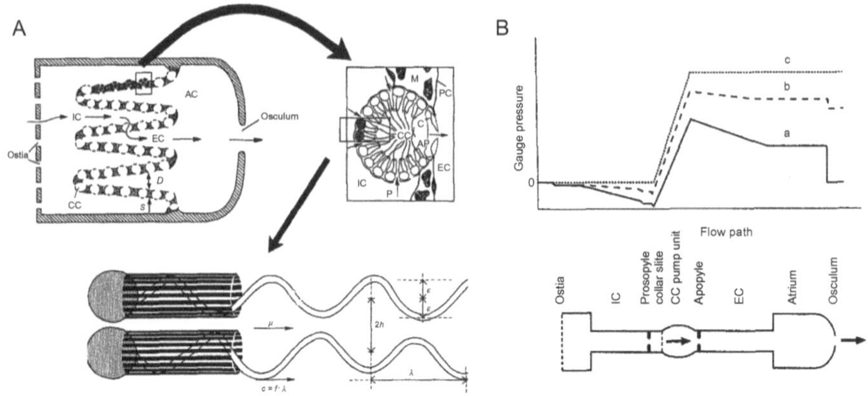

Figure 1.3 (A) Diagram illustrating the basic sponge design with pumps (choanocyte chambers) in parallel. Water enters through ostia to incurrent canals, through the choanocyte chambers and out excurrent canals to the osculum. (B) Schematic of the pressure head variation along a flow path from ostium to osculum, illustrating the pressure at low (a), medium (b) and high (c) resistance along the path. From Larsen and Riisgård, 1994.

3.2. Choanocyte structure

Choanocytes are very small cells (2–10 μm in diameter) with a flagellum that is surrounded by a 'collar' of microvilli, finger-like actin-filled extensions that arise from the cell membrane (Fig. 1.4). Bidder (1895) described choanocytes from Calcarea (*Sycon raphanus* and *Sycon compressa*) as being barrel-shaped cells, like 'full corn-sacks standing side by side in a granary', with a collar that is almost a perfect cylinder. Later, their structure and function were studied in live specimens and in ultrathin sections of spongillids using electron microscopy by Kilian (1952), Rasmont *et al.* (1958), and Rasmont (1959), and each account has confirmed the presence of common components including a fibrillar material that links microvilli of the collar.

The collar microvilli rise about 2/3 the height of the flagellum. Choanocytes may be squat or tall and may lie on a thin or a thick collagenous mesohyl, appearing in the former case to be suspended in the inhalant canal system, and in the latter to be embedded in the mesohyl (see Section 3.3). That choanocytes adhere to one another or to endopinacocytes with sealing junctions should be considered the typical situation because tissues fixed with ruthenium red showed no tracer passes between choanocytes or between pinacocytes and choanocytes into the mesohyl (Adams, 2010).

Between choanocyte microvilli stretch fine strands of proteoglycan ('glycocalyx') mesh (Fig. 1.4G, H, I). Already in the late 1800s, Bidder (1895) saw in sections of osmium-fixed tissues a 'film of some other substance' uniting the collar microvilli; later in ultrathin sections of spongillids, Rasmont (1958) refers to 'bridge-like structures' between the microvilli. The mesh is

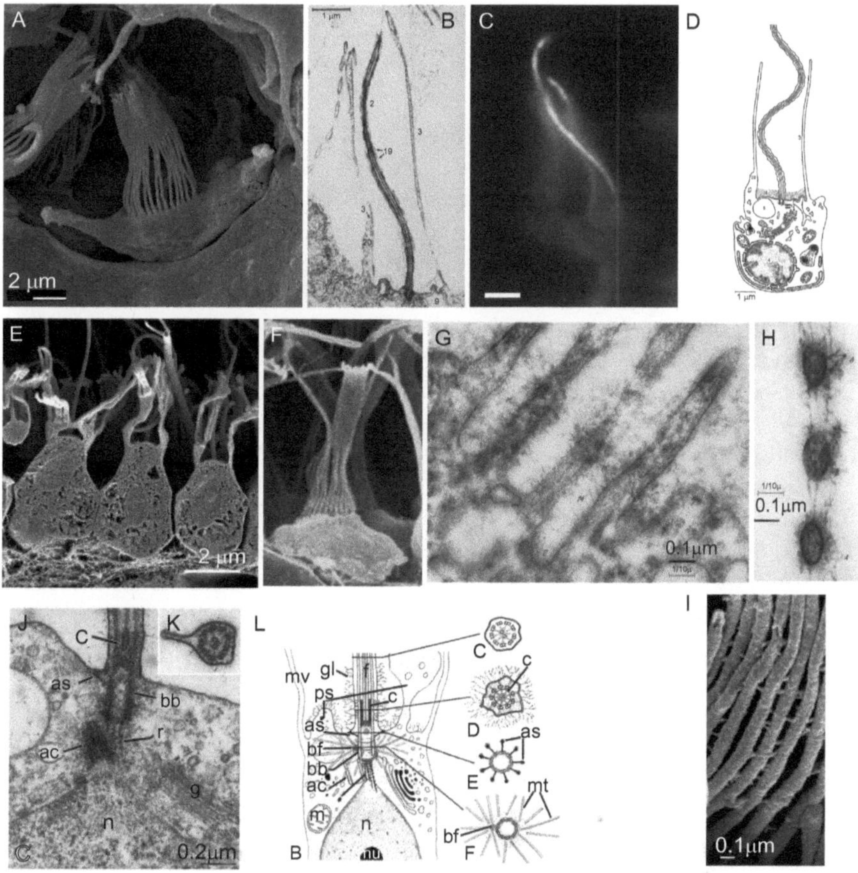

Figure 1.4 Choanocyte structure. (A) Choanocyte from *Ephydatia muelleri* (SEM); (B) Longitudinal section through the flagellum of *E. fluviatilis* TEM; (C) Fluorescence image of two choanocytes from *Ephydatia muelleri* showing actin labelled with Bodipy Fl phallacidin (green) and tubulin labelled with anti-alpha tubulin (red); nuclei are labelled with Hoechst 33342 (blue); (D) Diagram of a choanocyte from Brill, 1973); (E) Choanocytes in *Sycon coactum*, Calcarea; (F) Choanocyte from *Oscarella lobularis*, Homoscleromorph; (G–I) Fibrillar proteoglycan mesh between microvilli of *Spongilla lacustris*, G and H, and between microvilli of *Sycon coactum*, I. (J–L) Structure of the flagellum and basal body of a choanocyte in *Halisarca dujardini*. (A, C, E, Leys unpublished; B, D, Brill, 1973; F, Boury-Esnault et al., 1984; G, H, Fjerdingstad, 1961; J–L, from Gonobobleva and Maldonado, 2009). (For interpretation of the references to colour in this figure legend, the reader is referred to the online version of this chapter.)

better preserved by glutaraldehyde/osmium fixation and has subsequently been found both outside and inside the collar in many sponges. Electron micrographs show a mesh on the collar in *Spongilla lacustris* (demosponges, Fjerdingstad, 1961), *Sycon* (Calcarea, Ledger 1976 in Simpson 1984), and *Farrea occa* and *Rossella racovitzae* (hexactinellids, Reiswig and Mehl, 1991;

Koestler in Leys *et al.*, 2007). The mesh forms spaces of approximately 40 square nanometres, and importantly, this is the smallest space (and greatest surface area) through which the water actually passes in the sponge.

In freshwater demosponges, a layer of glycocalyx also forms what looks like a gasket—a structure filling in the space between neighbouring collars approximately 2–3 μm above the cell's surface (Weissenfels, 1992). In some marine demosponges (e.g. *Haliclona mollis*), a set of cells forms a layer similar to that gasket, sealing the spaces between each of the collars (Adams, 2010) (Fig. 1.5A–C); it is possible this layer of cells is what is referred to as Sollas' membrane (Bidder, 1895), an enigmatic structure whose existence has been difficult to confirm. This gasket is well known in glass sponges (Hexactinellida) where it is called the 'secondary reticulum', a layer of syncytial tissue that rises from the primary reticulum to surround the collar of each collar body (Mackie and Singla, 1983). The secondary reticulum may play a role in channelling flow through the collar, as inferred from a study that watched the passage of micorospheres through sponge fragments using video microscopy (Wyeth *et al.*, 1996; Wyeth, 1999). In *Ephydatia fluviatilis*, it is suggested to function as a 'one-way valve' to prevent back flow of water (Langenbruch and Weissenfels, 1987), so in both cases, perhaps it has a similar function.

3.3. Organization of choanocyte chambers—Terminology

There are some 50–200 choanocytes in demosponge chambers (Reiswig, 1975; Rasmont and Rozenfeld, 1981), 200–300 in a hexactinellid chamber (Leys et al., 2011), and up to 1000 in a calcareous syconoid chamber (*Sycon*, Leys, personal observation). Openings to choanocyte chambers are called 'pyles' (orifice Gr.), and, therefore, entrances are called prosopyles (L. forward of) and exits, apopyles (Gr. away from). Prosopyles can be formed by pseudopodial extensions of choanocytes arranged so as to create a hole

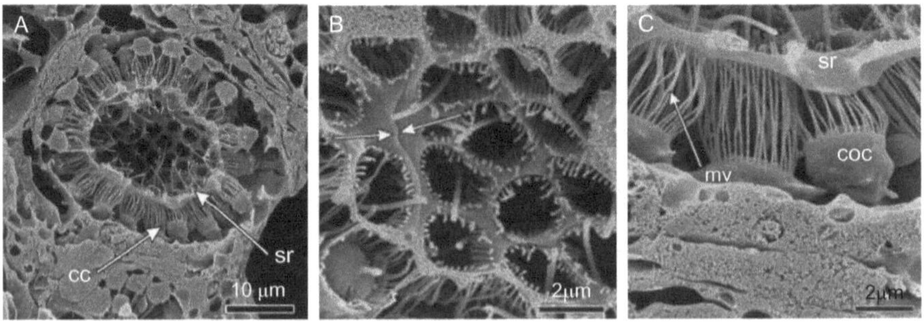

Figure 1.5 Gasket of cells surrounding the collars in choanocyte chambers of *Haliclona mollis*. (Adams, 2010). (A) Low and (B,C) high magnification views of the cells that envelop the collar microvilli.

(choanocytic prosopyles), or by pinacocytes that contact the choanocytes and form something like a porocyte (pinacocytic prosopyle). Apopyles are larger and, therefore, formed by several cells, called either 'cone' cells or apopylar cells (De Vos *et al.*, 1990), which are interpreted to arise from choanoblasts during early chamber formation (Weissenfels, 1981). Unlike choanocytes, however, which are flagellated, cone cells are ciliated and the cilium beats in a slow whip-like fashion away from the chamber (Fig. 1.6) (see Section 4.2). Some sponges lack cone cells and instead several endopinacocyte-like cells form a sieve plate at the exit of the chamber (e.g. *Tethya wilhelma* Nickel, personal communication; SPL, unpublished data).

3.4. Choanocyte function—Feeding

The action of the choanocyte flagellum in generating a low pressure to draw water through the collar is well described by Simpson (1984) from Van Tright (1919) and Kilian (1952). Though the basics are quite clear, exactly how the water moves through the collar and chamber is not actually known. For example, while it is understood that water follows a pressure gradient generated by the choanocyte, as noted above, it is suggested that there must be one-way valves to prevent water from flushing the other way under changes in external pressure (Vogel, 1978). Though potential one-way valves (apopylar cells and a glycocalyx mesh between the collars) have been identified by Langenbruch and Weissenfels (1987), it would be difficult to test how they work. Also, whether all water must pass through the collar filter or whether there are bypass canals is likewise not known. Some authors suggest that bypass structures exist, for example, to compensate for pressure changes that occur during rhythmic contractions (Nickel *et al.*, 2006), and other authors, using corrosion casts to study the aquiferous system, also find links between incurrent and excurrent canals (Bavestrello

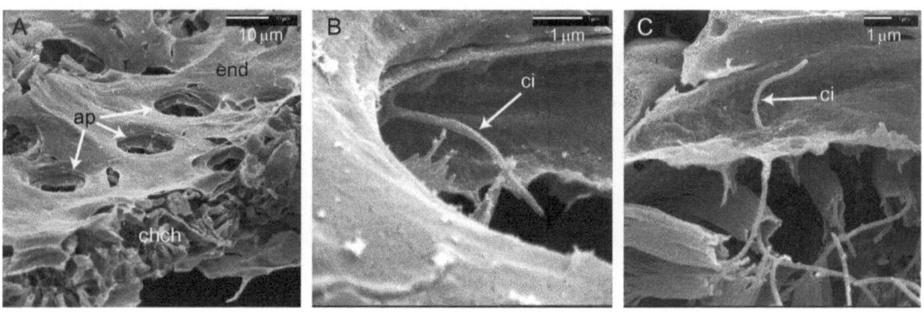

Figure 1.6 Apopyles and apopylar (cone) cells. (A) Several apopyles (ap) open into an excurrent canal in *Ephydatia muelleri* (end, endopinacoderm; chch, choanocyte chambers). (B, C) Apopylar cells with cilia (ci) which are approximately 4 µm long and show the characteristic bending of ciliary motion. (Leys, unpublished).

et al., 1988). However, since using corrosion casts may cause tearing or rupture of the canal system, doubt remains as to whether such bypasses exist, especially when it is considered that these would generate leakage in the pressure system that the choanocyte filter depends upon. Feeding studies which show greater than 95% retention of bacteria < 1 μm by the sponge do not support the presence of bypass canals in all sponges (Pile *et al.*, 1996; Yahel *et al.*, 2003, 2007).

Food capture occurs at choanocytes, prosopyles, and incurrent canal epithelia (Kilian, 1952; Weissenfels, 1976; Imsiecke, 1993; Leys and Eerkes-Medrano, 2006); even exopinacocytes have been shown capable of uptake of latex beads (Willenz and Van de Vyver, 1982), so epithelial cells in general must be capable of phagocytosis. Generally however, bacteria and other picoplankton are captured by pseudopodia at the choanocyte, and once in that cell, are transferred to amoebocytes for digestion and transfer of food to other cells in the sponge. The transfer of food between choanocyte and amoebocyte has been shown by video microscopy by Imsiecke (1993), and between amoebocyte and other cells in a suite of studies by Willenz and colleagues (Willenz and Van de Vyver, 1984, 1986; Willenz *et al.*, 1986).

3.5. Choanocyte differentiation and turnover

Choanocytes are one of the better-understood cells in sponges in terms of their origin, function, and differentiation potential. They generate flow, capture food, and can differentiate into other choanocytes, spermatocytes, and even oocytes. Other studies suggest that they can even become archaeocytes, and in this sense, they are a type of stem cell (Funayama *et al.*, 2010). In cases where sperm arise from choanocytes, usually whole populations of choanocytes in chambers undergo this modification, multiplying in number, losing the collar, packaging the nucleic material, and modifying the cell body to form spermatic cysts (Boury-Esnault and Jamieson, 1999; Maldonado and Riesgo, 2008). Therefore, choanocytes are at the same time dispensable cells and have the potential to re-differentiate into gametes.

In sponges hatched from gemmules or from aggregates of dissociated cells, choanocytes develop from archaeocytes (Buscema *et al.*, 1980), but during development from larvae, they arise from ciliated epithelial cells (Amano and Hori, 1996; Leys and Degnan, 2002). In each instance, the collar develops secondarily: when choanocytes develop from archaeocytes, they first develop the flagellum and then later the collar (Weissenfels, 1981); when choanocytes develop from ciliated epithelial cells of the larva, the cilium is withdrawn into the cell (engulfed), its axoneme is disassembled into tubulin dimers, and then it is reassembled as a flagellum and the collar is formed (Leys and Degnan, 2002) (Fig. 1.7). In larvae that already have choanocytes (or in glass sponges, collar bodies rather than choanocytes), the collar is lost at metamorphosis and is only formed again after the choanocytes of the small existing chambers move to new positions and form new chambers

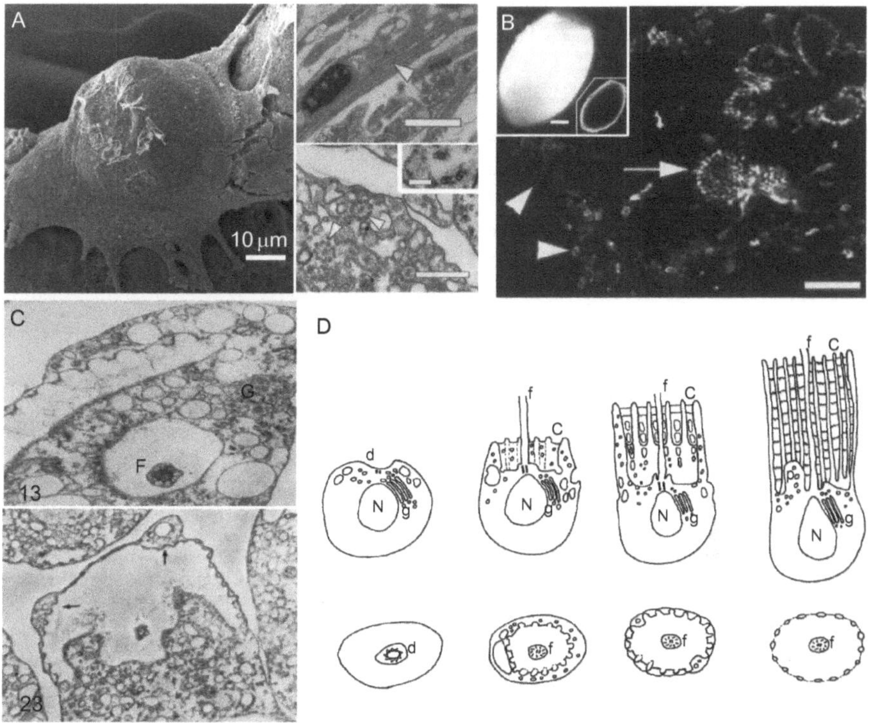

Figure 1.7 Differentiation of choanocytes. (A) Loss of cilia from the metamorphosing larva of *Amphimedon queenslandica* and their disassembly in cells of the post-larva. (B) Fluorescently labelled ciliated cells of the larva transdifferentiate into choanocytes (arrows) of the juvenile sponge. (C, D) Collar formation in *Tetilla serica*: a tube becomes gradually separated into microvilli. (A, B, Leys and Degnan, 2002; C, D, Watanabe, 1978b).

(Leys, personal observations). An excellent account of collar formation comes from Watanabe (1978a) who showed that in *Tetilla serica*, new choanocytes form the collar as a membranous tube (essentially a pseudopodial extension) which, once it reaches a 2–3 μm in height, divides into the fingers that become microvilli (Fig. 1.7). Choanocytes can, therefore, dedifferentiate and re-differentiate easily—they are fairly plastic.

Funayama et al. (2005a) took advantage of the feeding behaviour of choanocytes to find a molecular marker for the choanocyte cell lineage. They fed the sponge fluorescent plastic microspheres (latex beads), and once these were captured by the sponge, they dissociated the sponge tissue and via fluorescent cell sorting isolated the cell fraction that contained the fluorescent beads. A gene that was expressed highly in mRNA isolated from that fraction was *annexin,* and via *in situ* hybridization its expression in choanocytes and choanocyte-committed archaeocytes was confirmed (Funayama et al., 2005a). *Annexin* has many roles in cell adhesion, cell motility, endocytosis, and vesicle transport, and so its specific expression in choanocytes may reflect the role of those cells in nutrient uptake and transport (Funayama et al., 2005a).

The high turnover rate of choanocytes has been known for some time, and in one instance, it has been suggested that spent choanocytes are sloughed off several times a day (De Goeij *et al.*, 2009). Measurements of the choanocyte cell cycle range from 20 h for *Hymeniacidon synapium* (Shore, 1971) to 15 h for *Baikalospongia bacillifera* (Efremova and Efremov, 1979) and 5.4 h for *Halisarca caerulea* (De Goeij *et al.*, 2009). The latter estimate based on the number of BrdU positive cells in sections of labelled sponges may be on the short side; mammalian intestinal epithelial cells turnover every 3–60 days (Karam, 1999). Regardless of the precise rate, clearly, choanocytes are replaced frequently compared to other cells in the sponge body, and this has important implications. Firstly, it means there is a population of stem cells that replaces them and this appears to be archaeocytes committed to the choanocyte lineage. In contrast to pinacocytes, choanocytes are continually dividing, as can be shown by BrdU staining of dividing nuclei (De Goeij *et al.*, 2009; Funayama *et al.*, 2010) and the labelling of archaeocytes that are committed to be choanocytes as well as choanocytes themselves with the stem cell marker Piwi (Funayama *et al.*, 2010). Secondly, it implies that either choanocytes are cheap to make or that pumping is a costly activity and that these cells are destroyed or are easily damaged. The small size of choanocytes might reflect the need to be able to replace them rapidly at a minimal expense. Rasmont and Rozenfeld (1981) determined that based on the volumes of each cell type, in *E. fluviatilis*, a single archaeocyte could divide into at least 30 choanocytes.

Determining the cost of production of a choanocyte would greatly help us understand the exact cost of pumping for the sponge. It has been estimated that the cost of pumping amounts to 1% of metabolism (Riisgård *et al.*, 1993; Larsen and Riisgård, 1994). However, recent measurements of oxygen consumption in a sponge that went from not pumping to a maximum pumping rate suggest that the cost of pumping is up to 30% of metabolism (Hadas *et al.*, 2008), and a new study of filtration in glass sponges (Leys et al., 2011) suggests this may be due to the resistance generated by the fine dimensions of the glycocalyx mesh on the collar, which was not taken into account earlier. More data on the dimensions of passages through sponges, and of oxygen consumption in a variety of sponges, will greatly help refine our understanding of energy expenditure and the cost of filtration for sponges.

3.6. Control over flow

Sponges can control the flow through their aquiferous system, but they do this in different ways. In glass sponges (Hexactinellida), the flagella stop moving ('arrest') when the sponge is irritated by sediment or there is damage to the tissue. This happens by the spread of an action potential through the syncytial tissues (see below); the flagella only start to beat again after the membrane potential has reset. Other sponges are cellular and as far as we know are not able to spread action potentials and changes in flow never occur within the timeframe that would suggest electrical activity. In those sponges, the flagella beat without arrest for hours on end or until 'conditions are unfavourable' (Grant, 1825).

Grant was probably one of the first to observe this incessant pumping of water and gives a nice account of using a microscope and the reflection of candle light to watch the flow pour from the osculum for up to 5 h without change in rate (Grant, 1825). Cellular sponges can contract their canals and close their ostia and oscula, so flow can be slowed and stopped gradually, but in general, it is understood that the flagella of choanocytes—as do cilia—beat relentlessly, even if each flagellum might beat at a slightly different rate than the others. Observations of flagella beating in intact living tissues (damaged dissociated choanocytes have a highly variable beating pattern) indicate that the flagella beat in sinusoidal waves in a single plane and each choanocyte flagellum in a chamber beats in a slightly different plane to the others (Jones, 1964; Kilian and Wintermann-Kilian, 1979; Wyeth *et al.*, 1996; Wyeth, 1999). Jones (1964) notes that in the calcareous sponge, *Leucosolenia*, the beat varies in frequency but is usually too fast to be measured; the same is true for freshwater demosponges (Kilian and Wintermann-Kilian, 1979); the beat eventually slows and stops after some hours (Kilian, 1952; Jones, 1964). According to Kilian and Wintermann-Kilian (1979), however, in *E. fluviatilis*, some flagella do occasionally stop beating. While slowing and arrest after several hours of observation can be attributed to damage (light, tearing, lack of oxygen or flow under the microscope slide), it is unclear what the mechanism of arrest of an individual flagellum would be.

So far, only glass sponges (hexactinellids) are known to be able to arrest the flagella in an instant (<30 s), and they can do this because they have syncytial tissue which, upon a mechanical or electrical stimulus to the sponge (a strong prod, or damage to the tissue), propagates electrical signals throughout the entire sponge body (Leys and Mackie, 1997). The action potential would cause calcium to enter the collar bodies, and the rise in intracellular calcium would arrest the flagellum (Leys *et al.*, 1999). The flagella beat resumes only once calcium is pumped out of the cell body. In demosponges, homoscleromorphs, and calcareous sponges, no electrical (rapid) signalling mechanism is known. If flagella were to stop, as they do in hexactinellids, this implies that there must be control over calcium levels in individual choanocytes and it is unclear how that would happen except during contraction of the choanosome. All cellular sponges have been known to be able to contract their tissues slowly in response to poor water quality, mechanical or chemical irritation, or as a spontaneous rhythm (e.g. Parker, 1910; McNair, 1923; Prosser *et al.*, 1962; Emson, 1966; Prosser, 1967; De Vos and Van de Vyver, 1981;

Weissenfels, 1990; Nickel, 2004; Elliott and Leys, 2007), and it has now also been shown that the lyssacine hexactinellid *Oopsacas minuta* can also carry out very slow contractions of the entire body (Nickel, 2010). Contractions are thought to be coordinated by calcium waves that travel through the tissues because contraction amplitude is reduced in the absence of calcium and abolished in the absence of calcium and magnesium (Elliott and Leys, 2010). In other animals, calcium waves are propagated by a repeated cycle in which a trigger (e.g. mechanical irritation) causes a rise in intracellular calcium $[Ca^{2+}]_i$ which causes release of molecules (e.g. glutamate, NO, ATP, or cAMP; Elliott and Leys, 2007; Elliott and Leys, 2010) that bind receptors on neighbouring cells triggering a rise in $[Ca^{2+}]_i$ in those cells, and so on until all the cells in a region have experienced a calcium rise (reviewed in Nickel, 2010). During a strong contraction, even whole choanocyte chambers contract and choanocyte collars and flagella are so compressed that they project out of the exit opening (apopyle) of the chamber (Elliott, 2009); at that point, the motion of the flagella is also noted to have stopped (Weissenfels, 1990). Water may also be slowed from moving through canals by the contraction of sphincters formed by cells in canals. Reiswig (1971) showed that reduced oscula diameter correlated with reduced pumping velocity in *Tethya crypta*, and constriction of atrial canals corresponded with cessation of pumping in *Verongia gigantea*.

Control over the flow may be necessary to prevent collapse of the canals due to unequal pressure in the incurrent and excurrent canal system. If the incurrent canals clog, then the choanocyte pumps will be unable to bring water into the chambers, or if the osculum is torn or any part of the excurrent or incurrent canal system is opened by wounding, the pressure drop across the choanocytes in the chamber presumably will be lost. The sponge must have mechanisms to sense fine changes in the pressure across the entire system and adjust the dimensions of canals by contractions or remodelling. To repair the system, changes must occur in the space of minutes. We might, therefore, expect sensory cells in different regions of the aquiferous system, as well as the ability of the sponge tissues, to have positional information to determine which regions are incurrent and which are excurrent so as to be able to consistently remodel according to this directional body plan.

4. The Pinacoderm Epithelium

4.1. Pinacoderm description and overview of function

The main function of an epithelium is compartmentalization, to control the secretion and absorption of water and the transport of chemicals, proteins and ions between body regions. Epithelia also detect and ward off infection (Van de Vyver, 1970; Müller and Müller, 2003), and provide structural support (Teragawa, 1986), especially when out of water as would be the case for intertidal sponges. In Hexactinellida, epithelia are syncytial: a single unbroken 1-μm-thin sheet of tissue called the trabecular reticulum covers the entire sponge. As these were described in some detail previously (Leys *et al.*, 2007), they will not be covered here. In 'cellular' sponges (demosponges, calcareous, and homoscleromorph sponges), epithelia are formed by a single layer of squamous (flat) cells that adhere to one another with a short overlap of their edges (Fig. 1.8). The slight nature of this epithelium coupled with the idea that cells can move readily into and out of the epithelium has led to the view that sponge epithelia are tenuously held together; but imaging of sponge epithelial cells *in vivo*, physiological experiments on their sealing and secretory ability, and studies of the molecules

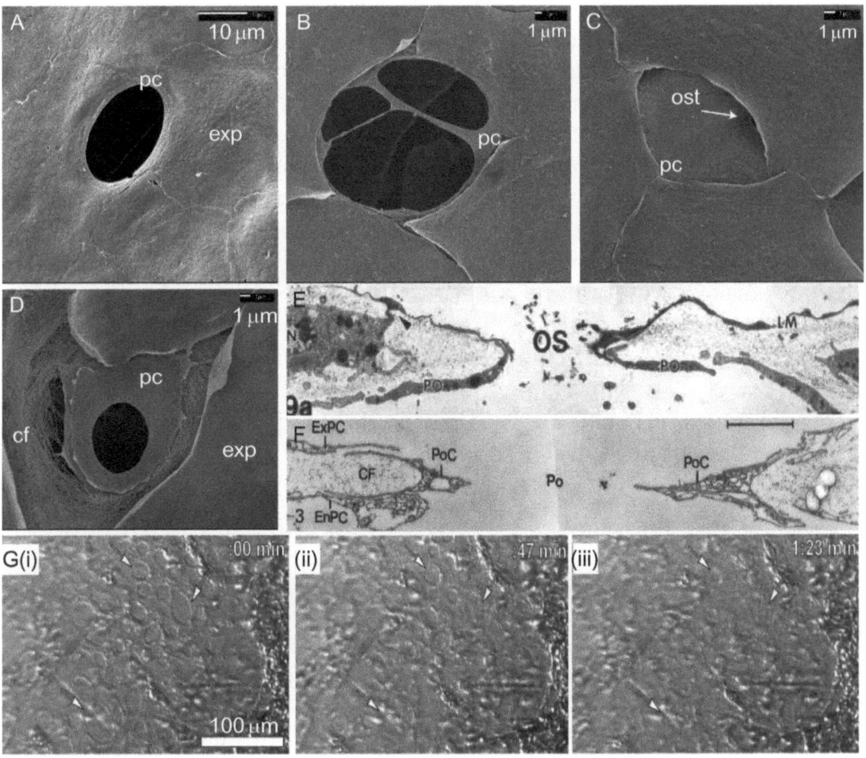

Figure 1.8 Dynamics of opening and closing of ostia. (A–D) Porocytes in *Ephydatia muelleri* showing A, B, open and C, closed and largely hidden under the exopinacocytes. (D) Contraction of exopinacocytes during fixation reveals the collagen layer on which pinacocytes lie and to which pinacocytes attach (pc, porocyte; exp, exopinacocyte; ost, postium; cf, collagen fibers). (E, F) Cross-sections through porocytes show a similar structure in E, *Dysidea etheria* and F, *Haliclona elegans*. (G, i–iii) video images of the ostia closing in the dermal membrane of *Ephydatia muelleri*. (A–D, Leys, unpublished; E, Teragawa, 1986; F, Langenbruch and Scalera-Liaci, 1986; G, Elliott and Leys, 2007).

present in sponges shows that despite their apparent simplicity, established epithelia in sponges can and do form effective barriers (Adams *et al.*, 2010).

4.2. Pinacocytes—Terminology

The sponge surface is formed by a flat cell type called a pinacocyte (literally 'pavement' cell and pinacoderm, a pavement epithelium). A good description of the pinacoderm is given by Simpson (1984) and De Vos *et al.* (1991). Pinacocytes are thin (often 1 μm thick except at the nucleus) broad cells (up to 20 μm in diameter) that form a squamous sheet. In many sponges, pinacocytes are T-shaped, with the nucleus in a cell body that projects down into the collagenous middle layer. Pinacoderms line the sponge surface and canals and also line the internal cavity and canals of the larvae of freshwater sponges.

At least six types of pinacocytes can be identified by morphological or positional differences. Pinacocytes forming the outer surface of the sponge are exopinacocytes, those on internal surfaces (on canals or lacunae) are endopinacocytes, and those that form the base of the sponge and provide the attachment to a surface and which can secrete a massive calcareous skeleton (e.g. Gilis, 2011) are distinguished as basopinacocytes. Furthermore, the polarity of the aquiferous system defines the prefixes given to pinacocytes before (pro) and after (ap) the feeding chambers, hence prosopinacocytes and apopinacocytes (Boury-Esnault and Rützler, 1997). As more gene expression data become available, with genes labelling particular regions of a sponge, it is increasingly important to use terminology that accurately defines location. Differences might be expected, for example, in the populations of potassium channels or glutamate receptors on incurrent and excurrent canal epithelia.

4.3. Cilia and flagella—Function and location in the sponge

An interesting and little discussed feature of sponge pinacocytes is that they can be ciliated (Boury-Esnault and Rützler, 1997); Simpson refers to the cilia as flagella (Simpson, 1984, p. 52). There is little (if any) difference in the ultrastructure of cilia and flagella and often both organelles are referred to by a single term (usually cilium) (Lindemann and Lesich, 2010). However, cilia and flagella have fundamentally different patterns of movement—cilia beat in a whip-like manner with active and recovery strokes (or are non-motile, sensory, organelles), whereas flagella beat in sinusoidal waves, and this has been traced to differences in the domains of proteins involved in forming the axonome (Linck, 1973). Sponges are one of the few animals that have somatic cells (choanocytes) with flagella; flagella are also found in the flame cells of Platyhelminthes, but it seems that most other metazoans have cilia rather than flagella on somatic cells. Choanocytes can differentiate into sperm, which also possess flagella. In contrast, the sponge larval epithelium is formed by ciliated cells, which beat with a whip-like motion in metachronal waves, like the ciliated epithelia of other invertebrate larvae; at metamorphosis of the larva, however, these ciliated cells dedifferentiate, resorb or lose the cilium, and migrate inwards, where they re-differentiate into choanocytes with a flagellum and a collar (Amano and Hori, 1996; Leys and Degnan, 2002). Not all sponge larvae retain their ciliated cells for use as choanocytes. In *E. fluviatilis*, the ciliated epithelial cells are engulfed by amoeboid cells which then spread out to form the new pinacocytes (Wielsputz and Saller, 1990), but since choanocytes are already formed in freshwater sponge larvae presumably, the ciliated epithelial cells are no longer necessary to make new choanocytes at metamorphosis. It is interesting that only freshwater sponges and hexactinellids have choanocyte chambers in the larva; in both cases, these are ready to be used once the osculum has formed after metamorphosis. The two types of sponges share little in common except perhaps living in low nutrient habitats. If nutrients are limited, perhaps it is more advantageous to be equipped to feed immediately upon metamorphosis rather than use up energy to reshuffle epithelial cells and organelles into new locations.

As the sponge forms new canals after metamorphosis, ciliated pinacocytes arise on newly formed canal epithelia (SPL personal observation). Therefore, pinacocytes must be able to differentiate cilia, but whether all pinacocytes can do this, or whether certain archaeocytes are committed to the ciliated lineage and differentiate into ciliated pinacocytes is not yet known. In demosponges, endopinacocytes in the canals may have one cilium or a pair of cilia, for example, in *Hippospongia communis* and *Tethya aurantium* (Simpson, 1984, p. 52; Vacelet *et al.*, 1989; Boury-Esnault *et al.*, 1990, figs. 23 and 24) or *Ephydatia muelleri* (Elliott, 2009), but all endopinacocytes that form the inner lining of the osculum are ciliated (in all sponges observed so far) with one to four cilia per pinacocyte (Ludeman, 2010; Leys, personal observation). Cilia also occur on the apopylar cells which form a 'gasket'-like exit to chambers (see Section 3.3). Many authors refer to these as flagella, but in *E. muelleri*, where chambers and canals can be observed live in thin areas around the periphery of the animals, while the flagella of choanocytes beat in sinusoidal waves, the organelles on the apopylar cells beat like cilia in a whip-like manner (SPL personal observation). Furthermore, using the fluorescent vital dye FM 1-43, which is taken up by channels in the cilium membrane, and using phase contrast microscopy to watch their motion cilia in the canals and osculum can be located and watched, and these do not 'beat' at all but simply vibrate as if in the flow of water (SPL and D. Ludemann, unpublished observations).

In homoscleromorph sponges, all endopinacocytes are ciliated, each with one cilium, and in most species so are exopinacocytes (Boury-Esnault *et al.*, 1984; Maldonado and Riesgo, 2007). But no study has examined whether these are motile or non-motile, nor if motile, how they beat. In Calcarea, no ciliated pinacocytes have ever been reported, and no cilia have been found on any part of the syncytial trabecular reticula or dermal membranes of hexactinellids.

4.4. Pinacoderm: Role in sealing and osmoregulation

In his studies of contraction mechanisms in sponges, Prosser (1967) found that marine sponges have an ionic gradient across the tissues, inward sodium, and outward potassium, such that he anticipated the membrane potential would be -30 to -65 mV. Freshwater sponge pinacocytes have a low osmolarity (12–15 MOsM) and possess contractile vacuoles which actively excrete water to the outside as external solute concentration falls (Brauer, 1975). These data together suggest that for both marine and freshwater sponges, pinacocytes differ in ion content from the external environment (salt or freshwater) and the pinacoderm controls the ion content of the mesohyl.

Sponge pinacoderms are thin and elastic and are coated with a proteoglycan matrix which has so far prevented intracellular recordings of true membrane potential. But it is still possible to determine the ability of the pinacoderm to seal by studying the transport of ions across the whole epithelium in culture. Adams and colleagues cultured sponge tissues in small cups with a permeable membrane bottom and recorded the ionic resistance across the sponge epithelium (Adams *et al.*, 2010). Using miniature cell cultures, they simultaneously recorded current in the medium above and below the tissue. Cultures that formed a confluent (continuous) sheet covering all of the membrane in the cup had a transepithelial resistance similar to that of epithelia in other metazoans ($>2\ \mathrm{M}\Omega$). Furthermore, for a culture with high resistance, there was no passage of a small molecular weight tracer across the tissue; if the tissue in the cup was torn, then the tracer was able to pass across the tissues (Adams *et al.*, 2010). The authors considered the resistance was generated across the basopinacoderm (the tissue that was directly in contact with the cup membrane), but the endopinacoderm would also contribute to the resistance to passage of ions across the fine mesohyl and between pinacocytes. Therefore, it can be concluded that the pinacoderm (baso- and endopinacoderm at least, and likely also the exopinacoderm) provides good resistance to the passage of ions and controls the passage of small molecules across it.

In the context of being able to control cell–cell signalling via the release of small molecules, a sealed and stable pinacoderm makes sense. Images of the exopinacoderm labelled with a vital fluorescent dye (FM 1–43) also show that contacts between neighbouring pinacocytes remain stable for at least 1 h (Leys *et al.*, 2009; Adams, 2010). So why is it thought that sponge epithelia are transient? Statements about the looseness of sponge pinacocytes appear to stem from observations of highly mobile cells filmed at the periphery of explants of sponge tissue (e.g. Bond, 1992), and from sections and scanning electron micrographs of fixed tissues that show cells not adherent to one another at the choanocyte chambers (e.g. fig. 6 in Simpson, 1984), and the supposed entry of bacteria through endopinacocytes into the mesohyl (Weissenfels, 1976). Whereas the motility of cells in explants is to be expected because these cells are actively seeking new unexploited areas to establish tissues in, the direct openings to the mesohyl from incurrent canals are not (Weissenfels, 1976, 1983). The images showing openings to those areas were subsequently acknowledged to have been caused by damage (Langenbruch and Weissenfels, 1987). Tissues that are poorly preserved often reveal openings to the mesohyl, but careful preservation and handling shows the cells in contact. The fact that sponge epithelia require such careful handling in order to preserve well, however, may be taken as some indication of the tenuous adhesion of sponge junctional molecules.

4.5. Cell adhesion and cell junctions

Sponges are almost best known for their ability to be disassociated—usually by being forced through fine mesh—and for the individual cells to re-associate in a species-specific manner (Wilson, 1907). Carbohydrate molecules, and specifically large sulphated proteoglycans originally known as 'aggregation factors', carry out species-specific adhesion (Fernandez-Busquets and Burger, 2003; Carvalho de Souza *et al.*, 2009; Vilanova *et al.*, 2009). The adhesion occurs at moieties on glycan fingers that extend from massive proteoglycan molecules attached loosely to the cell surface. Both the attachment to the cell and the species-specific recognition with other cells are calcium dependent. Carbohydrates are unusual cell-adhesion molecules, not commonly found in other phyla, although perhaps this mechanism of adhesion has not been adequately explored in other metazoans (Fernandez-Busquets and Burger, 2003), but they have a remarkable capacity for adaptation, and are able after one interaction to modify the types of ligands so as to recognize second encounters, that is they are polymorphic (Fernandez-Busquets and Burger, 1999). This type of adhesion has been considered a potential 'first' adhesion between early unicellular animals (Fernandez-Busquets and Burger, 1999; Vilanova *et al.*, 2009).

4.5.1. Molecules in sponge cell junctions—Overview

From the genome of the demosponge *A. queenslandica*, we know that sponges have molecules that are involved in cell polarity, adhesion and sealing, and in attachment to and communication with the extracellular matrix (Fahey and Degnan, 2010). Transcriptome data from several other sponges are also beginning to provide a rich picture of potential epithelial genes in sponges. Therefore, in addition to the carbohydrate type adhesion described above which is involved in species-specific aggregation, sponges produce membrane-associated proteins that are used in cell–cell adhesion. Junctions in sponges differ from those in other invertebrates and vertebrates, but the details of position of the junction, molecules, and proteins involved in each junction type are only just being examined (e.g. Leys and Riesgo, 2011).

4.5.2. Adherens junctions

Adherens junctions hold cells together; they usually occur on one side of a sealing junction and their presence around the apex of the cell is considered an indication of tissue stability. While many authors refer to adhering junctions in sponges as desmosomes, strictly speaking desmosomes (macula adherentes) have desmosomal cadherins (desmogleins, desmocollins, and plakophilins) and plakoglobins, which are innovations of vertebrates. Invertebrates usually have zonula adherens, a belt-form junction, but in sponges, adherens junctions are usually spot-form junctions, appearing as dense plaques of actin between cells (Bagby, 1965; Pavans de Ceccatty, 1981; Boury-Esnault *et al.*, 2003; Elliott and Leys, 2007). Focal adhesions (spot desmosomes) have also been found on the basal epithelial layer of *E. muelleri* (Pavans de Ceccatty, 1986) and perhaps are the only real desmosomal adhesive junction sponges have. However, since none of the proteins that make up any of these adherens junctions have been studied, the true nature of sponge adherens junctions remains to be explored.

The principal component of adherens junctions is *cadherin*, a transmembrane molecule with extracellular cadherin (EC) repeats, EGF, and LAMG repeats (altogether called Fat cadherins). Sponges have 17 cadherins (Fahey and Degnan, 2010), and one of these (first reported by Abedin and King, 2008) is a classical Fat cadherin, which is usually associated with adherens junctions. In choanoflagellates, *cadherins* have binding sites for tyrosine phosphorylation and are thought to be involved in cell signalling (Abedin and King, 2008). In metazoans (including sponges), *cadherins* also have a conserved cytoplasmic domain, which has a binding site for *β-catenin* that is involved in signalling via the Wnt pathway. Therefore, adhesion and cell–cell signalling likely went hand-in-hand in the earliest metazoans. One *cadherin* in *A. queenslandica* has a *hedgehog*-like domain (Adamska et al., 2007b), and it is suggested that its expression around the pigmented posterior pole of the sponge larva reflects a role in signalling during the development of polarity in that sponge (Adamska et al., 2007b, 2010) (see Section 7).

Other adhesion molecules present in the sponge genome belong to the MAGUK family of transmembrane proteins (Sakaraya *et al.*, 2007; de Mendoza *et al.*, 2010; Fahey and Degnan, 2010). There are 10 types of MAGUKs (de Mendoza *et al.*, 2010), some of which generate the scaffold that anchors post-synaptic molecules in their place in nerve cells and others that generate tight junctions in vertebrates. Sponges have neither nerves nor tight junctions, but given the presence of members of the DLG, MPP, and MAGI super families of MAGUK proteins in sponges, we might expect these to be involved in cell–cell adhesion and/or signalling at cell junctions.

In situ hybridization of MAGI RNA in 'primorphs' (small sponges grown from dissociated and reaggregated cells) suggested expression in both the dermal membrane (exopinacoderm) and the choanoderm of sponge epithelia (Adell *et al.*, 2004). Similar hybridization experiments with DLG and several of the post-synaptic scaffolding genes in the *A. queenslandica* larva suggest a varied expression with some label in the inner cell mass, some in the sub-epithelial layer, and some specifically in a type of globular cell (also called a mucous cell) that populates the larval epithelium (Sakaraya *et al.*, 2007).

4.5.3. Occluding junctions

Occluding junctions in sponge epithelia are both 'parallel membrane' junctions (Green and Bergquist, 1979, 1982; Pavans de Ceccatty, 1981; Adams, 2010) and septate junctions (tight junctions are considered a chordate innovation). The former, in which the plasma membranes of two cells lie parallel to one another for about 500 nm with only 10–20 nm separation, occlude the passage of small molecular weight tracers (^{3}H-inulin, 5KD, and ruthenium red, 0.85KD) (Adams *et al.*, 2010). In these junctions, faint septae can sometimes be seen in transmission electron micrographs, but generally the occluding region occurs either at the parallel junctions or at regions adjacent to them in which the membrane of one cell appears merged (fused) or in tight contact with the other cell. Septate junctions have been reported in all sponges (see review by Leys *et al.*, 2009), but only in calcareous sponges are the septae very clear with 10 nm spacing between cells and between septae (Ledger, 1975; Green and Bergquist, 1979).

Septate junctions have been best characterized in *Drosophila* where two types are known. Pleated sheet SJs (pSJ) are found in the epidermis, salivary gland, and photoreceptors, while smooth SJ (sSJ) occur in the mid-gut and malpighian tubules. The ultrastructure of each shows that pSJ have distinct septae and sSJ lack them. Both junctions have the molecules neuroglian and contactin and both molecules are needed for barrier function, but only pSJ have *neurexin IV* (a contactin-associated protein, or Caspr) (Baumgartner *et al.*, 1996; Genova and Fehon, 2003; Faivre-Sarrailh *et al.*, 2004), so this molecule seems to be associated with the presence of septae but is not necessary for barrier function (Baumgartner *et al.*, 1996; Baumann, 2001). sSJ are instead characterized by *fasciclin III*, *ankyrin*, and *α,β-spectrin*.

Although the composition of the sponge septate and parallel membrane junctions is not yet known, the *A. queenslandica* genome does not seem to have clear orthologs of *neuroglian*, *neurexin IV*, *fibronectin III*, or *contactin* (Fahey and Degnan, 2010). *Neurexin IV*, however, was reported in *Oscarella carmela*, a homoscleromorph (Nichols *et al.*, 2006), and *A. queenslandica* does have a *claudin* gene ortholog which aligns well with sequences from two other sponges as well as *Drosophila*, nematode, and human sequences (Leys and Riesgo, 2011). *Claudins* are known from tight junctions, but in the fly, these proteins are localized to septate junctions and have been shown to play a role in occlusion (Asano *et al.*, 2003). *Claudins* are also present in the genome of *Hydra* (Chapman *et al.*, 2010) and recent work using GFP-claudins shows labelling around the apex of each cell (Salvenmoser, 2011). It is, therefore, quite possible that the demosponge *claudin* molecules may also function in occlusion, and that in sponge parallel and/or septate junctions, perhaps *claudins* rather than *neurexins* perform the sealing role (Leys and Riesgo, 2011). To better understand how the sponge junction works, antibody localization and gene knockdown is needed, as well as further more detailed exploration of other sponge genomes as they become sequenced.

Other molecules found in the *A. queenslandica* genome and which may be involved in establishing polarity of the epithelium, or in localizing junctions to their polarized position in the epithelium, include *Scrb*, *Par*, *Crumbs*, and *Patj* (Fahey and Degnan, 2010). In other invertebrates, *Scrb* and *Par* are localized to septate junctions. Given the very narrow region which sponge epithelial cells make contact with one-another, it is conceivable that these molecules are also present at the sponge junctions, but testing this will require antibody localization using immunogold to confirm the position at the ultrastructural level.

4.6. The basement membrane: Differences among sponge epithelia

The integrity of the epithelium is achieved not only by adhesion between cells but also by the support of the extracellular matrix below. Depending on the location in the sponge body, epithelia have some sort of basal support provided by collagen, and most also have an upper cuticle. There are typically two components to the basement membrane—the lamina reticularis (a sheet of collagens I, II, and V), and the basal lamina (a 50–100 nm layer of type IV collagen together with laminins and proteoglycans such as nidogen/entactin and perlecan) that lies immediately below, and follows every contour of, the cell membrane. Electron micrographs frequently show a dense mat of collagen underlying the sponge epithelium, which may be the equivalent of a lamina reticularis. In electron micrographs of homoscleromorph sponges, there is also a fine basal lamina below both the pinacoderm and choanoderm and antibodies to type IV collagen raised from that sponge recognize a layer that is beneath the epithelium (Boute *et al.*, 1996). Recently, type IV collagen that corresponds well to alpha and beta chains has also been found in the calcareous sponge *Sycon coactum* and another homoscleromorph, *Corticium candelabrum* (Leys and Riesgo, 2011). The finding of type IV collagen in *Sycon* is all the more unusual since electron micrographs do not show a basement membrane structure in any calcareous sponge.

The *A. queenslandica* genome does not have either type IV collagen or the proteoglycans nidogen or perlecan, and though there are several laminin-related genes, only one (*Lam* γ-like) has a domain architecture sufficiently comparable to laminins from bilaterians to be considered proper laminins (Fahey and Degnan, 2010, 2012). Type IV collagen is necessary for basal lamina formation since its globular heads link to form a mesh that is attached to the cell surface by laminins. Although no other demosponge genomes are available, searches of other sponge transcriptomes and use of degenerate PCR has not found conventional type IV collagen in other demosponges. However, it has been shown that demosponges have a related form of this molecule called 'spongin short chain collagen' (SSCC), which shares very similar NC1a and NC1b domains with type IV collagen and possesses four of the six conserved cysteine molecules typical of type IV collagen (Aouacheria *et al.*, 2006). SSCC is also predicted to have the same three-dimensional structure and therefore to be capable of forming the same interactions with the epithelium as type IV collagen (Aouacheria *et al.*, 2006). SSCC is expressed by cells of the basal pinacoderm (Exposito and Garrone, 1990; Exposito *et al.*, 1991). Further support for the idea that the three-dimensional structure of SSCC may be similar to type IV collagen comes from a study by De Goeij *et al.* (2009) who showed that polyclonal antibodies to type IV collagen from rabbit also recognize a layer below the choanoderm in the demosponge *H. caerulea.* It is possible that these antibodies recognize a conserved 3D structure in the protein despite what would be considered to be a little conserved amino acid sequence.

4.7. Pinacoderm function: Biomineralization

Biomineralization in sponges—the production of the skeleton—is the subject of several excellent reviews (Uriz *et al.*, 2003; Uriz, 2006; Sethmann and Wörheide, 2008; Wang *et al.*, 2010) and is also covered by Wang and coauthors in this issue. Most secretion of mineral and organic skeletons in sponges takes place in the mesohyl, but here, we alert the reader to two instances in which the pinacoderm functions in biomineralization. The first is the observation that some spicules are formed by pinacocytes in the homoscleromorph sponge *Corticium candelabrum* (Maldonado and Riesgo, 2007). This observation may not be so unusual if pinacocytes in homoscleromorphs give rise to sclerocytes as they are said to do in Calcarea (Simpson, 1984, p. 147), and also to microsclerocytes as they are said to do in some demosponges (Simpson, 1984, p. 178). Maldonado and Riesgo (2007), however, do not think this is the case in *C. candelabrum* and therefore this points to an unusual secretory ability of the pinacoderm in that homoscleromorph sponge. The epithelium also plays a role in biomineralization in hypercalcified sponges, both demosponges (e.g. *Acanthochaetetes wellsi*, *Astrosclera willeyana*) and Calcarea (e.g. *Petrobiona massiliana*). The precise mechanism of mineral uptake, transport and secretion is not yet fully understood, but histology, histochemistry, and many types of energy X-ray analyses of the basopinacoderm from both demosponges and calcareous sponges suggest that minerals (magnesium and/or calcium) are sequestered into vesicles in basopinacocytes and secreted as an amorphous material on the basal side of the epithelium (Reitner and Gautret, 1996; Wörheide, 1998; Gilis, 2011; Gilis *et al.*, 2011). Polymerization (crystallization) of the material below the basopinacoderm is promoted by the prior or concurrent secretion of an organic matrix, which is shown by staining with alcian blue, ruthenium red and acridine orange, to contain highly sulphated or acidic polysaccharides (e.g. glycosaminoglycans) (Gilis, 2011). Together, these materials allow for a biologically controlled crystallization of the minerals moulding them into a basal skeleton and incorporating into it spicules also produced by the sponge. In *Astrosclera,* biomineralization has also been shown to involve a type of mesohyl cell, a large vacuolar cell (LVC) (Wörheide, 1998), which is found in the choanosome and ectosome above the calcified skeleton. Demonstration that these cells express the enzyme carbonic anhydrase required for catalyzing HCO_3^- (Jackson *et al.*, 2007) suggests that LVCs are involved in uptake of calcium and its transport to basopinacocytes where secretion of the matrix occurs.

4.8. Pinacoderm development

The origin of pinacocytes is, after more than a century of study, still unclear. Earlier research using a variety of models, such as dissociated tissue forming reaggregates (Wilson and Penney, 1930), dissociated larvae forming juvenile sponges (Borojevic and Lévi, 1964, 1965), and explants of adult sponges (Simpson, 1963; Bagby, 1972), all from different species, concluded that archaeocytes differentiate into collencytes which become pinacocytes. Galtsoff (1925) who was the first to use aggregates for this, thought that pinacocytes remained differentiated upon dissociation of a sponge, and simply 'sorted' out to reform the pinacoderm. Wilson and Penney (1930) concluded that a type of cell called a 'grey cell' with grey granular inclusions, formed the pinacocytes. Bagby (1972) determined that since there were no grey cells in *Microciona* archaeocytes must become the pinacocytes. The only 'live cell' study of this problem was by Simpson (1963), who showed that in explants pinacocytes did not divide to form the pinacoderm, but rather they arose from mesohyl cells which differentiated and moved into the pinacoderm. Study of the metamorphosing larva of *E. fluviatilis* supports the view that archaeocytes migrate up into and form the pinacoderm of the juvenile sponge (Boury-Esnault, 1976; Wielsputz and Saller, 1990). So in sum, new pinacocytes arise by differentiation of archaeocytes from the mesohyl.

Curiously, a recent study of sponges differentiated from gemmules of *Suberites domuncula* suggests the opposite that the expression of genes in the pinacoderm triggers the differentiation of pinacocytes into sclerocytes and mesohyl cells (Schroder *et al.*, 2004). After stimulation of canal formation using increased temperature and Fe^{3+}, pinacocytes were the first cells to express *noggin* and *silicatein*, and subsequently, these genes were expressed in cells in the mesohyl (Schroder *et al.*, 2004). The authors concluded that a subpopulation of pinacocytes is able to differentiate into other cell types; but it is also possible that genes expressed by pinacocytes could induce the differentiation of cells in the mesohyl into other cell types. Whether cells from the pinacoderm can differentiate into other cells might be studied by using a fluorescent cell-tracker such as CMFDA which incorporates only into cells it has direct contact with; therefore, using a bath application mesohyl cells would not be labelled. So far, two studies have used labelling of new nuclei using 5-bromo-2′-deoxyruridine (BrDU) to identify stem cells, one in adult specimens of the crevice-dwelling marine sponge *Halisarca* (De Goeij *et al.*, 2009) and the other in the freshwater sponge *E. fluviatilis* differentiating from gemmules (Funayama *et al.*, 2010). Although both groups found incorporation of the label into choanocytes and archaeocytes within the 10-h time frame of exposure to the drug, neither found the label in pinacocytes.

In sum, pinacocytes do not appear to divide, although this does not mean they are terminally differentiated. It is possible that some pinacocytes can differentiate into other cell types, but it is perhaps more likely that they trigger/stimulate the differentiation of mesohyl cells into other cell types. The latter role is important in the formation of the canals and differentiation of the sponge body plan. Pinacocytes are, if not the first, certainly among the first cells to differentiate in the early development of a sponge from a larva or from a gemmule and during aggregation after dissociation of the adult. Pinacocytes line the incipient canal epithelia and are responsible for controlling ion exchange across the epithelium into the mesohyl.

5. The Aquiferous System

5.1. Differentiation of porocytes and canals

Ostia—incurrent openings into the canal system—are formed by porocytes, either singly, by having a large ‘hole’ in the middle of the cell, or possibly by several meeting up around an ostium to form the pore (Simpson, 1984). Evidence for the latter, however, comes from electron micrographs that show several pinacocytes converging on an ostium in a freshwater sponge exopinacoderm. Given that the porocyte lies below the exopinacocytes and between the two pinacoderm epithelia, when the pore is open, the cell body may actually be completely hidden by the two pinacoderms (Fig. 1.8).

Porocytes have been seen to form in as little as 10 min in *E. fluviatilis* (Weissenfels, 1980) and within 10 h in *Leucosolenia variabilis* (Jones, 1961). The former observation might be explained by a porocyte having been hidden by pinacocytes and opening to reveal itself since videos of opening and closing ostia show this sort of time frame (Elliott and Leys, 2007)

(Fig. 1.8), but the latter suggests that—if cell division takes 10–20 h—they probably arise from mesohyl cells, rather than dividing from other pinacocytes. Oscula are also thought to arise initially from a single porocyte (Weissenfels, 1980), but how they coordinate with other porocytes to form a larger osculum is still unclear. In freshwater sponges hatched from gemmules, the region in which the osculum will differentiate appears first as a bulge, which according to Brien (1932) is caused by pressure from water pushed by the beat of newly formed chambers; but one wonders how that pressure is dissipated if there is still no vent or opening that is the osculum. Nevertheless, the movement of fluid over the tissues to generate signals for differentiation of tissues is a common theme in animal development, even in, for example, the nodal cilia.

Much of our understanding of canal formation comes from the observations by Wintermann (1951) and Mergner (1964; 1966) and later by Weissenfels (1981), all of whom found that the excurrent canal system arises first, chambers plug into the canals, and the incurrent canal system joins up last (and is most plastic). The canals themselves arise from lacunae, small epithelia-lined cavities within the sponge mesohyl, which move around, often in a clockwise centripetal manner, until they merge with others to form canals (Leys, personal observation). Hans Mergner first experimented with the role of the osculum in canal formation and found that if he removed the osculum (from a small freshwater sponge hatched from gemmules), it remained differentiated and formed a small ball of exo and endopinacoderm, which he called the 'oscula ball'. He found that if he inserted this oscula ball anywhere in the sponge canal system it caused a new osculum to form in that spot (Mergner, 1964), and surmised that a substance ('Induktionstoffes') from the osculum tissue caused the differentiation of canal epithelia and formation of an osculum.

Recent experiments support the idea that osculum sends out a factor controlling canal formation. Windsor and Leys (2010) used Mergner's (1964) experimental approach to remove oscula from the freshwater sponge *E. muelleri* and place them on the surface of a different sponge, one hatched from the same batch of gemmules. They found that the transplanted osculum not only attached to the surface of the sponge but also linked up with the excurrent canal system and in fact drew the entire aquiferous system towards it, causing the original osculum to cease to function (Windsor and Leys, 2010). This seemed to confirm Mergner's (1966) idea that a substance secreted by the oscula tissues induces canal formation, competing with gradients secreted by the original host osculum. An alternative hypothesis, however, is that the secreted factor stimulates regeneration, and regeneration triggers the release of factors that result in oscula formation. Windsor and Leys (2010) proposed that the secreted factor is Wnt protein because soaking the developing sponge in Wnt activators caused a phenotype of multiple oscula and disruption of the aquiferous system.

5.2. Role of *Wnt* in canal differentiation and polarity in sponges

To date, *Wnt* pathway genes have been isolated from at least 10 different sponges—*Amphimedon queenslandica* (Adamska *et al.*, 2010; Srivastava *et al.*, 2010), *Suberites domuncula* and *Lubomirskia baicalensis* (Adell *et al.*, 2007), *Ephydatia muelleri* (Windsor and Leys, 2010), *Oscarella lobularis* (Lapébie *et al.*, 2009), and *Sycon ciliatum* (Adamska, unpublished); *S. coactum, Pseudocorticium jarrei, Spongilla lacustris* and *Petrosia ficiformis* (Ana Riesgo, Pamela Windsor and Sally Leys, unpublished). Expression of *Wnt* genes has been studied in larvae, young adults and adult sponge tissue, but not in the different developmental stages (larvae, postlarva and adult) of any one species. Our understanding the role of *Wnt* in sponge development and body plan formation is therefore still patchy.

The most complete studies to date of *Wnt* gene and ligand expression have been carried out in the *A. queenslandica* sponge larva (Adamska *et al.*, 2010) where it has been shown that one of the three *Wnt* genes (*WntA*) has a polarized expression at the posterior pole of the larva together with pigment cells that form a ring around the posterior pole. *WntA* expression is juxtaposed with expression of *TGFβ*, which labels the entire anterior portion of the larval ciliated epithelium (Adamska *et al.*, 2010). Some *Wnt* ligands also have a polarized expression (e.g. LRP) while the expression of others is localized more to the outer, middle or inner layers of the larva (Adamska *et al.*, 2010). Since LiCl experiments have so far not been successful in *A. queenslandica*, it is unknown what function Wnt has except for a probable role in guiding anterior and posterior axis formation in the larva.

In the cnidarians *Hydra* and *Nematostella*, artificial activation of the *Wnt* pathway causes the formation of ectopic tentacles and multiplication of the 'head' end (Broun *et al.*, 2005), and gradients of *Wnt* gene expression mark the anterior to posterior poles of the polyp (Kusserow *et al.*, 2005). To determine whether the sponge canal system might develop in a polarized manner by gradients of *Wnt* gene expression, Windsor and Leys (2010) hatched sponges in a growth medium with drugs that activate the *Wnt* pathway (lithium chloride and alsterpaullone) and found that with both drugs the sponges had many more oscula than normal sponges, and that the aquiferous system was completely malformed and non-functional. Furthermore, in the treated sponges, there were either fewer ostia and porocytes, or they were completely absent. A similar approach was used in *Oscarella lobularis* by Lapebie *et al.* (2009). Incubation in azakenpaullone caused the sponge to form many more ostia; oscula were not studied. These results suggest that secreted Wnt protein may be Mergner's 'induktionstoffe' (Mergner 1966) which polarizes the epithelia during development to determine where ostia and incurrent canals or oscula and excurrent canals form. In *E. fluviatilis*, there is also evidence that the *Wnt* pathway may be involved independently in radial organization of the skeleton because sponges grown in BIO or azakenpaullone have a disorganized skeleton (Nakata and Funayama, 2011). If production of the skeleton were necessary for proper organization of aquiferous system, these results would make sense together, but since it is also known that normal sponges form in the absence of materials to construct spicules (Weissenfels and Langenbruch, 1985), the simplest explanation is that the *Wnt* pathway has multiple roles in sponge development.

Other genes that may have a role in canal formation include the homeobox genes *Iroquois* and *NK* class. Both show expression around canals or on the outer epithelium (Perović *et al.*, 2003; Gazave *et al.*, 2008, see Section 7), but experimental approaches will be needed to determine what role these genes might have in actual canal formation.

6. Epithelia as Sensory and Contractile Tissues

6.1. Overview of sensory and coordinating tissues

Sponges lack nerves in the conventional sense; that is, they do not possess a defined cell type that transmits signals using electrical impulses that travel along the cell membrane and communicate to other cells via a chemical synapse. This does not mean, however, that sponges do not send either electrical or chemical signals, but rather that the speed at which the signals propagate through the sponge is slower than conventional electrical signalling. Furthermore, the means by which signals propagate is not necessarily by electrical impulses, as generally is the case in other animals. Glass sponges propagate electrical signals that shut down their feeding current (Leys and Mackie, 1997; Leys *et al.*, 1999; reviewed in Leys and Meech, 2006; Leys *et al.*, 2007), but so far, there is no evidence of electrical signalling in cellular sponges. However, at least one species (*Hymedesmia*) is capable of dropping a tissue flap over its ostial pore fields in under 1 s (Reiswig, 1979), which indicates that localized, rapid (perhaps electrical) signalling may occur in some species. Signals that are not electrical impulses would presumably be waves of calcium (propagated, localized increases in calcium concentration that spread through the cell), triggered by stimuli to the cell membrane, either via receptors on the cell surface or via channels in the cell membrane. Signals could propagate through cells of the epithelium or cells of the mesohyl, but in recent years, the epithelium has become recognized as the primary conduit of coordination in sponges.

In the 1970s, it was already known that signal transduction largely occurred through the epithelia of all sponges, although thick walled sponges were thought to have single contractile cells or bundles of cells in the mesohyl which were considered to play a role in contraction (Pavans de Ceccatty, 1974a,b). Viewed live by cine-film or time-lapse digital microscopy, amoebocytes in the mesohyl of thin-walled sponges are very active, some of the cells move in a highly directional manner, others travel almost in circles (Pavans de Ceccatty, 1969, 1976; Adams, 2010). Collencytes leave a trail of collagen in their wake, sclerocytes sit stationary and take minutes to hours to develop a single spicule, lophocytes cluster around spicules, cells with inclusions appear stationary, archaeocytes are sluggish and tend to stay near choanocyte chambers.

It is not possible to film cell activity in thick walled sponges, and certainly mesohyl cells may not be equally active in all sponges, or perhaps the activity of cells in the mesohyl depends on the thickness or density of collagen. Ultrastructural studies in the 1960s and 1970s showed cells stretching across the mesohyl and in contact with epithelia at either end, interactions that are suggestive of cell–cell communication (Pavans de Ceccatty *et al.*, 1970; Pavans de Ceccatty, 1974a,b, 1986). The presence of actin microfilaments in these cells was taken to indicate they might be contractile; therefore, they were termed myocytes (Bagby, 1965, 1970). The general view was that where sponge tissues were thin, it was more likely that pinacocytes were contractile, while in thick tissues, it was more likely that mesohyl cells (myocytes) were contractile. In very thick walled sponges, sphincters around canals contract more rapidly than the whole sponge can and reduce the diameter of canals; the contractile cells in sphincters are pinacocytes. In recent years, pinacocytes have come to be considered the primary contractile cell (Nickel *et al.*, 2006; Nickel, 2010; Nickel *et al.*, 2011), and actinocyte substituted as a more generic term (Boury-Esnault and Rützler, 1997), even though strictly speaking all cells would be expected to have actin. Although it is not possible to view mesohyl cells in thick sponges *in vivo*, estimates from area projections of contracted and relaxed sponges using microCT scanning suggest that even where the mesohyl is substantial, the epithelium is the primary contractile and conductive pathway (Nickel *et al.*, 2011). Nevertheless, even in thin-walled sponges like *Ephydatia* in which periodic contractions have been studied by time-lapse imaging since the mid-20th century (Pavans de Ceccatty *et al.*, 1960; Pavans de Ceccatty, 1969; De Vos and Van de Vyver, 1981; Weissenfels, 1990), it is still unclear whether both pinacoderm and mesohyl cells are involved in contractions, but it is likely that the epithelium is the principal conduit for signal propagation through the sponge.

6.2. Molecules involved in coordination and signal transduction

One of the most central questions in the evolution of animals is *when did nervous tissues first arise*? Because sponges show propagated signals that translate into slow but coordinated behaviour, they appear to bridge the 'gap' between non-neural and neuronal states. Thus, it would seem that genomics and transcriptomics promise to shed much light on the topic of neuron origins by revealing what aspects of ion channels, transmitters and synapses are present in extant species. Molecules identified to date, however, seem to simply support the current idea that sponges have an 'almost nervous system' of the type discussed by Pavans de Ceccatty (1974a,b, 1979). Nonetheless, it is not yet clear if sponges never fully went down the path to conventional nerves, or if they have in fact lost some of the key molecules involved in conventional nervous systems. For example, the *Amphimedon* sponge genome has genes for calcium and potassium channels but no sodium channels (Liebeskind *et al.*, 2011), yet sodium channels seem to be present in choanoflagellates, the unicellular protistan lineage sister to the Metazoa. The absence of sodium channels in sponges may make sense given that physiological experiments on glass sponges (Hexactinellida) suggest they use calcium/potassium impulses (depolarization of the membrane by calcium influx and repolarization by potassium) rather than sodium/potassium as do most other animals (Leys *et al.*, 1999). Calcium impulses are temperature sensitive and have been put forward as one of the reasons that glass sponges are restricted in their range to deep, cooler waters (Leys *et al.*, 2003). The presence of a functional sodium channel in choanoflagellates, however, prompts speculation that sponges may have lost aspects of a conventional nervous system. The *Amphimedon* genome has both inward rectifying (K_{ir}) and putative votage gated K^+ channels (Sakaraya *et al.*, 2007; Alie and Manuel, 2010), and expression of K_{ir} channels in the frog oocyte system has shown that sponge K_{ir} channels are strong rectifiers (Tompkins-MacDonald *et al.*, 2009), which indicates that the membrane potential can be quickly returned to normal should it be depolarized. These data suggest that *Amphimedon* cells might normally depolarize, as if for electrical signalling.

Most research has focused on seeking evidence for synapses or synaptic molecules in sponges. There is now abundant evidence that many molecules known from the post-synaptic scaffold (PSD) are present in *Amphimedon* (Sakaraya *et al.*, 2007; Alie and Manuel, 2010; Srivastava *et al.*, 2010), but no work has convincingly shown the presence of a synapse. *In situ* hybridization of *PSD* genes from *Amphimedon* (*GKAP*, *GRIP* and *HOMER*) is confusing, showing potential expression in a globular cell in the sponge epithelium, but also in many other regions of the sponge larva (Sakaraya *et al.*, 2007). Can these all be post-synaptic signaling sites, even in the supposed receptor cell (the globular cell)? Perhaps the best morphological evidence for a synaptic-like structure is still found in the electron micrographs of Pavans de Ceccatty (1959). Because of the lack of recognizable landmarks in a sponge, gene expression data are very difficult to interpret. What is needed now is a merging of the two techniques by making antibodies to the proteins coded by the genes involved in the post-synaptic density for visualization by immunogold electron microscopy to determine whether sponges in fact possess a discrete 'pseudo-synapse' as often suggested, or whether these proteins are distributed more widely around the sponge cells.

More progress has been made recently by physiological studies of sponge responsiveness. Molecules found to be present in the *Amphimedon* sponge genome do appear to have a specific role in signaling along sponge epithelia. The range of studies that have demonstrated the responsiveness of sponges to different stimuli was recently reviewed by Nickel (2010). Since the early 1900s, chemicals as well as mechanical and electrical stimuli were applied to sponges to determine the speed at which they could respond (Parker, 1910; Emson, 1966; Prosser, 1967; Ellwanger and Nickel, 2006). Rates of contraction were fastest in the osculum (McNair, 1923), but all regions of the sponge have now been shown to contract periodically (Nickel, 2010).

Since the *A. queenslandica* genome contains several GABA receptors, at least seven metabotropic glutamate receptors, serotonin, and a host of other transmitters known from neuronal systems (Sakaraya *et al.*, 2007; Srivastava *et al.*, 2010), it is expected that demosponges are fully capable of responsiveness and signaling. In recent years, it has been shown that contractions can be triggered by bath application of glycine, cAMP and the neurotransmitters adrenaline and serotonin, while nitric oxide appears to modulate contraction intensity (Ellwanger *et al.*, 2004; Ellwanger and Nickel, 2006). Glutamate seems to be particularly important. It is able not only to trigger small contractions, but at the right concentration it also initiates the propagation of a coordinated inflation and contraction (I/C) behaviour, termed a 'sneeze' because clogging by dye particles also triggers this behaviour, which through several such I/C events manages to rid the chamber of the particles (Elliott and Leys, 2007). It has also been shown that GABA acts antagonistically to glutamate to prevent sneezes, which importantly shows that even without conventional nerves, sponges have and use glutaminergic signalling systems (Elliott and Leys, 2010). Antibodies to GABA receptors recognize pinacocytes in *Chondrosia reniformis* (Ramoino *et al.*, 2007), but given that those antibodies were not raised against sponge molecules, the possibility of cross-reactivity with other cell surface molecules cannot be ruled out. Use of antibodies raised to specific sequences of glutamate, GABA and other receptor and signalling molecules from the sponges themselves will help determine whether sponge incurrent and excurrent endopinacoderms have different receptor populations, or whether all pinacoderms present a uniform population of receptors.

6.3. Gene expression as an indicator of sensory epithelia

Two sets of genes have been suggested as indicators of sensory cells and tissues in sponges. An *Antennapedia* (*ANTP*) *NK* class homeobox transcription factor (*HomoNK*) has been isolated from *Oscarella lobularis* and, based on phylogenetic analysis, likened to *NK6* and seven genes which in vertebrates are expressed in neurosensory cells. In *O. lobularis HomoNK* is expressed in the choanocytes of the chambers in both reproductive and non-reproductive adult tissue, and these, therefore, were tentatively suggested to have a sensory function (Gazave *et al.*, 2008). Choanocytes were also suggested as potential mechanosensory cells by Jacobs *et al.* (2007), but functional data to support this idea still needs to be obtained. *A. queenslandica* larvae also express genes of the basic Helix loop Helix (*bHLH*) family (Simionato *et al.*, 2007), and *bHLH* as well as the *Notch* receptor and *Delta* ligand are expressed in globular cells of the outer epithelium (Richards *et al.*, 2008; see also Section 7.4). Phylogenetic analysis shows *A. queenslandica bHLH* is ancestral to the *Atonal/neurogenin* family, and functional studies demonstrate that the sponge gene can substitute for the native gene to trigger neuronal differentiation in *Xenopus* as well as development of sensory bristles in *Drosophila*. Given this finding, the authors speculate that the globular cells themselves might be a type of sensory cell in the sponge larval epithelium (Richards *et al.*, 2008). However, it should perhaps be considered that the globular cells play a role in causing the differentiation of other, neighbouring cells—perhaps the flask cell—into sensory cells in the epithelium. It would be useful to find a marker for sensory cells in sponges in general and to be able to test this hypothesis using knock down by RNAi.

7. Tissue Formation During Sponge Development

7.1. Overview of embryogenesis and larval morphogenesis in sponges

Mechanisms of sexual reproduction and larval development are highly varied in sponges (see Leys and Ereskovsky, 2006). Gametes are formed from archaeocytes or choanocytes, and whereas sperm are released into the water column, in many cases, oocytes are retained and fertilized in the parent sponge developing into a swimming larva that is released. In fewer known instances, both eggs and sperm are released into the water column, and in these cases, fertilization is usually (but not always) external. The different types of development and different types of larvae are reviewed elsewhere (Ereskovsky, 2004, 2005; Leys and Ereskovsky, 2006), but the most studied and probably the best understood modes of development involve brooded larvae. See Ereskovsky (2005) and Leys and Ereskovsky (2006) for a more comprehensive review of sponge development.

Eight types of larvae have been described in sponges (Maldonado and Bergquist, 2002), and these fall into four categories: parenchymella, coeloblastula (hollow blastula), amphiblastula, and trichimella. There are other unusual types of larvae which differ in construction of cell layers, skeleton, polarity, regionalization, for example, the direct developer of *T. serica* (Watanabe, 1978a,b), the solid larva of *Chondrosia* (Lévi and Lévi, 1976), the armoured larva of *Alectona* (Vacelet, 1999), and while these lack specific names, they are each very different in morphology from the other four categories.

Sponges develop through cleavage to form a solid or hollow blastula. Cellular differentiation begins after approximately the 64 cell stage to form micromeres and macromeres which sort, often by cellular migration, to position micromeres on the outside and macromeres on the inside, or by centrifugal movement of all cells to generate a hollow larva. Other mechanisms of sorting include inversion of a hollow blastula in syconoid Calcaronea. The result is a layered and/or polarized larva. For the parenchymella, embryogenesis results in a layered larva in which there is an outer pseudostratified ciliated epithelium (the nuclei lie at different heights in the epithelium), an inner cell mass with amoeboid cells and often also with spicules, and a thin sub-epithelial layer of cells that lie perpendicular to the other layers thereby forming a sort of belt or girdle around the larva. The coeloblastula is a hollow larva that has a single layer of ciliated cells, and within that layer some other cells either at the middle of the larva (e.g. the cinctoblastula of homoscleromorphs) or at the posterior pole (e.g. calcinean calcareous sponge larvae). The amphiblastula is also formed by a single layer of cells, but only the anterior half is ciliated, while the posterior half large has large globular cells. Amphiblastulae larvae also have some maternal cells which migrate in from the parent mesohyl at the time of inversion. The trichimella larva is somewhat unusual in being syncytial. Blastomeres fuse during early development to form the future trabecular reticulum and that tissue envelops cellular regions. The trichimella has precociously formed flagellated chambers and multiciliated cells whose cilia pierce the smooth syncytial outer tissue.

The sorting of cellular material that occurs during sponge embryogenesis has been likened to gastrulation (Leys and Degnan, 2002) although by definition gastrulation involves the formation of the feeding epithelium, and in sponges, this is, therefore, a two-step process, the first involving positioning the cells that will do the feeding (that is those that will generate the feeding current and do much of the particle capture) into a tissue, and the second involving moving those cells to a region in which they can feed; the latter step usually occurs at metamorphosis (Leys, 2004).

7.2. Regulatory genes in development

The 'genetic toolkit' present in sponge genomes that is presumably responsible for the organization of sponge development is largely absent from other non-metazoan genomes including choanoflagellates. This finding supports the notion that ontogenetic development originated at the base of the metazoans and that sponge genetics will play an important role in our understanding of metazoan development in general (Srivastava *et al.*, 2010), and sponge developmental processes in particular. Two major groups of gene families are generally considered when discussing the genetic programming of animal development and include the molecules involved in signal transduction pathways and the host of factors that influence transcription of developmentally important genes.

Generally speaking there are seven major cell–cell signalling pathways that control the developmental programmes of bilaterian animals: *Wnt*, *TGF-β*, *Hedgehog*, *receptor tyrosine kinase*, *nuclear receptor*, *Notch*, and *JAK/STAT* (e.g. reviewed in Barolo and Posakony, 2002). Sponge genomes are known to possess ligands and receptors for the majority of these pathways. The components of arguably the most prominent developmental control pathways, *Wnt*, *TGF-beta* and *Notch-delta* appear to be complete in *A. queenslandica* (Adamska *et al.*, 2011), while important components of the *Hedgehog*, *receptor tyrosine kinase*, *nuclear receptors* and *JAK/STAT* signaling pathways are also found in Porifera (Müller, 2003; Wiens *et al.*, 2003; Larroux *et al.*, 2006; Nichols *et al.*, 2006; Adamska *et al.*, 2007b).

Many transcription factor genes that play important roles in developmental processes in animals are present in sponges. Here, we will confine our discussion to those transcription factors and signal pathway genes for which we have some expression or functional data in tissue formation or function. A more complete description of classes of sponge transcription factors can be found elsewhere (e.g. Larroux *et al.*, 2008; Degnan *et al.*, 2009; Srivastava *et al.*, 2010).

7.3. Gene expression during early embryogenesis

Little is presently known about the earliest molecular factors that set the developmental programme in fertilized eggs leading to proper cleavage and blastula formation in sponges. This is partially due to the difficulty of obtaining eggs, sperm, and/or early embryos in most species. A host of genes are known to be expressed in cleavage and/or blastula stage embryos of *A. queenslandica*, and these include components of the *Wnt* signal transduction pathway: *WntA*, *Frizzled* (*FzdA*), *Secreted Frizzled Related Proteins* (*SFRPA, C, D*), *Lrp5/6*, *GSK3*, *APC*, *Axin*, *Dishevelled*, *β-catenin*, *Tcf* and *Groucho*. In most cases, the expression patterns of these genes appear in subsets of micromeres distributed throughout the embryo (including the follicle envelope) with some genes demonstrating apparently higher expression levels than others. *Secreted Frizzled Related Protein*, *SFRPD*, however, seems to be a marker for the follicle membrane surrounding embryos and is not expressed in the embryo proper. Interestingly, another *Secreted Frizzled Related Protein*, *SFRPC*, is expressed only in the inner part of the blastula embryo (and is excluded from the follicle layer). *Lrp5/6* is also expressed in the follicle layer of early embryos and does not become expressed in the embryo until movements of gastrulation (Adamska *et al.*, 2010). Another signalling pathway gene, *TGF-β*, is expressed in cleavage stage embryos of *A. queenslandica* in small micromeres that are distributed uniformly throughout the embryo; expression is subsequently confined to the outer region of the blastula (Adamska *et al.*, 2007a).

Several transcription factors are known to be expressed in *A. queenslandica* embryos: these include *PaxB*, *POU*, *Lim3*, *Bsh*, *Prox2*, *NK2*, *Sox* (B, C and F), *Fox* (J and L1), *Tbx* (A and B), *Mef2*, *HNF4*, *Ets* (Larroux *et al.*, 2006) and *NF-κB* (Gauthier and Degnan, 2008). Two of those genes, *Bsh* and *NF-κB*, are expressed in subsets of micromeres at cleavage; *NF-κB* expression, however, is strongest in granular cells of the post-cleavage embryo (Larroux *et al.*, 2006; Gauthier and Degnan, 2008). Given the fact that most of the genes studied thus far in embryos are from one species (*A. queenslandica*) and that most reveal expression patterns in 'subsets of micromeres', it is difficult to draw any conclusions about the role of these molecules in early development. For two of these transcription factors, *PaxB* and *Six1/2*, expression in embryos and larvae has been partially evaluated in the demosponge *Chalinula loosanoffi*. *PaxB* expression levels during early developmental stages (embryos through to metamorphosing larvae) are very low. *PaxB* transcripts could not be detected by *in situ* hybridization in pre-release larvae for this species (Hill *et al.*, 2010). These data are consistent with what has been observed for *A. queenslandica PaxB* (Larroux *et al.*, 2006). *Six1/2* transcripts are evident throughout all developmental stages. In blastula/early gastrula stage embryos, *Six1/2* expression is restricted to the outer epithelial layer (Hill *et al.*, 2010). Interestingly, *Six1/2* expression is found in the choanosome of adult tissue in this species. These data might fit well with what is known of demosponge larval metamorphosis where the ciliated outer epithelial cells of the larvae end up migrating inward to form choanocytes (e.g. Leys and Degnan, 2002). In this scenario, *Six1/2* could be a marker of that process. More detailed analysis of gene expression during sponge embryogenesis from more species as well as experimental manipulation of gene expression at early developmental stages will be necessary.

7.4. Gene expression during gastrulation and formation of larval layers

In *A. queenslandica*, *Wnt* expression patterns during gastrulation and formation of larvae indicate a role in anterior–posterior axial polarity as *Wnt*-expressing cells become progressively confined to the posterior pole during gastrulation. The expression of *Wnt* later coincides with the region where pigment cells will form a spot and ring at the posterior end of the larvae (Adamska *et al.*, 2007a). Other members of the *Wnt* pathway are also expressed during gastrulation and larval development; this work is detailed in Adamska *et al.* (2010). The expression of the genes encoding *Wnt* pathway receptors, cytoplasmic domains and nuclear components of the pathway discussed earlier (e.g. *Frizzled*, *GSK3*, *β-catenin*) becomes localized to the outer layer of the early gastrula stage embryo, and it is suggested that the *Wnt* pathway is activated in a gradient from posterior to anterior (Adamska *et al.*, 2010, see also Section 4.2). It is also likely that the *Wnt* pathway plays a role in pigment cell migration because the pigmented ring marking the posterior of the *A. queenslandica* larvae becomes established at the same stages at which *Wnt* is expressed at the posterior pole. Functional and/or larval manipulation experiments would help support this hypothesis.

At 'gastrulation' (migration of cells to form the two layers of the larva) of *A. queenslandica*, *TGF-β* is expressed in the outer layer with expression domains at both poles of the larva. Later in development, however, *TGF-β* expression becomes restricted to an anterior pole domain suggesting a role for this pathway in establishing axial polarity (antero-posterior) along with the *Wnts* (Adamska *et al.*, 2007a). *Hedgehog* (*Hh*) family molecules are also found in sponges, however, the *Hh* protein (named *Hedgling* in *A. queenslandica*) lacks important features of true metazoan *Hh* genes (Adamska et al., 2007b). Nonetheless, *Hedgling* is expressed in a sharp narrow band around the developing pigment spot and ring in *A. queenslandica* larvae and the expression of *Wnt*, *TGF-β*, and *Hedgling* molecules form a series of concentric, partially overlapping domains around the forming ring, which might indicate an ancient co-expression cassette (Adamska *et al.*, 2011) along with their probable roles in the establishment of axial polarity.

In *A. queenslandica*, the *Notch* receptor and the *Delta* ligand are co-expressed in the outer layer of the developing larva, and subsequently in cells of the larval subepithelia. Later, *Notch* remains expressed in the sub-epithelial layer and the *Delta* ligand becomes highly expressed in globular cells that are located beyond the subepithelia (Richards *et al.*, 2008). Richards *et al.* (2008) propose that *Notch-Delta* signalling plays a role in globular cell genesis and that the globular cells of these larvae may be related to eumetazoan neurons. Along with *Notch* and *Delta*, expression of the transcription factor *bHLH* is evident in the outer layer of the gastrulating embryo in subpopulations of cells (before the subepithelial layer forms). At the early pigment ring stage, *bHLH* is restricted to the subepithelial layer along with *Notch* and *Delta* and remains expressed in this layer along with Notch even later in larval development (Richards *et al.*, 2008). Interestingly, *A. queenslandica bHLH* demonstrates proneural activities in both *Xenopus* and *Drosophila* when the sponge gene is over-expressed in these organisms. Like *Xenopus neurogenin*, *A. queenslandica bHLH* can induce ectopic expression of *Xenopus N-tubulin* and *Delta* and can activate *Xenopus Notch*. In flies, *A. queenslandica bHLH* can induce formation of ectopic sensory bristles (Richards *et al.*, 2008). These experiments indicate a conservation of important amino acid residues related to *bHLH* function as found in genes such as *atonal* and *neurogenin* in bilaterians. These data support the idea that a functional *Notch-Delta* signalling pathway was also likely to have been present in Urmetazoa (Gazave *et al.*, 2009). However, these authors' notion that expression of *Delta, Notch* and *bHLH* during globular cell genesis in *A. queenslandica* is suggestive of bilaterian neurogenesis needs considerable further support.

Two other transcription factors have been evaluated during *A. queenslandica* gastrulation and larval development. *NF-kB* shows post-cleavage expression broadly, but especially in large granulated cells. Later, during cell differentiation and migration, these granular cells migrate to the outer layer and the expression of *NF-kB* persists in these cells in the outer and subepithelial layers. Expression is also observed in the so-called 'flask' cells that are ciliated and interspersed through the columnar epithelia. By late larval stages, expression of *NF-kB* is confined to flask cells of the subepithelial layer along with some expression observed in the inner cell mass (Gauthier and Degnan, 2008). The role that this transcription factor plays in sponge development is unclear, but the expression patterns are dynamic and in some cases cell specific, indicating that this could also serve as a good marker for flask cells during late larval development. It should be mentioned that the repeated expression of genes in 'flask' (globular) cells could reflect a 'stem cell' like property of these cells, but the possibility that those cells are 'sticky' and collect mRNA transcripts indiscriminately must also be considered.

The transcription factor *A. queenslandica Bsh* is expressed during gastrulation in a subset of cells in the outer layer and the inner cell mass. Later, however, during gastrulation, the expression of this gene is restricted to cells of the inner cell mass. Close inspection reveals that at least some of the cells within the inner cell mass that express *Bsh* are the larval sclerocytes (Larroux *et al.*, 2006). *Bar* (*Bar-/Bsh*) expression has also been studied in *Halichondria* where developmental regulation was observed during larval development (Hill *et al.*, 2004), and in this species, *Bar* expression is also restricted to the inner cell mass of larvae (Holstien *et al.*, 2010) and associated with sclerocytes (Hill, pers. obs.). It would be interesting to use functional analysis to determine if sponge *Bar* genes play roles in the specification of both larval and/or adult sclerocytes.

7.5. Developmental gene expression in larvae

In addition to the molecules discussed above, a few other developmental genes have been evaluated for expression in mature sponge larvae. In *A. queenslandica,* the *BZIP* transcription factor is expressed in two populations of cells of the inner cell mass. One pattern of expression is observed in a subset of cells that is directly adjacent to the boundary of the inner cell mass and subepithelial layer (making a ring pattern around the cell division). The other set of cells that strongly express *BZIP* are in the very centre of the inner cell mass (Larroux *et al.*, 2006). This finding indicates that there may be some cell specification/differentiation within the inner cell mass of the larvae. The metazoan-specific transcription factor *HNF4* is also expressed in *A. queenslandica* larvae where it is restricted to the outer epithelial layer and localized in small ciliated columnar cells (not in the larger epithelial cells like the mucous or flask cells, and also not around the pigment ring) (Larroux *et al.*, 2006). Thus, the aforementioned transcription factors do not have overlapping expression in mature larvae and at the least could be used along with *Bar* as markers for the inner cell mass and outer epithelial layers.

The *T-box* family of transcription factors has multiple representatives in Porifera (Adell *et al.*, 2003; Manuel *et al.*, 2004; Larroux *et al.*, 2006, 2008). In *Halichondria*, larval expression patterns for five family members have been evaluated (Holstien *et al.*, 2010). For two of the *T-box* transcription factors (*Tbx4/5* and *TbxPor*), expression profiles in free-swimming larvae reveal potential involvement in axial polarity. *TbxPor* is expressed at both the anterior and posterior poles. At the posterior pole, expression is evident and forms a gradient that seems to be strongest in the most posterior cells, whereas in the anterior region, expression is observed in distinct cells along the subepithelial layer and extends around the lateral sides of the larvae (possibly in 'flask' cells of the subepithelium). *Tbx4/5* also shows expression concentrated at both poles of the larvae. Given that some *T-box* transcription factors are known to be downstream targets of *Wnt* signaling it will be interesting to determine whether or not this regulatory connection and function (i.e. establishment/ maintenance of axial polarity) is utilized in sponges.

7.6. Gene expression patterns in juvenile and adult sponge tissues and cells

As is the case in other metazoans, many of the genes expressed during early development are also utilized in forming and maintaining the tissues of the adult sponge. Many of the transcription factors discussed above are also expressed during various stages of metamorphosing, juvenile and adult sponges (e.g. Seimiya *et al.*, 1997; Coutinho *et al.*, 2003; Adell and Müller, 2004; Hill *et al.*, 2004, 2010; Larroux *et al.*, 2006). However, little is known about their tissue-specific expression patterns or the roles they play in shaping adult sponge body plans.

In the demosponge *S. domuncula*, the homeobox transcription factor *Iroquois* is expressed in cells adjacent to canals (probably endopinacocytes) that form the epithelium of the canal structure. This gene is also turned on during the formation of canals in developing sponge aggregates (primorphs) (Perović *et al.*, 2003). In *Chalinula loosanoffi*, the transcription factors *PaxB* and *Six1/2* are also transcribed in pinacodermal cells (i.e. pinacodermal lining of the atrial opening of the oscular chimney) as well as in the immediately adjacent choanosome (Hill *et al.*, 2010). In *E. muelleri*, *PaxB* and *Six1/2* are both expressed in multiple cell types, and most notably in endopinacocytes of the canals. *PaxB* and *Six1/2* are also both expressed in cells or small groups of cells that are at the junction between choanocyte chambers. Given the placement, shape and occurrence of these cells, it is likely that they are archeocyte cells that are specified to develop into choanocytes. Very low levels of *PaxB* and *Six1/2* are found in fully formed choanocyte cells (April Hill, personal observation). Recent data indicate that when *PaxB* or *Six1/2* expression is knocked-down by RNAi, the resulting sponges have poorly formed canals with missing pinacocytes (Hill, personal observation). Thus, it seems likely that one role for *Iroquois*, *PaxB* and *Six1/2* may be in the development of the endopinacocytes of adult sponges.

In the homoscleromorph sponge, *Oscarella lobularis*, an *NK* homeobox transcription factor is expressed strictly in the choanocyte cells of the chambers in both reproductive and non-reproductive adult tissue (Gazave *et al.*, 2008). In the freshwater sponge *E. fluviatilis*, two additional genes are also known to be expressed in choanocytes. While these genes are not formally transcription factors or signal transduction pathway genes, they merit mention here. The *annexin* gene is expressed in choanocyte-committed archeocytes and chamber-forming choanocytes (Funayama et al., 2005a; see also Section 3.5). Additionally, the stem cell marker *Piwi* is expressed in archeocytes and choanocytes of *Ephydatia* and studies from this group suggest that choanocytes might represent ancestral-type stem cells (Funayama *et al.*, 2010; see also Section 5.4).

8. The Immune System

8.1. Molecules with a potential role in the immune response in sponges

A host of genes that are involved in the immune system function in bilaterians are present in sponge genomes. Importantly, these include genes that play roles in cell adhesion and possibly allorecognition (i.e. aggregation factors; see Müller, 2003), as well as a host of metazoan specific factors such as molecules in the *Toll-like receptor* (*TLR*) and *interleukin1 receptor* (*IL-1R*) pathways, *macrophage expressed gene 1* (*MPEG1*), *interferon regulatory factor-like proteins* and *MDA-5-like* RNA helicases regulated upon exposure to LPS in primorphs (Schröder *et al.*, 2003). Thus, it may be that these lectin proteins are involved in sponge bacterial defense.

One of the most important and best-characterized sets of pathways involved in innate immune defense are the *TLR* and *interleukin 1* (*IL-1R*) receptor pathways. First isolated and reported for sponges in *S. domuncula*, *TLR* transcripts are expressed constitutively in epithelial layers of the sponge surface (Wiens *et al.*, 2007). In an earlier study, Wiens *et al.* (2005) found that *myeloid differentiation factor 88* (*MyD88*), a protein containing a Toll/interleukin-1 receptor domain and a death domain, is up-regulated when sponges are exposed to LPS. Another gene of the macrophage-expressed protein class is also expressed highly at the surface in the sponges after LPS treatment. These data indicate that sponges may have a system in place to recognize Gram-negative bacteria (Wiens *et al.*, 2005). Additional work showed that *caspase* gene product is increased in *S. domuncula* during treatment with LPS, and that when recombinant *TLR* was added, it could prevent the induced expression of *caspase* (Wiens *et al.*, 2007). These data support the hypothesis that an active *TLR* pathway is utilized in sponges. Recent sequencing of the *A. queenslandica* genome revealed that although a conventional *TLR* or *IL-1R* receptor is not present, there are two highly-related receptors that contain *IL-1R-like* immunoglobulins (*Ig*) extracellular domains and an intracellular *TLR-like* (*Toll/interleukin 1*) receptor/resistance domain (called *IgTlRs*) (Gauthier *et al.*, 2010). Most of the other members of the conserved pathway (e.g. *Tollip, MyD88, Pellino*) are also present in this demosponge (Gauthier *et al.*, 2010). The authors of the latter study correctly point out that the *Toll* pathway is also known to play important roles in animal development and their study clearly demonstrates expression of *IgTlR, MyD88, Tollip* and *Pellino* orthologs during larval morphogenesis. Interestingly, *IgTlR, Tollip* and *Pellino* are expressed in the globular cells of the larvae (Nehyba *et al.*, 2009; Gauthier *et al.*, 2010; Srivastava *et al.*, 2010).

One possible marker in sponges of cells involved in innate immunity is the *lectin* gene. In *E. fluviatilis*, this gene is not expressed early in sponge development, but is expressed at late stages in large, differentiated cells that are distributed throughout the sponge body and localized beneath epithelial cells (Funayama et al., 2005b). The *E. fluviatilis lectin* gene is most similar to a *lectin* gene isolated in *S. domuncula* called *GBP* (*galactose-binding protein*). The *S. domuncula GBP* gene is able to bind to lipopolysaccharides (like the bacterial endotoxin LPS) and can inhibit bacterial growth (Schröder *et al.*, 2003). Indeed, *S. domuncula GBP* is up-regulated upon exposure to LPS in primorphs (Schröder *et al.*, 2003). Thus it may be that these *lectin* proteins are involved in sponge bacterial defense. Studies determining whether or not the *TLR* and *interleukin 1* receptor pathways play roles in immunity as well as development in *A. queenslandica* will help clarify the function of these pathways in innate immunity.

9. Conclusions

Whether or not sponges represent the oldest extant phylogenetic lineage, they are clearly a crucial group of ancient non-bilaterian animals whose structure and physiology will shed light on the evolution, development and molecular genetics of all animals. Here, we try to revisit the comparative morphology and physiology of sponges in light of a growing body of data that reveals the molecular machinery underlying cell and tissue function in these animals. Important advances have been made in understanding the sponge epithelium, cell adhesion and junctions, differentiation of the aquiferous system, coordination and contraction, as well as possible aspects of a sensory system. A host of signaling pathway components and other genetic factors have also been uncovered leading to a more complete assessment of the molecules that play roles in sponge development and function. However, it is clear that there is much more work to be done on every level (molecular, cellular, physiological) before we can understand more fully the origin of multicellularity, the evolution of tissues, signal pathways, body polarity and coordination systems in the sponges and, indeed, across the animals.

ACKNOWLEDGEMENTS

We thank P. Willenz, A. Riesgo, M. Hill, N. Boury-Esnault and R. Brown for comments. This work was supported by grants from the Natural Sciences and Engineering Research Council (Canada) (SPL) and from the National Science Foundation and the Jeffress Memorial Trust (USA) (ALH).

REFERENCES

Abedin, M., and King, N. (2008). The premetazoan ancestry of cadherins. *Science* **319,** 946–948.

Adams, E. D. M. (2010). Physiology and morphology of epithelia in the freshwater demosponge, *Spongilla lacustris* MSc thesis, University of Alberta.

Adams, E. D. M., Goss, G., and Leys, S. (2010). Freshwater sponges have functional, sealing epithelia with high transepithelial resistance and negative transepithelial potential. *Public Library of Science One* **5,** e15040.

Adamska, M., Degnan, S. M., Green, K. M., Adamski, M., Craigie, A., Larroux, C., and Degnan, B. M. (2007a). Wnt and TGF-β expression in the sponge *Amphimedon queenslandica* and the origin of metazoan embryonic patterning. *Public Library of Science One* **2,** e1031.

Adamska, M., Matus, D. Q., Adamski, M., Green, K., Rokhsar, D. S., Martindale, M. Q., and Degnan, B. M. (2007b). The evolutionary origin of hedghog proteins. *Current Biology* **17,** R836–R837.

Adamska, M., Larroux, C., Adamski, M., Green, K., Lovas, E., Koop, D., Richards, G. S., Zwafink, C., and Degnan, B. M. (2010). Structure and expression of conserved Wnt pathway components in the demosponge *Amphimedon queenslandica. Evolution & Development* **12,** 494–518.

Adamska, M., Degnan, B. M., Green, K., and Zwafink, C. (2011). What sponges can tell us about the evolution of developmental processes. *Zoology* **114,** 1–10.

Adell, T., and Müller, W. E. G. (2004). Isolation and characterization of five Fox (Forkhead) genes from the sponge *Suberites domuncula. Gene* **334,** 35–46.

Adell, T., Grebenjuk, V. A., Wiens, M., and Müller, W. E. G. (2003). Isolation and characterization of two T-box genes from sponges, the phylogenetically oldest metazoan taxon. *Development Genes and Evolution* **213,** 421–434.

Adell, T., Gamulin, V., Perovic-Ottstadt, S., Wiens, M., Korzhev, M., Muller, I. M., and Muller, W. E. G. (2004). Evolution of metazoan cell junction proteins: The scaffold protein MAGI and the transmembrane receptor tetraspanin in the demosponge *Suberites domuncula*. *Journal of Molecular Evolution* **59,** 41–50.

Adell, T., Thakur, A., and Muller, W. (2007). Isolation and characterization of Wnt pathway-related genes from Porifera. *Cell Biology International* **31,** 939–949.

Alié, A., and Manuel, M. (2010). The backbone of the post-synaptic density originated in a unicellular ancestor of choanoflagellates and metazoans. *BMC Evolutionary Biology* **10,** 34.

Amano, S., and Hori, I. (1996). Transdifferentiation of larval flagellated cells to choanocytes in the metamorphosis of the demosponge *Haliclona permollis*. *The Biological Bulletin* **190,** 161–172.

Aouacheria, A., Geourjon, C., Aghajari, N., Navratil, V., Deléage, G., Lethias, C., and Exposito, J.-Y. (2006). Insights into early extracellular matrix evolution: Spongin short chain collagen-related proteins are homologous to basement membrane Type IV collagens and form a novel family widely distributed in invertebrates. *Molecular Biology and Evolution* **23,** 2288–2302.

Asano, A., Asano, K., Sasaki, H., Furuse, M., and Tsukita, S. (2003). Claudins in *Caenorhabditis elegans*: Their distribution and barrier function in the epithelium. *Current biology: CB* **13,** 1042–1046.

Bagby, R. M. (1965). The fine structure of myocytes in the sponges *Microciona prolifera* (Ellis and Solander) and *Tedania ignis* (Duchassaing and Michelotti). *Journal of Morphology* **118,** 167–182.

Bagby, R. M. (1970). The fine structure of pinacocytes in the marine sponge *Microciona prolifera* (Ellis and Solander). *Cell and Tissue Research* **105,** 579–594.

Bagby, R. M. (1972). Formation and differentiation of the upper pinacoderm in reaggregation masses of the sponge *Microciona prolifera* (Ellis and Solander). *The Journal of Experimental Zoology* **180,** 217–225.

Barolo, S., and Posakony, J. W. (2002). Three habits of highly effective signaling pathways: Principles of transcriptional control by developmental cell signaling. *Genes & Development* **16,** 1167–1181.

Baumann, O. (2001). Posterior midgut epithelial cells differ in their organization of the membrane skeleton from other *Drosophila* epithelia. *Experimental Cell Research* **270,** 176–187.

Baumgartner, S., Littleton, J. T., Broadie, K., Bhat, M. A., Harbecke, R., Lengyel, J. A., Chiquet-Ehrismann, R., Prokop, A., and Bellen, H. J. (1996). A *Drosophila* neurexin is required for septate junction and blood-nerve barrier formation and function. *Cell* **87,** 1059–1068.

Bavestrello, G., Burlando, B., and Sarà, M. (1988). The architecture of the canal systems of *Petrosia ficiformis* and *Chondrosia reniformis* studied by corrosion casts (Porifera, Demospongiae). *Zoomorphology* **108,** 161–166.

Bidder, G. (1895). The collar cells of Heterocoela. *The Quarterly Journal of Microscopical Science* **38,** 9–43.

Bond, C. (1992). Continuous cell movements rearrange anatomical structures in intact sponges. *The Journal of Experimental Zoology* **263,** 284–302.

Borojevic, R., and Lévi, C. (1964). Etude au microscope électronique des cellules de l'éponge *Ophlitaspongia seriata* (Grant) au cours de la réorganisation après dissociation. *Zeitschrift für Zellforschung und Mikroskopische Anatomie* **64,** 708–725.

Borojevic, R., and Lévi, C. (1965). Morphogenèse expérimentale d'une éponge à partir de cellules de la larve nageante dissociée. *Zeitschrift für Zellforschung* **68,** 57–69.

Boury-Esnault, N. (1976). Morphogenèse experimentale des papilles inhalantes de l'éponge *Polymastia mamillaris* (Muller). *Archives de zoologie expérimentale et génèrale* **117,** 181–196.

Boury-Esnault, N. (2006). Systematics and evolution of Demospongiae. *Canadian Journal of Zoology* **84,** 205–224.

Boury-Esnault, N., and Jamieson, B. G. M. (1999). Porifera. In "Reproductive Biology of Invertebrates" (B. G. M. Jamieson, ed.), pp. 1–20. Oxford and IBH Publishing Co. Pvt. Ltd, New Delhi, Vol. 9.

Boury-Esnault, N., and Rützler, K. (1997). Thesaurus of Sponge Morphology. Smithsonian Institution, Washington.

Boury-Esnault, N., De Vos, L., Donadey, C., and Vacelet, J. (1984). Comparative study of the choanosome of Porifera: The Homoscleromorpha. *Journal of Morphology* **180,** 3–17.

Boury-Esnault, N., De Vos, L., Donadey, C.Vacelet, J. *et al.* (1990). Ultrastructure of choanosome and sponge classification. In "New Perspectives in Sponge Biology" (K. Rutzler, ed.), pp. 237–244. Smithsonian Institution Press, Washington, DC.

Boury-Esnault, N., Ereskovsky, A., Bézac, C., and Tokina, D. (2003). Larval development in the Homoscleromorpha (Porifera, Demospongiae). *Invertebrate Biology* **122,** 187–202.

Boute, N., Exposito, J. Y., Boury-Esnault, N., Vacelet, J., Nor, N., Miyazaki, K., Yoshizato, K., and Garrone, R. (1996). Type IV collagen in sponges, the missing link in basement membrane ubiquity. *Biologie Cellulaire* **88,** 37–44.

Brauer, E. B. (1975). Osmoregulation in the fresh water sponge, *Spongilla lacustris*. *The Journal of Experimental Zoology* **192,** 181–192.

Brien, P. (1932). Contribution a l'étude de la régénération naturelle chez les spongillidae. *Spongilla lacustis* (L.), *Ephydatia fluviatilis* (L.). *Archives Zoologie Experimentale et Generale* **74,** 461–506.

Brill, B. (1973). Untersuchungen zur Ultrastruktur der Choanocyte von Ephydatia fluviatilis L. *Zeitschrift für Zellforschung und mikroskopische Anatomie* **14,** 231–245.

Broun, M., Gee, L., Reinhardt, B., and Bode, H. R. (2005). Formation of the head organizer in hydra involves the canonical Wnt pathway. *Development* **132,** 2907–2916.

Buscema, M., De Sutter, D., and Van de Vyver, G. (1980). Ultrastructural study of differentiation processes during aggregation of purified sponge archaeocytes. *Wilhelm Roux's Archives* **188,** 45–53.

Carvalho de Souza, A., Ganchev, D., Snel, M., van der Eerden, J., Vliegenthart, J., and Kamerling, J. (2009). Adhesion forces in the self-recognition of oligosaccharide epitopes of the proteoglycan aggregation factor of the marine sponge *Microciona prolifera*. *Glycoconjugate Journal* **26,** 457–465.

Cavalcanti, F., and Klautau, M. (2011). Solenoid: A new aquiferous system to Porifera. *Zoomorphology* **130,** 255–260.

Chapman, J. A., Kirkness, E. F., Simakov, O., Hampson, S. E., Mitros, T., Weinmaier, T., Rattei, T., Balasubramanian, P. G., Borman, J., Busam, D., Disbennett, K.Pfannkoch, C. *et al.* (2010). The dynamic genome of *Hydra*. *Nature* **464,** 592–596.

Coutinho, C. C., Fonseca, R. N., Mansure, J. J. C., and Borojevic, R. (2003). Early steps in the evolution of multicellularity: Deep structural and functional homologies among homeobox genes in sponges and higher metazoans. *Mechanisms of Development* **120,** 429–440.

De Goeij, J. M., De Kluijver, A., Van Duyl, F. C., Vacelet, J., Wijffels, R. H., De Goeij, A. F. P. M., Cleutjens, J. P. M., and Schutte, B. (2009). Cell kinetics of the marine sponge *Halisarca caerulea* reveal rapid cell turnover and shedding. *The Journal of Experimental Biology* **212,** 3892–3900.

de Mendoza, A., Suga, H., and Ruiz-Trillo, I. (2010). Evolution of the MAGUK protein gene family in premetazoan lineages. *BMC Evolutionary Biology* **10,** 93.

De Vos, L., and Van de Vyver, G. (1981). Etude de la contraction spontanee chez l'eponge d'eau douce *Ephydatia fluviatilis* cultivee en vitro. *Annales de la societe Royale zoologique de Belgique* **111,** 21–31.

De Vos, L., Boury-Esnault, N., and Vacelet, J. (1990). The apopylar cell of sponges. In "New Perspectives in Sponge Biology" (K. Ruetzler, ed.), pp. 153–158. Smithsonian Institution Press, Washington, DC.

De Vos, L., Ruetzler, K., Boury-Esnault, N., Donadey, C., and Vacelet, J. (1991). Atlas of Sponge Morphology. Smithsonian Institution, Washington, DC.

Degnan, B. M., Vervoort, M., Larroux, C., and Richards, G. S. (2009). Early evolution of metazoan transcription factors. *Current Opinion in Genetics & Development* **19,** 591–599.

Dunn, C. W., Hejnol, A., Matus, D. Q., Pang, K., Browne, W. E., Smith, S. A., Seaver, E., Rouse, G. W., Obst, M., Edgecombe, G. D., Sorensen, M. V., Haddock, S. H. D., Schmidt-Rhaesa, A., Okusu, A., Mobjerg Kristensen, R., Wheeler, W. C., Martindale, M. Q., and Giribet, G. (2008). Broad phylogenomic sampling improves resolution of the animal tree of life. *Nature* **452,** 745–749.

Eerkes-Medrano, D., and Leys, S. P. (2006). Ultrastructure and embryonic development of a syconoid calcareous sponge. *Invertebrate Biology* **125,** 177–194.

Efremova, S. M., and Efremov, V. I. (1979). Proliferation cellulaire chez la larvae nageante de l'eponge d'eau douce: Baikalospongia bacillifera (Dybowski). *Colloque Internationaux du CNRS Biologie des spongiaires* **291,** 59–65.

Elliott, G. R. D. (2009). The cytology and physiology of coordinated behaviour in a freshwater sponge, *Ephydatia muelleri* PhD thesis, University of Alberta.

Elliott, G. R. D., and Leys, S. P. (2007). Coordinated contractions effectively expel water from the aquiferous system of a fresh water sponge. *The Journal of Experimental Biology* **210,** 3736–3748.

Elliott, G. R. D., and Leys, S. P. (2010). Evidence for glutamate, GABA and NO in coordinating behaviour in the sponge, *Ephydatia muelleri* (Demospongiae, Spongillidae). *The Journal of Experimental Biology* **213,** 2310–2321.

Ellwanger, K., and Nickel, M. (2006). Neuroactive substances specifically modulate rhythmic body contractions in the nerveless metazoon *Tethya wilhelma* (Demospongiae, Porifera). *Frontiers in Zoology* **3,** 710.1186/1742-9994-1183-1187.

Ellwanger, K., Brümmer, F., and Nickel, M. (2004). Glutamate, GABA and serotonin induce contractions in the sponge *Tethya wilhelma* (Porifera: Demospongiae). In "Jahrestagung der Deutschen Zoologischen Gesellschaft, Abstractband" (R. Kinzelbach, ed.), p. 157. Zoologisches Institut der Universität Rostock, Rostock.

Emson, R. H. (1966). The reactions of the sponge *Cliona celata* to applied stimuli. *Comparative Biochemistry and Physiology* **18,** 805–827.

Ereskovsky, A. V. (2004). Comparative embryology of Sponges and its application for poriferan phylogeny. In "Sponge Science in the New Millennium" (M. Pansini, R. Pronzato, G. Bavestrello and R. Manconi, eds), pp. 301–318. Bolletin dei Musei e Instituto Biologia, Universitaire de Genova, Genova 68.

Ereskovsky, A. V. (2005). Comparative Embryology of Sponges. Sanct-Petersburg University Press, St. Petersburg.

Exposito, J.-Y., and Garrone, R. (1990). Characterization of a fibrillar collagen gene in sponges reveals the early evolutionary appearance of two collagen gene families. *Proceedings of the National Academy of Sciences of the United States of America* **87,** 6669–6673.

Exposito, J.-Y., Le Guellec, D., Lu, Q., and Garrone, R. (1991). Short chain collagens in sponges are encoded by a family of closely related genes. *The Journal of Biological Chemistry* **266,** 21923–21928.

Fahey, B., and Degnan, B. M. (2010). Origin of animal epithelia: Insights from the sponge genome. *Evolution & Development* **12,** 601–617.

Fahey, B., and Degnan, B. M. (2012). Origin and Evolution of Laminin Gene Family Diversity. *Molecular Biology and Evolution*. doi:10.1093/molbev/mss060.

Faivre-Sarrailh, C., Banerjee, S., Li, J., Hortsch, M., Laval, M., and Bhat, M. A. (2004). Drosophila contactin, a homolog of vertebrate contactin, is required for septate junction organization and paracellular barrier function. *Development* **131,** 4931–4942.

Fernandez-Busquets, X., and Burger, M. M. (1999). Cell adhesion and histocompatibility in sponges. *Microscopy Research and Technique* **44,** 204–218.
Fernandez-Busquets, X., and Burger, M. M. (2003). Circular proteoglycans from sponges: First members of the spongican family. *Cellular and Molecular Life Sciences* **60,** 88–112.
Fjerdingstad, E. J. (1961). The ultrastructure of choanocyte collars in *Spongilla lacustris* (L.). *Zeitschrift für Zellforschung* **53,** 645–657.
Funayama, N., Nakatsukasa, M., Hayashi, T., and Agata, K. (2005a). Isolation of the choanocyte in the fresh water sponge, *Ephydatia fluviatilis* and its lineage marker, *Ef annexin*. *Development, Growth & Differentiation* **47,** 243–253.
Funayama, N., Nakatsukasa, M., Kuraku, S., Takechi, K., Dohi, M., Iwabe, N., Miyata, T., and Agata, K. (2005b). Isolation of Ef silicatein and Ef lectin as molecular markers of sclerocytes and cells involved in innate immunity in the freshwater sponge *Ephydatia fluviatilis*. *Zoological Science* **22,** 1113–1122.
Funayama, N., Nakatsukasa, M., Mohri, K., Masuda, Y., and Agata, K. (2010). Piwi expression in archeocytes and choanocytes in demosponges: Insights into the stem cell system in demosponges. *Evolution & Development* **12,** 275–287.
Galtsoff, P. S. (1925). Regeneration after dissociation (an experimental study on sponges) I. Behavior of dissociated cells of *Microciona prolifera* under normal and altered conditions. *Journal of Experimental Zoology* **42,** 183–221.
Gauthier, M., and Degnan, B. (2008). The transcription factor NF-κB in the demosponge *Amphimedon queenslandica* insights on the evolutionary origin of the Rel homology domain. *Development Genes and Evolution* **218,** 23–32.
Gauthier, M. E. A., Du Pasquier, L., and Degnan, B. M. (2010). The genome of the sponge *Amphimedon queenslandica* provides new perspectives into the origin of Toll-like and interleukin 1 receptor pathways. *Evolution & Development* **12,** 519–533.
Gazave, E., Lapébie, P., Renard, E., Vacelet, J., Rocher, C., Ereskovsky, A. V., Lavrov, D. V., and Borchiellini, C. (2010). Molecular Phylogeny Restores the Supra-Generic Subdivision of Homoscleromorph Sponges (Porifera, Homoscleromorpha). *Public Library of Science One* **5,** e14290.
Gazave, E., Lapébie, P., Renard, E., Bézac, C., Boury-Esnault, N., Vacelet, J., Pérez, T., Manuel, M., and Borchiellini, C. (2008). NK homeobox genes with choanocyte-specific expression in homoscleromorph sponges. *Development Genes and Evolution* **218,** 479–489.
Gazave, E., Lapébie, P., Richards, G., Brunet, F., Ereskovsky, A., Degnan, B., Borchiellini, C., Vervoort, M., and Renard, E. (2009). Origin and evolution of the Notch signalling pathway: An overview from eukaryotic genomes. *BMC Evolutionary Biology* **9,** 249.
Genova, J. L., and Fehon, R. G. (2003). Neuroglian, Gliotactin, and the Na+/K+ ATPase are essential for septate junction function in *Drosophila*. *The Journal of Cell Biology* **161,** 979–989.
Gilis. (2011). Biomineralization of Basal Skeletons in Recent Hypercalcified Sponges PhD thesis, Université Libre de Bruxelles.
Gilis, M., Grauby, O., Willenz, P., Dubois, P., Legras, L., Heresanu, V., and Barronnet, A. (2011). Multi-scale mineralogical characterization of the hypercalcified sponge *Petrobiona massiliana* (Calcarea, Calcaronea). *Journal of Structural Biology* **176,** 315–329.
Gonobobleva, E., and Maldonado, M. (2009). Choanocyte ultrastructure in *Halisarca dujardini* (Demospongiae, Halisarcida). *Journal of Morphology* **270,** 615–627.
Grant, R. (1825). Observations and experiments on the structure and functions of the sponge. *Edinburgh Philosophical Journal* **8,** 94–107.
Green, C. R., and Bergquist, P. R. (1979). Cell membrane specialisations in the Porifera. *Colloques Internationaux du CNRS Biologie des Spongiaires* **291,** 153–158.

Green, C. R., and Bergquist, P. R. (1982). Phylogenetic relationships within the invertebrata in relation to the structure of septate junctions and the development of occluding junctional types. *Journal of Cell Science* **53,** 279–305.
Hadas, E., Ilan, M., and Shpigel, M. (2008). Oxygen consumption by a coral reef sponge. *The Journal of Experimental Biology* **211,** 2185–2190.
Harcet, M., Roller, M., Ćetković, H., Perina, D., Wiens, M., Müller, W. E. G., and Vlahoviček, K. (2010). Demosponge EST sequencing reveals a complex genetic toolkit of the simplest metazoans. *Molecular Biology and Evolution* **27,** 2747–2756.
Hill, A., Tetrault, J., and Hill, M. (2004). Isolation and expression analysis of a poriferan Antp-class Bar-/Bsh-like homeobox gene. *Development Genes and Evolution* **214,** 515–523.
Hill, A., Boll, W., Ries, C., Warner, L., Osswalt, M., Hill, M., and Noll, M. (2010). Origin of pax and six gene families in sponges: Single PaxB and Six1/2 orthologs in *Chalinula loosanoffi*. *Developmental Biology* **343,** 106–123.
Holstien, K., Rivera, A., Windsor, P., Ding, S., Leys, S., Hill, M., and Hill, A. (2010). Expansion, diversification, and expression of T-box family genes in Porifera. *Development Genes and Evolution* **220,** 251–262.
Hyman, L. H. (1940). The Invertebrates. McGraw-Hill, New York.
Imsiecke, G. (1993). Ingestion, digestion, and egestion in *Spongilla lacustris* (Porifera, Spongillidae) after pulse feeding with *Chlamydomonas reinhardtii*. *Zoomorphology* **113,** 233–244.
Jackson, D. J., Macis, L., Reitner, J., Degnan, B. M., and Worheide, G. (2007). Sponge Paleogenomics Reveals an Ancient Role for Carbonic Anhydrase in Skeletogenesis. *Science* **316,** 1893–1895.
Jacobs, D., Nakanishi, N., Yuan, D., Camara, A., Nichols, S., and Hartenstein, V. (2007). Evolution of sensory structures in basal metazoa. *Integrative and Comparative Biology* **47,** 712–723.
Jones, W. C. (1961). Properties of the wall of *Leucosolenia variabilis*. *The Quarterly Journal of Microscopical Science* **s3-102,** 543–550.
Jones, W. C. (1964). Photographic records of living oscular tubes of *leucosolenia variabilis* I. The choanoderm boundary, the choanocytes and the pore arrangement. *Journal of the Marine Biological Assocation of the United Kingdom* **44,** 67–85.
Karam, S. (1999). Lineage commitment and maturation of epithelial cells in the gut. *Frontiers in Bioscience* **4,** d286–d298.
Kilian, E. F. (1952). Wasserströmung und Nahrungsaufnahme beim Süsswasserschwamm *Ephydatia fluviatilis*. *Zeitschrift für vergleichende Physiology* **34,** 407–447.
Kilian, E. F., and Wintermann-Kilian, G. (1979). Mouvement cellulaire et contraction chez *Spongilla lacustris* et *Ephydatia fluviatilis*. *Colloques internationaux du CNRS* **291,** 137–143.
Kusserow, A., Pang, K., Sturm, C., Hrouda, M., Lentfer, J., Schmidt, H. A., Technau, U., von Haeseler, A., Hobmayer, B., Martindale, M. Q., and Holstein, T. W. (2005). Unexpected complexity of the *Wnt* gene family in a sea anemone. *Nature* **433,** 156–160.
Langenbruch, P. F. (1983). Body structure of marine sponges I. Arrangement of the flagellated chambers in the canals system of *Reniera* sp.. *Marine Biology* **75,** 319–325.
Langenbruch, P. F. (1988). Body structure of marine sponges V. Structure of choanocyte chambers in some Mediterranean and Caribbean haplosclerid sponges (Porifera). *Zoomorphology* **108,** 13–21.
Langenbruch, P.-F., and Jones, W. C. (1990). Body structure of marine sponges. VI. Choanocyte chamber structure in the Haplosclerida (Porifera, Demospongiae) and its relevance to the phylogenesis of the group. *Journal of Morphology* **204,** 1–8.
Langenbruch, P. F., and Scalera-Liaci, L. (1986). Body structure of marine sponges IV. Aquiferous system and choanocyte chambers in *Haliclona elegans* (Porifera, Demospongiae). *Zoomorphology* **106,** 205–211.

Langenbruch, P. F., and Weissenfels, N. (1987). Canal systems and choanocyte chambers in freshwater sponges (Porifera, Spongillidae). *Zoomorphology* **107,** 11–16.
Langenbruch, P. F., Simpson, T. L., and Scalera-Liaci, L. (1985). Body structure of marine sponges III. The structure of choanocyte chambers in *Petrosia ficiformis* (Porifera, Demospongiae). *Zoomorphology* **105,** 383–387.
Lapébie, P., Gazave, E., Ereskovsky, A., Derelle, R., Bézac, C., Renard, E., Houliston, E., and Borchiellini, C. (2009). WNT/β-Catenin signalling and epithelial patterning in the homoscleromorph sponge *Oscarella. Public Library of Science One* **4,** e5823.
Larroux, C., Fahey, B., Liubicich, D., Hinman, V., Gauthier, M., Gongora, M., Green, K., Woerheide, G., Leys, S. P., and Degnan, B. M. (2006). Developmental expression of transcription factor genes in a demosponge: Insights into the origin of metazoan multicellularity. *Evolution & Development* **8,** 150–173.
Larroux, C., Luke, G. N., Koopman, P., Rokhsar, D. S., Shimeld, S. M., and Degnan, B. M. (2008). Genesis and expansion of metazoan transcription factor gene classes. *Molecular Biology and Evolution* **25,** 980–996.
Larsen, P. S., and Riisgård, H. U. (1994). The sponge pump. *Journal of Theoretical Biology* **168,** 53–63.
Ledger, P. W. (1975). Septate junctions in the calcareous sponge *Sycon ciliatum. Tissue & Cell* **7,** 13–18.
Lévi, C., and Lévi, P. (1976). Embryogénèse de *Chondrosia reniformis* (Nardo), démosponge ovipare, et transmission des bactéries symbiotiques. *Annales des Sciences Naturelles, Zoologie et Biolologie animale* **18,** 367–380.
Leys, S. P. (1999). The choanosome of hexactinellid sponges. *Invertebrate Biology* **118,** 221–235.
Leys, S. P. (2004). Gastrulation in sponges. In "Gastrulation: From Cells to Embryo" (C. Stern, ed.), pp. 23–31. Cold Spring Harbor Press, New York.
Leys, S. P., and Degnan, B. M. (2002). Embryogenesis and metamorphosis in a haplosclerid demosponge: Gastrulation and transdifferentiation of larval ciliated cells to choanocytes. *Invertebrate Biology* **121,** 171–189.
Leys, S., and Eerkes-Medrano, D. (2006). Feeding in a calcareous sponge: Particle uptake by pseudopodia. *The Biological Bulletin* **211,** 157–171.
Leys, S. P., and Ereskovsky, A. (2006). Embryogenesis and larval differentiation in sponges. *Canadian Journal of Zoology* **84,** 262–287.
Leys, S. P., and Mackie, G. O. (1997). Electrical recording from a glass sponge. *Nature* **387,** 29–30.
Leys, S. P., and Meech, R. W. (2006). Physiology of coordination in sponges. *Canadian Journal of Zoology* **84,** 288–306.
Leys, S., and Riesgo, A. (2011). Epithelia: An evolutionary novelty of metazoans. *Journal of Experimental Zoology (Molecular and Developmental Evolution).* doi:10.1002/jez.b.21442.
Leys, S. P., Mackie, G. O., and Meech, R. W. (1999). Impulse conduction in a sponge. *The Journal of Experimental Biology* **202,** 1139–1150.
Leys, S. P., Mackie, G., and Meech, R. W. (2003). Effect of temperature on impulse conduction in the sponge *Rhabdocalyptus dawsoni.* Proceedings of the Physiological Society, Cambridge.
Leys, S. P., Mackie, G. O., and Reiswig, H. (2007). The biology of glass sponges. *Advances in Marine Biology* **52,** 1–145.
Leys, S. P., Nichols, S. A., and Adams, E. D. M. (2009). Epithelia and integration in sponges. *Integrative and Comparative Biology* **49,** 167–177.
Leys, S., Yahel, G., Reidenbach, M. A., Shavit, U., Tunnicliffe, V., and Reiswig, H. (2011). The sponge pump: The role of current induced flow in the design of the sponge body plan. *Public Library of Science One* **6,** (12), e27787.

Liebeskind, B. J., Hillis, D. M., and Zakon, H. H. (2011). Evolution of sodium channels predates the origin of nervous systems in animals. *Proceedings of the National Academy of Sciences* **108,** 9154–9159.
Linck, R. W. (1973). Chemical and structural differences between cilia and flagella from the lamellibranch mollusc, *Aequipecten irradians. Journal of Cell Science* **12,** 951–981.
Lindemann, C. B., and Lesich, K. A. (2010). Flagellar and ciliary beating: The proven and the possible. *Journal of Cell Science* **123,** 519–528.
Ludeman, D. (2010). Origin of Sensory Cells within the Metazoa: Do Sponges have Sensory Cilia? Honours thesis, University of Alberta.
Maldonado, M., and Bergquist, P. R. (2002). Phylum Porifera. In "Atlas of Marine Invertebrate Larvae" (C. M. Young, M. A. Sewell and M. E. Rice, eds), pp. 21–50. Academic Press, San Diego.
Maldonado, M., and Riesgo, A. (2007). Intra-epithelial spicules in a homosclerophorid sponge. *Cell and Tissue Research* **328,** 639–650.
Maldonado, M., and Riesgo, A. (2008). Reproductive output in a Mediterranean population of the homosclerophorid *Corticium candelabrum* (Porifera, Demospongiae), with notes on the ultrastructure and behavior of the larva. *Marine Ecology* **29,** 298–316.
Maldonado, M., and Young, C. (1998). Limits on the bathymetric distribution of keratose sponges: A field test in deep water. *Marine Ecology Progress Series* **174,** 123–139.
Manuel, M. (2009). Early evolution of symmetry and polarity in metazoan body plans. *Comptes Rendus Biologies* **332,** 184–209.
Manuel, M., Borojevic, R., Boury-Esnault, N., and Vacelet, J. (2002). Class Calcarea Bowerbank, 1864. (J. N. A. Hooper and R. W. M. Van Soest, eds), Systema Porifera: A Guide to the Classification of Sponges, pp. 1103–1110. Kluwer Academic/Plenum Publishers, New York, vol. 2.
Manuel, M., Le Parco, Y., and Borchiellini, C. (2004). Comparative analysis of Brachyury T-domains, with the characterization of two new sequences, from a hexactinellid and a calcisponge. *Gene* **340,** 291–301.
McNair, G. T. (1923). Motor reactions of the fresh-water sponge *Ephydatia fluviatilis. The Biological Bulletin* **44,** 153–166.
Mergner, H. (1964). Uber die inducktion neuer oscularrohre bei *Ephydatia fluviatilis. Roux's Archives of Developmental Biology* **155,** 9–128.
Mergner, H. (1966). Zum Nachweis der Artspezifitat des Induktionsstoffes bei Oscularrohr-neubildungen von Spongilliden. *Verhandlungen der Deutschen Zoologischen Gesellschaft* **30,** 522–564 (supplementband).
Müller, W. E. G. (2003). The origin of metazoan complexity: Porifera as integrated animals. *Integrative and Comparative Biology* **43,** 3–10.
Müller, W. E. G., and Müller, I. M. (2003). Origin of the metazoan immune system: Identification of the molecules and their functions in sponges. *Integrative and Comparative Biology* **43,** 281–292.
Nakata, Y. and Funayama, N. (2011). Investigation of the molecular mechanism that recruit spicules to their holding-up points. In: Bosch T, Holstein TW, editors. Searching for Eve: basal metazoans and the evolution of multicellular complexity Tutzing, Germany.
Nehyba, J., Hrdličková, R., and Bose, H. R. (2009). Dynamic evolution of immune system regulators: The history of the interferon regulatory factor family. *Molecular Biology and Evolution* **26,** 2539–2550.
Nichols, S. A., Dirks, W., Pearse, J. S., and King, N. (2006). Early evolution of animal cell signaling and adhesion genes. *Proceedings of the National Academy of Science of United States of America* **103,** 12451–12456.
Nickel, M. (2004). Kinetics and rhythm of body contractions in the sponge *Tethya wilhelma* (Porifera: Demospongiae). *The Journal of Experimental Biology* **207,** 4515–4524.
Nickel, M. (2010). Evolutionary emergence of synaptic nervous systems: What can we learn from the non-synaptic, nerveless Porifera? *Invertebrate Biology* **129,** 1–16.

Nickel, M., Donath, T., Schweikert, M., and Beckmann, F. (2006). Functional morphology of *Tethya* species (Porifera): 1. Quantitative 3D-analysis of *Tethya wilhelma* by synchrotron radiation based X-ray microtomography. *Zoomorphology* **125,** 209–223.

Nickel, M., Scheer, C., Hammel, J. U., Herzen, J., and Beckmann, F. (2011). The contractile sponge epithelium sensu lato—Body contraction of the demosponge Tethya wilhelma is mediated by the pinacoderm. *The Journal of Experimental Biology* **214,** 1692–1698.

Parker, G. H. (1910). The reactions of sponges with a consideration of the origin of the nervous system. *The Journal of Experimental Zoology* **8,** 765–805.

Pavans de Ceccatty, M. (1959). Les structures cellulaires de type nerveux chez *Hippospongia communis* LMK. *Annales des Sciences Naturelles, Zoologie et Biolologie animale* **12,** 105–112.

Pavans de Ceccatty, M. (1969). Les systèmes des activites motrices, spontanées et provoquées des Éponges. *Comptes Rendus de l'Académie de Sciences Paris, Sciences de la vie* **269,** 596–599.

Pavans de Ceccatty, M. (1974a). Coordination in sponges. The foundations of integration. *American Zoologist* **14,** 895–903.

Pavans de Ceccatty, M. (1974b). The origin of the integrative systems: A change in view derived from research on coelentrates and sponges. *Perspectives in Biology and Medicine* **17,** 379–390.

Pavans de Ceccatty, M. (1976). Cellular movements and pathways of coordination in the sponges. *Bulletin de la Societe de Zoologie Francaise* **100,** 154.

Pavans de Ceccatty, M. (1979). Cell correlations and integration in Sponges. *Colloques Internationaux du CNRS Biologie des Sponiaires* **291,** 123–135.

Pavans de Ceccatty, M. (1981). Demonstration of actin filaments in sponge cells. *Cell Biology International Reports* **5,** 945–952.

Pavans de Ceccatty, M. (1986). Cytoskeletal organization and tissue patterns of epithelia in the sponge *Ephydatia muelleri. Journal of Morphology* **189,** 45–65.

Pavans de Ceccatty, M., Cargouil, M., and Coraboeuf, E. (1960). Les reactions motrices de l'eponge *Tethya lyncurium* (Lmk.) a quelques stimulations experimentales (I). *Vie et Milieu* **11,** 594–600.

Pavans de Ceccatty, M., Thiney, Y., and Garrone, R. (1970). Les bases ultrastructurales des communications intercellulaires dans les oscules de quelques éponges. *Zoological Society of London* **25,** 449–466.

Perović, S., Schroder, H. C., Sudek, S., Grebenjuk, V. A., Batel, R., Štifanic, M., Müller, I. M., and Müller, W. E. G. (2003). Expression of one sponge *Iroquois* homeobox gene in primorphs from *Suberites domuncula* during canal formation. *Evolution & Development* **5,** 240–250.

Pile, A. J., Patterson, M. R., and Witman, J. D. (1996). *In situ* grazing on plankton < 10um by the boreal sponge *Mycale lingua. Marine Ecology Progress Series* **141,** 95–102.

Prosser, C. L. (1967). Ionic analysis and effects of ions on contractions of sponge tissues. *Zeit vergleichliche Physiologie* **54,** 109–120.

Prosser, C. L., Nagai, T., and Nystrom, R. A. (1962). Oscular contractions in sponges. *Comparative Biochemistry and Physiology* **6,** 69–74.

Ramoino, P., Gallus, L., Paluzzi, S., Raiteri, L., Diaspro, A., Fato, M., Bonanno, G., Tagliafierro, G., Ferretti, C., and Manconi, R. (2007). The GABAergic-like system in the marine demosponge *Chondrilla nucula. Microscopy Research and Technique* **70,** 944–951.

Rasmont, R., Bouillon, J., Castiaux, P., and Vandermeerssche, G. (1958). Ultra-structure of the choanocyte collar cells in freshwater sponges. *Nature* **181,** 58–59.

Rasmont, R., and Rozenfeld, F. (1981). Étude micro-cinematographique de la formation des chambres choanocytaires chez une éponge d'eau douce. *Annales de la Societe Royale Zoologique de Belge* **111,** 33–44.

Rasmont, R., Bouillon, J., Castiaux, P., and Vandermeerssche, G. (1958). Ultra-structure of the choanocyte collar cells in freshwater sponges. *Nature* **181,** 58–59.

Reiswig, H. M. (1971). *In situ* pumping activities of tropical demospongiae. *Marine Biology* **9,** 38–50.
Reiswig, H. M. (1975). The aquiferous systems of three marine demospongiae. *Journal of Morphology* **145,** 493–502.
Reiswig, H. M. (1979). A new sponge with rapid contraction systems. In "Annual Meeting of the Canadian Society of Zoology. Programme abstracts, pp.83a Laval University, Quebec."
Reiswig, H. M., and Mehl, D. (1991). Tissue organization of *Farrea occa* (Porifera, Hexactinellida). *Zoomorphology* **110,** 301–311.
Reitner, J., and Gautret, P. (1996). Skeletal formation in the modern but ultraconservative chaetetid sponge *Spirastrella (Acanthochaetetes) wellsi* (Demospongiae, Porifera). *Facies* **34,** 193–208.
Richards, G. S., Simionata, E., Perron, M., Adamska, M., Vervoort, M., and Degnan, B. M. (2008). Sponge genes provide new insight into the evolutionary origin of the neurogenic circuit. *Current Biology* **18,** 1156–1161.
Riisgård, H., Thomassen, S., Jakobsen, H., Weeks, J., and Larsen, P. (1993). Suspension feeding in marine sponges *Halichondria panicea* and *Haliclona urceolus*: Effects of temperature on filtration rate and energy cost of pumping. *Marine Ecology Progress Series* **96,** 177–188.
Ruppert, E. E., Fox, R. S., and Barnes, R. D. (2004). Invertebrate Zoology: A Functional Evolutionary Approach. Thomson Brooks/Cole, Belmont, USA.
Sakaraya, O., Armstrong, K. A., Adamska, M., Adamski, M., Wang, I., Tidor, B., Degnan, B. M., Oakley, T. H., and Kosik, K. S. (2007). A post-synaptic scaffold at the origin of the animal kingdom. *Public Library of Science One* **2,** e506. doi: 510.1371/journal.pone.0000506.
Salvenmoser, W. (2011). New aspects of cellular ultrastructure in Hydra. In: Bosch T, Holstein TW, editors. Searching for Eve: basal metazoans and the evolution of multicellular complexity Tutzing. Germany .
Schröder, H. C., Ushijima, H., Krasko, A., Gamulin, V., Thakur, N. L., Diehl-Seifert, B., Müller, I. M., and Müller, W. E. G. (2003). Emergence and disappearance of an immune molecule, an antimicrobial lectin, in basal metazoa. *The Journal of Biological Chemistry* **278,** 32810–32817.
Schroder, H. C., Perovic-Ottstadt, S., Wiens, M., Batel, R., Muller, I. M., and Muller, W. E. G. (2004). Differentiation capacity of epithelial cells in the sponge *Suberites domuncula*. *Cell and Tissue Research* **316,** 271–280.
Seimiya, M., Watanabe, Y., and Kurosawa, Y. (1997). Identification of POU-class homeobox genes in a freshwater sponge and the specific expression of these genes during differentiation. *European Journal of Biochemistry* **243,** 27–31.
Sethmann, I., and Wörheide, G. (2008). Structure and composition of calcareous sponge spicules: A review and comparison to structurally related biominerals. *Micron* **39,** 209–228.
Shore, R. E. (1971). Growth and renewal studies of the choanocyte population in *Hymeniacidon sinapium* (Porifera: Demospongiae) using colcemid and 3-H thymidine. *The Journal of Experimental Zoology* **177,** 359–363.
Simionato, E., Ledent, V., Richards, G., Thomas-Chollier, M., Kerner, P., Coornaert, D., Degnan, B., and Vervoort, M. (2007). Origin and diversification of the basic helix-loop-helix gene family in metazoans: Insights from comparative genomics. *BMC Evolutionary Biology* **7,** 33.
Simpson, T. L. (1963). The biology of the marine sponge *Microciona prolifera* (Ellis and Solander) I. A study of cellular function and differentiation. *Journal of Experimental Zoology* **154,** 135–151.
Simpson, T. L. (1984). The Cell Biology of Sponges. Springer Verlag, New York.

Srivastava, M., Simakov, O., Chapman, J., Fahey, B., Gauthier, M., Mitros, T., Richards, G. S., Conaco, C., Dacre, M., Hellsten, U., Larroux, C.Putnam, N. *et al.* (2010). The *Amphimedon queenslandica* genome and the evolution of animal complexity. *Nature* **466,** 720–727.

Teragawa, C. K. (1986). Sponge dermal membrane morphology: Histology of cell-mediated particle transport during skeletal growth. *Journal of Morphology* **190,** 335–347.

Tompkins-MacDonald, G. J., Gallin, W. J., Sakarya, O., Degnan, B., Leys, S. P., and Boland, L. M. (2009). Expression of a poriferan potassium channel: Insights into the evolution of ion channels in metazoans. *The Journal of Experimental Biology* **212,** 761–767.

Uriz, M.-J. (2006). Mineral skeletogenesis in sponges. *Canadian Journal of Zoology* **84,** 322–356.

Uriz, M. J., Turon, X., Becerro, M. A., and Agell, G. (2003). Siliceous spicules and skeleton frameworks in sponges: Origin, diversity, ultrastructural patterns, and biological functions. *Microscopy Research and Technique* **62,** 279–299.

Vacelet, J. (1999). Planktonic armoured propagules of the excavating sponge *Alectona* (Porifera: Demospongiae) are larvae: Evidence from *Alectona wallichii* and *A. mesatlantica* sp. nov. *Memoirs of the Queensland Museum* **44,** 627–642.

Vacelet, J., and Boury-Esnault, N. (1995). Carnivorous sponges. *Nature* **373,** 333–335.

Vacelet, J., Boury-Esnault, N., De Vos, L., and Donadey, C. (1989). Comparative study of the choanosome of Porifera: II. The Keratose sponges. *Journal of Morphology* **201,** 119–129.

Van de Vyver, G. (1970). La non confluence intraspécifique chez les spongiaires et la notion d'individu. *Annales d'embryologie et de morphogènese* **3,** 251–262.

Van Tright, H. (1919). Contribution to the physiology of the fresh-water sponges (Spongillidae). *Tijdschrift voor Diergeneeskunde* **17,** 1–220.

Vilanova, E., Coutinho, C. C., and Mourão, P. A. S. (2009). Sulfated polysaccharides from marine sponges (Porifera): An ancestor cell–cell adhesion event based on the carbohydrate–carbohydrate interaction. *Glycobiology* **19,** 860–867.

Vogel, S. (1978). Evidence for one-way valves in the water-flow system of sponges. *The Journal of Experimental Biology* **76,** 137–148.

Wang, X., Hu, S., Gan, L., Wiens, M., and Müller, W. E. G. (2010). Sponges (Porifera) as living metazoan witnesses from the Neoproterozoic: Biomineralization and the concept of their evolutionary success. *Terra Nova* **22,** 1–11.

Watanabe, Y. (1978a). Structure and formation of the collar in choanocytes of *Tetilla serica* (Lebwohl), Demosponge. *Development, Growth & Differentiation* **20,** 70–91.

Watanabe, Y. (1978b). The Development of Two Species of *Tetilla* (Demosponge). Natural Science Report, Ochanomizu University, vol. 29, pp. 71–106.

Weissenfels, N. (1976). Bau und Funktion des Süsswasserschwamms *Ephydatia fluviatilis* L. (Porifera). III. Nahrungsaufnahme, Verdauung und Defäkation. *Zoomorphologie* **85,** 73–88.

Weissenfels, N. (1980). Bau und Funktion des Süsswasserschwamms *Ephydatia fluviatilis* L. (Porifera). VII. Die Porocyten. *Zoomorphologie* **95,** 27–40.

Weissenfels, N. (1981). Bau und Funktion des Süsswasserschwamms *Ephydatia fluviatilits* L. (Porifera) VIII. die Entstehung und Entwicklung der Kragengeisselkammern und ihre Verbindung mit dem ausfuhrenden Kanalsystem. *Zoomorphology* **98,** 35–45.

Weissenfels, N. (1983). Bau und Funktion des Subwasserschwamms *Ephydatia fluviatilis* (Porifera) X. Der Nachweis des offenen Mesenchyms durch Verfutterung von Backerhefe (*Saccharomyces cerevisiae*). *Zoomorphology* **103,** 15–23.

Weissenfels, N. (1990). Condensation rhythm of fresh-water sponges (Spongillidae, Porifera). *European Journal of Cell Biology* **53,** 373–383.

Weissenfels, N. (1992). The filtration apparatus for food collection in freshwater sponges (Porifera, Spongillidae). *Zoomorphology* **112,** 51–55.

Weissenfels, N., and Langenbruch, P. (1985). Langzeitkulturen von Subwasserschwammen (Porifera, Spongillidae) unter Laborbedingungen. *Zoomorphology* **105,** 12–15.

Weisz, J., Lindquist, N., and Martens, C. (2008). Do associated microbial abundances impact marine demosponge pumping rates and tissue densities? *Oecologia* **155,** 367–376.

Wielsputz, C., and Saller, U. (1990). The metamorphosis of the parenchymula-larva of *Ephydatia fluviatilis* (Porifera, Spongillidae). *Zoomorphology* **109,** 173–177.

Wiens, M., Batel, R., Korzhev, M., and Müller, W. E. G. (2003). Retinoid X receptor and retinoic acid response in the marine sponge *Suberites domuncula. The Journal of Experimental Biology* **206,** 3261–3271.

Wiens, M., Korzhev, M., Krasko, A., Thakur, N. L., Perović-Ottstadt, S., Breter, H. J., Ushijima, H., Diehl-Seifert, B., Müller, I. M., and Müller, W. E. G. (2005). Innate immune defense of the sponge *Suberites domuncula* against bacteria involves a MyD88-dependent signaling pathway. *The Journal of Biological Chemistry* **280,** 27949–27959.

Wiens, M., Korzhev, M., Perović-Ottstadt, S., Luthringer, B., Brandt, D., Klein, S., and Müller, W. E. G. (2007). Toll-Like receptors are part of the innate immune defense system of sponges (Demospongiae: Porifera). *Molecular Biology and Evolution* **24,** 792–804.

Willenz, P., and Van de Vyver, G. (1982). Endocytosis of latex beads by the exopinacoderm in the fresh water sponge *Ephydatia fluviatilis*: An *in vitro* and *in situ* study in SEM and TEM. *Journal of Ultrastructure Research* **79,** 294–306.

Willenz, P., and Van de Vyver, G. (1984). Ultrastructural localization of lysosomal digestion in the fresh water sponge *Ephydatia fluviatilis. Journal of Ultrastructure Research* **87,** 13–22.

Willenz, P., and Van de Vyver, G. (1986). Ultrastructural evidence of extruding exocytosis of residual bodies in the freshwater sponge *Ephydatia fluviatilis. Journal of Morphology* **190,** 307–318.

Willenz, P., Vray, B., Maillard, M. P., and Van de Vyver, G. (1986). A quantitative study of the retention of radioactively labeled *E. coli* by the freshwater sponge *Ephydatia fluviatilis. Physiological and Biochemical Zoology* **59,** 495–504.

Wilson, H. V. (1907). On some phenomena of coalescence and regeneration in sponges. *The Journal of Experimental Zoology* **5,** 245–258.

Wilson, H. V., and Penney, J. T. (1930). The regeneration of sponges (*Microciona*) from dissociated cells. *The Journal of Experimental Zoology* **56,** 72–134.

Windsor, P. J., and Leys, S. P. (2010). Wnt signaling and induction in the sponge aquiferous system: Evidence for an ancient origin of the organizer. *Evolution & Development* **12,** 484–493.

Wintermann, G. (1951). Entwicklungsphysiologische Untersuchungen an Süsswasserschwämmen. *Zoologische Jahrbuecher Abteilung fuer Anatomie und Ontogenie der Tiere* **71,** 428–486.

Wörheide, G. (1998). The reef cave dwelling ultraconservative corraline demosponge *Astrosclera willeyana* Lister 1900 from the Indo-Pacific. *Facies* **38,** 1–88.

Wyeth, R. C. (1999). Video and electron microscopy of particle feeding in sandwich cultures of the hexactinellid sponge *Rhabdocalyptus dawsoni. Invertebrate Biology* **118,** 236–242.

Wyeth, R. C., Leys, S. P., and Mackie, G. O. (1996). Use of sandwich cultures for the study of feeding in the hexactinellid sponge *Rhabdocalyptus dawsoni* (Lambe, 1892). *Acta Zoologica* **77,** 227–232.

Yahel, G., Sharp, J. H., Marie, D., Hase, C., and Genin, A. (2003). In situ feeding and element removal in the symbiont-bearing sponge *Theonella swinhoei*: Bulk DOC is the major source for carbon. *Limnology and Oceanography* **48,** 141–149.

Yahel, G., Whitney, F., Reiswig, H. M., Eerkes-Medrano, D. I., and Leys, S. P. (2007). In situ feeding and metabolism of glass sponges (Hexactinellida, Porifera) studied in a deep temperate fjord with a remotely operated submersible. *Limnology and Oceanography* **52,** 428–440.

CHAPTER TWO

Sponge–Microbe Symbioses: Recent Advances and New Directions

Robert W. Thacker[1] and Christopher J. Freeman

Contents

Department of Biology, University of Alabama at Birmingham, Birmingham, Alabama, USA
[1]Corresponding author: Email: thacker@uab.edu

Advances in Marine Biology, Volume 62
ISSN 0065-2881, DOI: 10.1016/B978-0-12-394283-8.00002-3

Abstract

Sponges can host abundant and diverse communities of symbiotic microorganisms. In this chapter, we review recent work in the area of sponge–microbe symbioses, focusing on (1) the diversity of these associations, (2) host specificity, (3) modes of symbiont transmission, and (4) the positive and negative impacts of symbionts on their hosts. Over the past 4 years, numerous studies have catalogued the diversity of sponge–microbe symbioses, challenging previous hypotheses of a uniform, vertically transmitted microbial community and supporting a mixed model of symbiont community transmission. We emphasize the need for experimental manipulations of sponge–symbiont interactions coupled with advanced laboratory techniques to determine the identity of metabolically active microbial symbionts, to investigate the physiological processes underlying these interactions, and to elucidate whether symbionts act as mutualists, commensals, or parasites. The amazing diversity of these complex associations continues to offer critical insights into the evolution of symbiosis and the impacts of symbiotic microbes on nutrient cycling and other ecosystem functions.

Key Words: Porifera; sponge; microbial symbiosis; mutualism; parasitism; coevolution; cospeciation; host specificity; transmission; stable isotope analyses

1. Introduction

Symbioses between microbes and metazoans are extremely common in marine, freshwater, and terrestrial environments (Margulis and Fester, 1991). Genomic studies suggest that such interactions were likely present in the earliest metazoans (Taylor *et al.*, 2007c; Thomas *et al.*, 2010; Siegl *et al.*, 2011) and a wealth of data confirm that modern sponges can host diverse and abundant microbial communities that contain nearly all major evolutionary lineages of bacteria and archaea (Wilkinson, 1978a; Wilkinson *et al.*, 1981; Unson *et al.*, 1994; Preston *et al.*, 1996; Sarà *et al.*, 1998; Friedrich *et al.*, 1999; Webster *et al.*, 2001, 2004; Hentschel *et al.*, 2002; Taylor *et al.*, 2007b; Webster and Taylor, 2011). In response to a renewed scientific interest in understanding the ecology and evolution of these ancient symbioses, the field of sponge microbiology, which is concerned with investigating the microbial communities associated with Porifera, has experienced a recent surge of new research and produced a large number of surveys, reviews, and commentary articles (Lee *et al.*, 2001; Hentschel *et al.*, 2006; Taylor *et al.*, 2007a,b,c; Usher *et al.*, 2007; Usher, 2008; Webster and Blackall, 2009; Webster and Taylor, 2011). Indeed, scientific publications in the general field of microbiology cite sponge-specific research articles with a higher than expected frequency (Becerro, 2008).

Microbial symbionts can comprise over 40% of the total sponge–microbe "holobiont" volume and can benefit host sponges by providing supplemental nutrition (Vacelet and Donadey, 1977; Wilkinson, 1983; Arillo *et al.*, 1993; Yahel *et al.*, 2003; Freeman and Thacker, 2011), producing secondary metabolites (Flatt *et al.*, 2005), enhancing the rigidity of the sponge skeleton (Wilkinson *et al.*, 1981), and providing protection from ultraviolet solar radiation (Regoli *et al.*, 2000). A large number of ongoing research programmes continue to provide evidence that sponges host evolutionarily unique bacterial lineages that are distinct from the bacterial communities found in ambient seawater (Hentschel *et al.*, 2002; Olson and McCarthy, 2005; Hill *et al.*, 2006; Taylor *et al.*, 2007b). All of these studies suggest key roles for microbial symbionts in the evolutionary history and ecological success of sponges and indicate that these ancient symbioses might provide fundamental insights into the evolution and maintenance of animal–microbe interactions (Taylor *et al.*, 2007a,c).

Photosynthetic microbial symbionts are associated with a wide variety of coral reef invertebrates, as exemplified by coral–dinoflagellate associations (Stanley, 2006). Coral reef sponges most often host cyanobacteria as photosynthetic symbionts (Rutzler, 1990; Diaz *et al.*, 2007; Erwin and Thacker, 2007) but can also form associations with filamentous algae (Carballo and Ávila, 2004) and dinoflagellates (Schönberg and Loh, 2005; Hill *et al.*, 2011). In addition to contributing nutrients to host metabolism (Wilkinson and Fay, 1979; Arillo *et al.*, 1993), these symbionts are hypothesized to enhance the growth rates and competitive abilities of their hosts in benthic communities (Wilkinson, 1983, 1987; Regoli *et al.*, 2000; Flatt *et al.*, 2005; Thacker, 2005; Erwin and Thacker, 2008b). Moreover, these symbioses can represent major sources of primary productivity and nutrient cycling in marine ecosystems (Diaz and Rützler, 2001; Mohamed *et al.*, 2008; Fiore *et al.*, 2010). Although cyanobacterial symbionts can comprise 25–50% of a sponge's cellular volume (Rutzler, 1990), we still know relatively little about the metabolic exchanges and ecological interactions between sponges and their microbial symbionts (Taylor *et al.*, 2007a).

Although microbial communities are associated with nearly all sponges investigated to date, vigorous debate surrounds the ecological and evolutionary nature of these communities. Reviews of sponge–microbe symbioses are quick to report the incredible abundance of diverse microbes found within some sponges and to list the metabolic pathways that may confer a benefit to the host sponge (Webster and Taylor, 2011), but fundamental questions are often overwhelmed by these assessments of microbial diversity and thus remain unaddressed. For example, associated microbes might be permanent residents of a particular host sponge or might simply be present in the water column at the time of sample collection. Similarly, associated microbes could be vertically transmitted from parent to offspring but might also be acquired from the surrounding seawater. Although Hentschel *et al.* (2003) suggested that "real experimental evidence for symbiotic interactions is scarce" and Taylor *et al.* (2007b) stated that it was "startling how little is known about many fundamental aspects of sponge symbiont biology", few studies have examined these interactions using a quantitative, experimental approach. Manipulative experiments can allow us to elucidate whether a symbiosis represents a mutualistic or a parasitic relationship, but such experiments have been conducted for only a handful of sponge species. In addition, studies identifying the nutritional benefit conferred to host sponges from microbial symbionts have only recently gained new interest, despite the fact that the seminal papers that provide the basis of such investigations were published over 30 years ago. Thus, our current state of knowledge in the field of sponge microbiology sets the stage to critically examine these and other topics, to revise many original hypotheses about these symbiotic interactions, and to propose new directions for improving and accelerating this area of research.

Our review was initiated by the organizing committee of the Eighth World Sponge Conference, held in Girona, Spain, in September 2010, with a charge of not only describing recent advances in the field of sponge microbiology but also describing new avenues of research in this field. We quantified increasing interest in sponge microbiology through the citation reports of the Science Citation Index Expanded database (Institute for Scientific Information, http://isiwebofknowledge.com). By searching the "topic" field for the query "((sponge or Porifera) and (bacteria* or microb*) and (symbio*))" from 1991 through May 2011, we obtained an estimate of the number of publications on sponge–microbe symbioses since 1991, which yielded a total of 271 papers (Fig. 2.1A). Almost half ($n = 134$, 49%) of these papers appeared after the 2006 International Sponge Symposium (Custódio *et al.*, 2007). Moreover, citations to these papers have increased exponentially (Fig. 2.1B), suggesting a growing recognition of this field of research by the broader scientific community. Since research in our laboratory focuses primarily on tropical taxa that host cyanobacterial symbionts, much of our review will consider examples drawn from these experimental systems. A secondary goal of this review will be to illustrate how our views of these fascinating symbioses have changed in recent years, from both ecological and evolutionary perspectives.

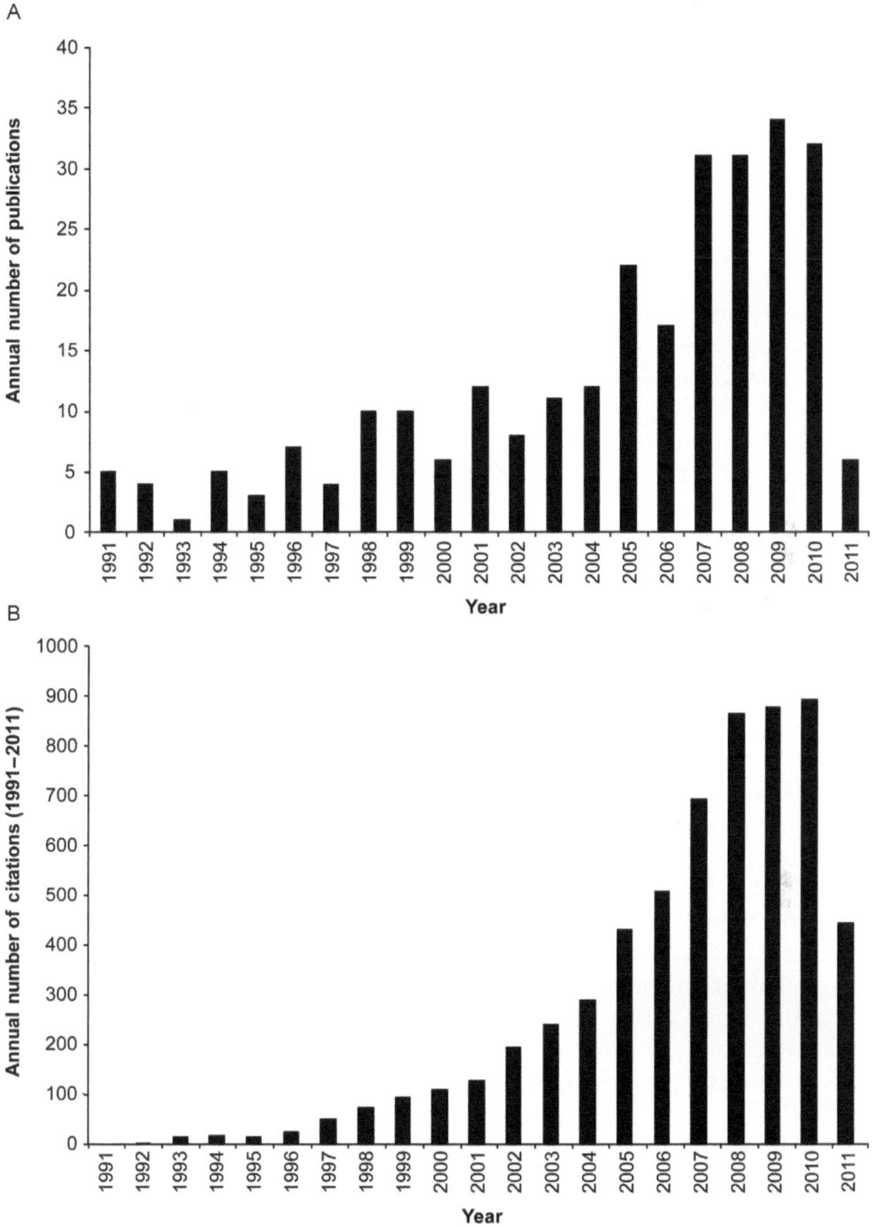

Figure 2.1 Increasing numbers of publications and citations in the field of sponge–microbe symbiosis, as measured through the citation reports of the Science Citation Index Expanded database (Institute for Scientific Information, isiwebofknowledge.com), searching the "topic" field for the query "((sponge or Porifera) and (bacteria★ or microb★) and (symbio★))". (A) Annual number of publications from 1991 through May 2011. (B) Annual number of citations from 1991 through May 2011. Note that 2011 is only partially represented.

2. Diversity of Sponge–Microbe Associations

2.1. Approaches to measuring diversity

Surveys that evaluate the diversity of sponge–microbe associations have tended to take either a microbial perspective or a host perspective. A microbial perspective usually evaluates the microbial community associated with one or a few hosts using culture-based or molecular-based techniques, identifying as many types of symbiotic microbes as possible. A host perspective generally samples many hosts within a particular geographic region and evaluates the presence/absence of a particular group of symbionts. Below, we will consider each of these perspectives, but first we will briefly review a variety of techniques used to measure the diversity of symbiotic microbial communities. The development of these techniques has mirrored the development of microbial community profiling methods in the broader field of environmental microbiology, moving from predominantly culture-based approaches to a greater emphasis on DNA sequencing and other molecular techniques (Nocker *et al.*, 2007; Gloor *et al.*, 2010).

The earliest studies of sponge microbiology used microscopic methods to document sponge–bacteria associations (Dosse, 1939; Bertrand and Vacelet, 1971; Sarà and Vacelet, 1973; Vacelet, 1975) along with traditional, culture-based approaches to quantify the diversity of cultivable bacteria associated with sponge hosts (Wilkinson, 1978b; Santavy and Colwell, 1990). Importantly, these researchers recognized the need to compare sponge-derived cultures with those derived from the surrounding seawater to address the issue of whether sponge-derived microbes simply reflect microbes present in the ambient environment at the time of sampling (Madri *et al.*, 1971; Wilkinson, 1978b; Santavy and Colwell, 1990). However, environmental microbiologists now recognize that the vast majority of microbial taxa have never been cultured (Amann *et al.*, 1995; Head *et al.*, 1998). This gross underestimation of microbial diversity is a major weakness of culture-based methodologies and has driven the development of culture-independent, molecular-based methodologies to assess both the diversity and species composition of microbial communities (Pace, 1997; Nocker *et al.*, 2007).

Traditionally, *16S* ribosomal RNA (rRNA) gene sequences are the most frequently targeted molecular marker for microbial characterization, yielding a large database for comparative analyses (Pace *et al.*, 1986; Gloor *et al.*, 2010). Caveats associated with the use of *16S* rRNA gene sequences include the presence of multiple copies of ribosomal operons in some microbial taxa (Stewart and Cavanaugh, 2007) and low levels of sequence divergence among closely related bacteria. For example, two marine *Prochlorococcus* species that exhibit >97% *16S* rRNA sequence identity differ considerably in genome size, gene number, and gene identity (Rocap *et al.*, 2003). These genetic differences correspond to ecological differences, as these species inhabit distinct ecological niches and display variable physiological capabilities (Moore *et al.*, 1998). Other molecular markers can be used to characterize closely related bacteria and increase fine-scale phylogenetic resolution, including the *16S–23S* rRNA internal transcribed spacer (ITS) region (Iteman *et al.*, 2000; Boyer *et al.*, 2001; Rocap *et al.*, 2002; Stewart and Cavanaugh, 2007).

Initial molecular-based approaches to measuring the diversity of sponge–microbe associations generated "clone libraries" of *16S* rRNA gene sequences derived from whole genomic DNA extracts of host sponges (e.g. Hentschel *et al.*, 2002; Hill *et al.*, 2006). Clone library-based approaches generally use PCR to amplify *16S* rRNA gene sequences from environmental samples; amplicons are then ligated to vectors, which are transformed into *Escherichia coli*, and grown in cultures. These cultures are plated and individual clones are sequenced, each of which contains a single *16S* rRNA sequence. This approach can provide a powerful description of both the abundance and sequence identity of microbes in a community but is also relatively time consuming and expensive. Consequently, many studies using clone libraries lack sufficient replication to test evolutionary and ecological hypotheses regarding host specificity and community similarity. The use of replicated clone libraries allows more sophisticated hypothesis testing, including explicit tests of phylogenetic community structure and analyses of community similarity (e.g. Martin, 2002; Erwin *et al.*, 2011a,b).

Microbial community fingerprinting approaches provide methods to decrease costs and increase sample sizes in ecological studies. Denaturing (or temperature) gradient gel electrophoresis (DGGE/TGGE) allows many samples to be screened simultaneously, but this technique can suffer from a lack of reproducibility (Panagiotis *et al.*, 2010). Banding patterns from DGGE can be quantified to differentiate microbial communities; moreover, individual bands can be cut from a gel and sequenced, but often these sequences do not provide enough information to permit the classification of taxa (Nocker *et al.*, 2007). Terminal restriction fragment length polymorphism (T-RFLP) profiling is extremely repeatable, and the polymorphisms observed can be referenced to a sequence database or a clone library to identify particular taxa (Clement *et al.*, 1998). For both D/TGGE and T-RFLP, the lack of sequence information can be overcome by pairing these methods with clone library sequences that characterize a community. Community profiling approaches have been particularly useful when applied to experimental manipulations of sponge–microbe symbioses (Isaacs *et al.*, 2009; Anderson *et al.*, 2010; López-Legentil *et al.*, 2010; Webster *et al.*, 2010b). Finally, the advent of next-generation sequencing technologies has led to the development of massively high-throughput community profiling methods. For example, the *V3, V5,* and *V6* regions of microbial *16S* rRNA can be targeted using Roche pyrosequencing or Illumina paired-end sequencing to generate millions of sequence reads (Sogin *et al.*, 2006; Gloor *et al.*, 2010). However, the high error rates of some next-generation methods can lead to incorrect taxonomic identifications and an overestimation of biodiversity (Reeder and Knight, 2009). Webster *et al.* (2010a) used a pyrosequencing approach to examine the diversity of *V6* sequences associated with three marine sponges, revealing a much greater estimate of diversity than previously suspected. Lee *et al.* (2011) also used pyrosequencing to examine three additional sponge species but reported lower symbiont diversity than Webster *et al.* (2010a). Each of these techniques has its own advantages and disadvantages, but all of these methods perform best with replicated sampling. Indeed, general ecological theory indicates that stronger inferences can be made when studies of community diversity are both spatially and temporally replicated (Dorazio *et al.*, 2006).

2.2. Microbial perspectives on the diversity of sponge–microbe associations

Most surveys of sponge–microbe associations consider a microbial perspective, reporting the diversity of microbes associated with a particular host or group of host sponges. These surveys have been the subject of multiple review articles in recent years (e.g. Taylor *et al.*, 2007b; Webster and Taylor, 2011). In general, studies published since the 2006 sponge symposium (Custódio *et al.*, 2007) continue to document the broad range of microbial taxa associated with sponges. For example, over 11,000 *16S* rRNA gene sequences from sponge-associated bacteria were compiled by Webster and Taylor (2011). These sequences include representatives of sixteen major lineages of bacteria, dominated by alpha-, beta-, gamma-, and deltaproteobacteria, as well as Acidobacteria, Actinobacteria, Bacteroidetes, Chloroflexi, Cyanobacteria, and Firmicutes, and also including the candidate phylum "Poribacteria". However, this compilation did not include numerous archaeal sequences reported from sponges (Taylor *et al.*, 2007b) or reports of additional microbial lineages from pyrosequencing reads acquired by Webster *et al.* (2010a), Lee *et al.* (2011), and Schmitt *et al.* (2011). Other recent studies have expanded our knowledge of sponge–microbe associations by exploring mesophotic (30–200 m), deep (>200 m), and polar habitats (e.g. Meyer and Kuever, 2008; Brück *et al.*, 2010; Nishijima *et al.*, 2010; Olson and Kellogg, 2010; reviewed by Webster and Taylor, 2011).

Two common themes arise from recent studies that consider either individual sponge hosts or a limited number of host species. Firstly, some sponge species host extremely diverse communities of symbionts, while other species host communities of much lower species richness.

For example, *Rhopaloeides odorabile* is reported to host nearly 3000 unique taxa (Webster *et al.*, 2010a), but *Agelas robusta* hosts only 32 unique OTUs (Sun *et al.*, 2010). Secondly, every study of sponge-associated microbial communities reports that some fraction of the observed diversity is likely to be represented by seawater-associated bacteria that were present in the sponge at the time of sample collection. Thus, the problem of distinguishing "resident" sponge symbionts from "transient" seawater bacteria is widely acknowledged, but it is currently unclear how much symbiont diversity can be assigned to each of these categories. We will consider these issues further in Section 3.

2.3. Host perspectives on the diversity of sponge–microbe associations

Several recent studies have examined the prevalence of sponge–microbe symbioses from the perspective of the sponge hosts by surveying a wide variety of hosts in a particular geographic area for the presence of a particular symbiont or group of symbionts. Most of these surveys have focused on photosynthetic cyanobacterial symbionts, since their presence/absence and abundance can be determined using microscopic techniques (e.g. TEM, SEM, fluorescence microscopy) and relatively simple chemical analyses for the presence of chlorophyll *a* (Wilkinson, 1983, 1987; Steindler *et al.*, 2002; Diaz *et al.*, 2007; Erwin and Thacker, 2007; Usher, 2008), as well as by pulse amplitude-modulated (PAM) fluorometry (Steindler *et al.*, 2002). Moreover, rates of photosynthesis and respiration can be readily compared among a variety of sponge hosts by measuring oxygen consumption in light and dark conditions (Wilkinson, 1983, 1987; Thacker *et al.*, 2007; Erwin and Thacker, 2008b).

Wilkinson (1987) compared the relative abundance of sponges hosting photosynthetic symbionts on reefs in the western Pacific (Great Barrier Reef, GBR) and the western Atlantic (Caribbean Sea), as well as the biomass percentage of sponges that displayed phototrophic nutrition (defined as a gross production to respiration ratio >1.5). Wilkinson concluded that phototrophic sponges were nearly absent on Caribbean reefs but comprised over 40% of the GBR sponge fauna. These results were surprising, since approximately one-third of the Caribbean sponges surveyed hosted photosynthetic symbionts, compared to 44% of the GBR sponges. Despite finding similar proportions of sponges hosting photosynthetic symbionts in the Caribbean and GBR faunas, Caribbean sponges were rarely measured as phototrophic. In contrast, surveys by Steindler *et al.* (2002) reported that 85% of intertidal and 64% of subtidal species in Zanzibar (West Indian Ocean) hosted photosynthetic symbionts, suggesting a much higher prevalence of these symbioses in other regions. Wilkinson (1987) hypothesized that the availability of particulate organic matter for heterotrophic filter feeding would drive the relative abundance of these symbioses, with a higher prevalence of photosymbionts in regions with less available POM.

Recent studies have questioned this hypothesis by documenting the occurrence of phototrophic sponges in regions with a high availability of POM. Surveys of sponges found in the Bocas del Toro Archipelago of Panama (Erwin and Thacker, 2007) revealed a similar proportion of sponges hosting photosynthetic symbionts as reported by Wilkinson (1987), with approximately one-third of the observed sponge species hosting photosynthetic symbionts; these species also accounted for approximately one-third of the observed abundance of sponges on coral reefs in this region. Despite the high observed diversity of sponge hosts, only two morphotypes of cyanobacteria dominated the Bocas del Toro sponge fauna: unicellular *Synechococcus spongiarum* and filamentous *Oscillatoria spongeliae* (Erwin and Thacker, 2007, 2008a,b). Measurements of gross production to respiration ratios revealed that several taxa considered heterotrophic by Wilkinson (1987) are, in fact, phototrophic when exposed to natural irradiances (Thacker *et al.*, 2007; Erwin and Thacker, 2008b). Since the levels of irradiance used by Wilkinson (1987) were less than 50% of natural irradiance, the methodology used in earlier studies appears to have underestimated the photosynthetic activity of many Caribbean sponge–cyanobacteria associations (Erwin and Thacker, 2008b).

Surveys of temperate reefs have revealed a prevalence of photosymbionts similar to that reported in tropical reefs. In New South Wales, Australia, over 65% of temperate reef sponges are estimated to host photosynthetic symbionts (Roberts *et al.*, 1999), while in Western Australia, 48% of sponge species and 63% of individuals contained photosymbionts (Lemloh *et al.*, 2009). The genetic sequences of these symbionts were closely related to those reported from tropical regions, including both the *S. spongiarum* and *O. spongeliae* clades (Lemloh *et al.*, 2009). The diversity of these two sponge-specific clades will be further discussed in subsequent sections.

Few studies have examined how the distribution of microbial symbionts varies among hosts across phylogenetic or taxonomic scales. Diaz *et al.* (2007) provided a review of sponges hosting symbiotic cyanobacteria, combining morphological evidence of cyanobacterial associations derived from light and electron microscopy with molecular evidence derived from *16S* rRNA gene sequencing. Over 100 sponge species were reported to host cyanobacteria, distributed across 36 families of Demospongiae and 3 families of Calcarea. Some taxonomic groups do appear to host cyanobacterial symbionts more frequently than others, including the orders Dictyoceratida (24 species), Haplosclerida (20 species), and Verongida (13 species). The majority of these species are distributed in just a few genera, including *Aplysina* (eight species) and *Xestospongia* (seven species). Clearly, similar studies are needed for the many diverse taxa of microbial symbionts to establish how symbiotic communities have changed during the evolution of sponges.

2.4. Is there a uniform community of bacterial symbionts?

The hypothesis that sponges host a common community of bacterial symbionts can be traced to Wilkinson's (1978b) study of cultured bacteria derived from three disparate sponge taxa (*Pericharax heteroraphis* [Calcarea, Clathrinida]; *Jaspis stellifera* [Demospongiae, Astrophorida]; and *Neofibularia irata* [Demospongiae, Poecilosclerida]). All three of these sponges hosted facultative anaerobic bacteria that were not found in the ambient water column, suggesting that diverse sponges could share similar microbial symbionts. The hypothesis of a "uniform community" received additional support from Hentschel *et al.*'s (2002) study of sponge-specific sequence clusters (SSSCs; see section 3, "Specificity") derived from clone libraries of *16S* ribosomal RNA gene sequences from *Aplysina aerophoba* (Demospongiae, Verongida), *Theonella swinhoei* (Demospongiae, Lithistida), and *R. odorabile* (Demospongiae, Dictyoceratida). In this study, 14 monophyletic SSSCs belonging to 7 bacterial divisions were recovered; 5 of these SSSCs were shared among these 3 host species, again suggesting that taxonomically diverse sponges host phylogenetically similar microbial communities (Hentschel *et al.*, 2002). Taylor *et al.* (2007b) concluded that their compilation of over 1500 *16S* rRNA gene sequences provided further support for this hypothesis by finding over 35 SSSCs. However, Webster *et al.* (2010a) obtained an exceptional diversity of symbiotic microbes using a deep sequencing approach and found many of the SSSCs reported by Taylor *et al.* (2007b) in seawater samples, questioning whether these clades are truly sponge specific. Surveys of sponge–microbe associations from a host perspective suggest that even common microbial symbionts are not found in all host sponges. For example, as discussed above, only about one-third of coral reef sponge species are estimated to host cyanobacterial symbionts; those sponges that do host cyanobacteria only host one of two possible genera (Erwin and Thacker, 2007).

Schmitt *et al.* (2011) compiled additional microbial community pyrosequencing data using the V3 region of the *16S* ribosomal RNA gene, from 32 sponge species. These authors divided the sponge microbiota into three groups: (1) a core community present in at least 22 of the 32 (70%) analysed sponges; (2) a variable community, whose members were present in between 2 and 22 species; and (3) a species-specific community, present in only a single sponge species. This study further divided each group into two other categories: plus-OTUs (which match to previously sponge-derived sequences) and minus-OTUs (which match to non-sponge-derived sequences). Schmitt *et al.* (2011) argue that the species-specific group is vertically transmitted but that a small core group of taxa are horizontally transmitted; however, this study did not assess samples of the ambient environment. This study agrees with those of Hentschel *et al.* (2002) and Taylor *et al.* (2007b), concluding that sponges "contain a uniform, sponge-specific bacterial community although each sponge species contains different bacterial species" (Schmitt *et al.*, 2011).

In contrast, Erwin *et al.* (2011b) reported unique bacterial lineages associated with sponges from the Gulf of Mexico; each sponge also shared a large number of microbial taxa with the ambient water column. These authors classified symbiotic microbial taxa as "specialist" or "generalist" symbionts based on host specificity. OTUs present only in sponges and exhibiting more than 2% sequence divergence from free-living bacteria reported in GenBank were considered specialists, while OTUs shared with seawater communities and/or closely related (>98% identity) to environmental (i.e. non-symbiont) DNA sequences in GenBank were considered generalist symbionts. Specialist symbionts were further classified based on their abundance in each community, with (1) "dominant symbionts" present in all samples from a host species, (2) "common symbionts" present in more than one sample from a host species and (3) 'rare symbionts' present in only one sample from a host species. For example, the sponge *Haliclona tubifera* hosted a single dominant specialist *gammaproteobacterium*; the remainder of *H. tubifera*'s symbiotic microbial community was not statistically different from seawater samples. Thus, some sponges do not fit the ideal of a "uniform microbial community" and instead host specialized symbionts that are phylogenetically distinct from previously proposed SSSCs.

The assertion that sponges host a "uniform community" has generated much debate among researchers, as it suggests that sponges have retained a common microbial community over millions of years of evolutionary change. This hypothesis also implies that microbial symbionts are predominately vertically transmitted from parents to offspring. Counterarguments to this hypothesis suggest that the observed bacterial communities simply reflect taxa that are abundant in the water column or that are enriched by a host sponge's filtration. In these cases, free-living seawater bacteria may adapt to exploit the ecological niches present within sponges, suggesting that bacterial colonization drives the evolution and ecology of these associations. These alternatives imply that symbiotic communities are predominately acquired from the surrounding water column (are horizontally transmitted). Below, we consider the specificity of these microbial communities and address these two modes of transmission.

Recent work by Erwin *et al.* (2011a) provides another interpretation of the presence of common microbial symbionts across diverse sponge hosts. Focusing their clone library analyses on three Mediterranean species of *Ircinia*, these investigators recovered 56 monophyletic sequence clusters, with 48% of these exclusively sponge associated. The sponge-associated clades frequently included symbionts previously reported from geographically and phylogenetically distant host taxa, suggesting that many of these symbionts are generalists that can inhabit a wide diversity of sponge hosts.

Despite the lack of host specificity found for most individual symbionts, statistical analyses revealed that each *Ircinia* species hosts a unique symbiotic community. Erwin *et al.* (2011a) reasoned that the patchy distributions of symbionts among sponge hosts could be structured by four non-exclusive processes: habitat-specific environmental factors; host-specific microenvironments; host–symbiont coevolution through vertical transmission; and periodic horizontal transmission. Interactions among these processes could form complex community structures that cannot be easily interpreted as exclusively vertically or horizontally transmitted.

3. Specificity of Sponge–Microbe Associations

Although symbioses may be generalized, with multiple potential hosts and symbiont partners, or specialized with a single host species and a single symbiont species, many symbioses fall along a continuum between these two extremes. The costs and benefits of generalization and specialization have been examined for a wide diversity of symbiotic interactions, including those between pollinators and plants (Thompson, 2005), rhizobia and legumes (Young *et al.*, 2003), ectomycorrhizae and plants (Bruns *et al.*, 2002), and dinoflagellates and corals (Santos *et al.*, 2004). For many of these interactions, a high degree of host specificity is associated with increased benefits to the host (Bruns *et al.*, 2002; Thompson, 2005). Alternatively, many coral species host different genotypes of dinoflagellates in different habitats, reflecting both environmental specialization and horizontal transmission of the symbionts (Rowan, 1998; Knowlton and Rohwer, 2003; Rowan, 2004). Sponge–symbiont associations may be most analogous to the microbial communities found in metazoan guts, since these associations contain a high diversity of both resident and transient microbial populations whose potential costs and benefits vary (Drake and Horn, 2007; Oh *et al.*, 2010).

3.1. Sponge-specific sequence clusters

As mentioned above, investigations of sponge–microbe associations often undertake an in-depth characterization of the microbial community associated with one or few sponge species, with many such studies compiled in recent reviews (Taylor *et al.*, 2007b; Webster and Taylor, 2011). This approach assesses the specificity of microbial symbionts by constructing phylogenetic trees using microbial *16S* rRNA sequences derived from the sponge host(s) of interest along with reference sequences obtained from GenBank. Ideally, such studies also obtain *16S* rRNA sequences from the target sponge's surrounding environment (including the water column, sediments, and other invertebrate hosts). When examining the constructed phylogeny, one hypothesizes that sequences that are reported exclusively from sponges can be considered an SSSC, and that this cluster represents a clade of microbes that are specifically adapted for life as sponge symbionts (Taylor *et al.*, 2007b).

However, there are several caveats to this approach. Firstly, if a particular sequence cluster also contains sequences of microbes living in the open water column, in the sediment, or associated with nearby invertebrates, does this pattern indicate that these microbes are horizontally transmitted among sponge hosts? Alternative explanations include: (1) these microbes are contaminants from the field or laboratory setting; or (2) these microbes are found in the water column and thus can be temporarily present in a sponge's canals. For example, the large-scale next-generation sequencing study by Webster *et al.* (2010a) found that sequences affiliated with 17 of 33 SSSCs were also present in samples of ambient seawater. These authors concluded that SSSCs not detected in seawater were likely to be vertically transmitted symbionts, while the rare observations of SSSCs found in seawater were likely to be horizontally transmitted symbionts. Alternatively, it is also possible that these sequences are derived from bacteria that transit host sponges without forming a stable symbiosis.

A second issue to consider is a lack of statistical rigour in evaluating hypotheses in studies focused on SSSCs. Although some SSSCs are clearly supported under any method of phylogenetic reconstruction, others are not. Approaches using SSSCs should endeavour to apply quantitative hypothesis testing of these clusters using methods designed to test associations between phylogeny and ecological traits (e.g. Ronquist, 2004; Kraft *et al.*, 2007; Hunt *et al.*, 2008). For example, one particularly powerful approach combines tests of phylogenetic lineage sorting (*P* tests) with analyses of molecular variation (*F* tests) to determine whether two sets of sequences are drawn from the same community (Fig. 2.2; Martin, 2002), but this approach has been applied to few studies of sponge–microbe symbioses (Taylor *et al.*, 2007b; Erwin and Thacker, 2008a; Erwin *et al.*, 2011a,b). Although Taylor *et al.* (2007b) used a *P* test to compare clone libraries across host sponge taxa, finding significant differences for all pairwise comparisons among host species, sponge-derived libraries were not compared to any libraries derived from the ambient environment, thereby confounding host taxa with their surrounding environments. Although the presence of SSSCs suggests that a particular microbe can form a stable, highly specific, and vertically transmitted symbiosis, additional methods of investigation are needed to determine whether or not this hypothesis is correct, as these sequence clusters could also represent horizontally transmitted symbionts, transient members of a community, or environmental contamination. A different way to assess this specificity is to not focus on individual host species, but instead, to examine the prevalence of a symbiont across many host species.

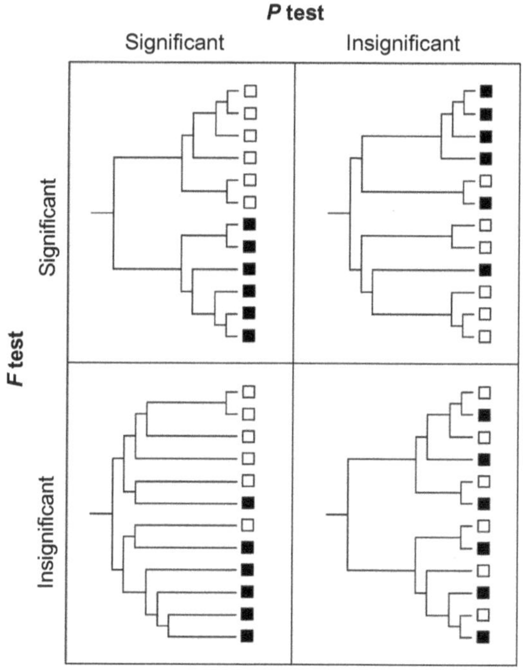

Figure 2.2 Examples of phylogenetic trees depicting patterns of relationships that would result in the four possible outcomes of two tests of phylogenetic differentiation. The open and solid squares represent two different communities from which sequences are obtained, for example, a host sponge and its surrounding environment. These trees are drawn with branch lengths proportional to the amounts of sequence evolution. Clear differentiation is evident if both F_{ST} and *P* tests are significant. By contrast, insignificance for both tests implies that the samples from two communities are drawn from the same pool of sequences. A significant F_{ST} test coupled with an insignificant *P* test implies that the tree contains several clades of closely related bacteria that are unique to one community or the other but that these clades are interspersed throughout the phylogenetic tree of all samples. Finally, a significant *P* test coupled with an insignificant F_{ST} test might reflect the existence of highly divergent lineages within each community (such that the within-community diversity approaches the total diversity) but indicates that there is significant covariation between community and phylogeny. Adapted from Martin (2002).

3.2. Comparisons of host and symbiont phylogenies

The specificity of host–symbiont associations is often examined by comparing phylogenies constructed for both the symbionts and their hosts (Ronquist, 1997; Page, 2003). This approach can be used to quantify whether symbiont clades are more specialized to host clades than expected by chance and to determine whether strict host–symbiont cospeciation occurs (Clark *et al.*, 2000; Thompson, 2005; Hosokawa *et al.*, 2006).

However, it is important to recognize that these phylogeny-based methods are only as reliable as the underlying phylogenetic information. To date, phylogenetic comparisons of sponge–microbe symbioses have focused on two major groups of cyanobacterial symbionts reported from marine sponges, filamentous *O. spongeliae* and unicellular *Candidatus* "*S. spongiarum*" (Steindler *et al.*, 2005; Diaz *et al.*, 2007). Following initial reports on the potential specificity and coevolution of these symbionts, the construction of additional phylogenies has changed our views of each of these interactions.

3.2.1. Case study: *S. spongiarum*

Unicellular, *S. spongiarum*-like symbionts were originally described as *Aphanocapsa feldmani* by investigators using electron microscopy to identify sponge symbionts (Rutzler, 1990; Usher *et al.*, 2006). Subsequent investigations based on molecular phylogenetic techniques placed these symbionts into the genus *Synechococcus* (Usher *et al.*, 2004; Thacker, 2005), and additional surveys revealed many common Caribbean hosts of *S. spongiarum*, including *Aplysina fulva*, *Neopetrosia subtriangularis*, and *Xestospongia muta* (Erwin and Thacker, 2007). Usher *et al.* (2001) provided evidence for the vertical transmission of these symbionts, observing the presence of cyanobacterial cells in eggs and sperm. Although vertical transmission suggests a high likelihood of strict cospeciation between hosts and symbionts, initial molecular phylogenetic analyses based on *16S* ribosomal RNA (rRNA) gene sequences suggested no specialization of these symbionts in particular host species (Hentschel *et al.*, 2002; Steindler *et al.*, 2005; Thacker, 2005).

The *16S* rRNA gene is contiguous with an ITS region that itself is contiguous with the gene that encodes the cyanobacterial *23S* rRNA subunit. For *Synechococcus* spp., the spacer region contains two transfer RNAs, *tRNA-Isoleucine* and *tRNA-Alanine* (Rocap *et al.*, 2002). The presence of these tRNAs can be used to amplify and sequence the *16S–23S* ITS regions from mixed cultures of cyanobacteria (Iteman *et al.*, 2000; Rocap *et al.*, 2002). Using this approach, Erwin and Thacker (2008a) found significant levels of ITS sequence divergence among clades of *S. spongiarum* that displayed nearly identical (>99% similarity) *16S* rRNA gene sequences, revealing a tremendous amount of cryptic diversity. This study also applied quantitative statistical techniques to examine the sorting of phylogenetic lineages of symbionts among host species and analyses of molecular variation to examine the interactive effects of geography and host phylogeny (Fig. 2.2; Martin, 2002). The genetic divergence of *S. spongiarum* clades showed statistically significant variation among host species, indicating a degree of host specialization and reflecting the previously documented vertical transmission of these symbionts. However, there was also a statistically significant geographic effect, suggesting that environmental conditions and horizontal transmission also influence the distribution of *S. spongiarum* among host species (Erwin and Thacker, 2008a).

3.2.2. Case study: *O. spongeliae*

The filamentous cyanobacterium *O. spongeliae* has been reported from a variety of Indo-Pacific and Caribbean sponges based on both morphological (Berthold *et al.*, 1982; Rutzler, 1990) and molecular (Thacker and Starnes, 2003; Ridley *et al.*, 2005; Thacker, 2005; Thacker *et al.*, 2007) evidence, with reported hosts including members of the dictyoceratid genera *Dysidea, Lamellodysidea, Lendenfeldia,* and *Phyllospongia,* as well as representatives of haplosclerid genera such as *Xestospongia* and *Haliclona. O. spongeliae* filaments are extracellular, approximately 10 μm wide, and range in length from 5 to 50 cells (Hinde *et al.*, 1994). Due to the diversity of chemical compounds reported from its hosts, *O. spongeliae* has been the subject of a tremendous number of natural products chemistry studies, which have suggested that the cyanobacterium produces many of the chemical compounds isolated from its host sponges (Unson and Faulkner, 1992; Unson *et al.*, 1994; Flowers *et al.*, 1998; Flatt *et al.*, 2005). Indeed, differences in chemistry among sponge collections have been attributed to differences in the symbiont genome rather than the host's genome or environmental influences (Flatt *et al.*, 2005).

Thacker and Starnes (2003) sequenced *16S* ribosomal RNA genes amplified from three species of Indo-Pacific sponges (in the genera *Dysidea* and *Lamellodysidea*) and their *O. spongeliae* symbionts. A comparison of the resulting phylogenies revealed complete host specificity, in which each sponge species hosts a unique clade (or phylotype) of *O. spongeliae.* However, the use of only three host taxa in these analyses prevented any conclusions regarding the likelihood of cospeciation. Additional sampling of Indo-Pacific hosts by Ridley *et al.* (2005) supported a hypothesis of strict cospeciation between hosts and symbionts but also provided qualitative evidence supporting one host-switching event among the Indo-Pacific hosts. However, this single event was not supported by a statistical comparison of host and symbiont phylogenies using Bayesian constraints (Page, 2003; Ridley *et al.*, 2005).

Subsequent faunal surveys revealed that *O. spongeliae* symbionts were also associated with several Caribbean sponges, including the dictyoceratid *Hyrtios violaceus* and two haplosclerid taxa, *Haliclona walentinae* and *Xestospongia bocatorensis* (Diaz *et al.*, 2007; Thacker *et al.*, 2007). The addition of the Caribbean taxa to the phylogeny of *O. spongeliae* did not support a hypothesis of strict cospeciation. If a thorectid ancestor were initially colonized by *O. spongeliae,* then two independent colonizations of Dysideidae and two independent colonizations of two different haplosclerid genera were needed to generate the observed phylogenetic pattern (Thacker *et al.*, 2007). The lack of similar filamentous *Oscillatoria* in other species of *Haliclona* and *Xestospongia* also supports a hypothesis of multiple colonizations of these hosts. Although we have recently discovered another Caribbean species of *Haliclona* that hosts *O. spongeliae* (Cristina Diaz and Robert Thacker, unpublished), our most recent dataset supports a hypothesis of independent colonization of these hosts by the symbionts and suggests that cospeciation events might only occur within genera (Thacker *et al.*, 2007). Additional data on the molecular systematics of the host and non-host sponges are clearly needed to quantify the contribution of sponge hosts to these patterns. Such work is currently underway, sponsored by the U.S. National Science Foundation as the Porifera Tree of Life project (www.portol.org).

3.3. Specificity is influenced by ecological interactions and dispersal

These case studies provide two examples of how our views of sponge–symbiont associations have changed as new evidence has been gathered. For *S. spongiarum*, the original concept of a homogeneous symbiont consisting of a single genotype has been replaced by evidence of multiple genotypes that display varying degrees of host and geographic specificity. For *O. spongeliae*, a hypothesis of strict cospeciation has been replaced by a hypothesis of multiple host colonization events, yet retaining evidence of strict specialization, since a single symbiont phylotype is associated with each host species. These observations suggest that different types of symbionts can have different degrees of host specificity, even within the same host sponges. The degree of host specificity may also be influenced by the ecological nature of a symbiosis; for example, whether it represents an obligate or facultative interaction and whether the outcome of the symbiosis is harmful or beneficial to the host.

Dispersal ability may play another key role in determining the extent of host specificity (Thompson, 2005). Symbionts able to disperse independently from their host may become less specialized, while symbionts that cannot disperse from their host may become more specialized. Law (1985) predicted that mutualistic endosymbionts would lack strong specificity to particular host species, as the symbionts would become so accommodating to their hosts that transfer among unrelated hosts would be favoured by selection. Therefore, early in the evolution of a mutualism, endosymbionts may evolve extreme specialization, but over time, host specificity may decrease as the endosymbionts become able to form associations with other hosts; however, this hypothesis relies on dispersal of symbionts among different host species (Thompson, 2005). Symbioses formed by *Synechococcus* and *Oscillatoria* may reflect different evolutionary stages, with *Oscillatoria* currently more specialized and *Synechococcus* more generalized. A reduced potential for dispersal by *Oscillatoria* may generate its extreme host specificity, while an ability to disperse among hosts in local environments could generate the patterns of specificity displayed by *Synechococcus*. These hypotheses are intertwined with our next topic, which considers the prevalence of vertical and horizontal transmission in sponge–microbe symbioses.

4. Modes of Symbiont Transmission

Symbionts can be transmitted from parents to offspring by two primary modes: vertical and horizontal transmission. Vertical transmission implies that symbionts are passed into either gametes or brooded embryos by a parent (Buchner, 1965; Mira and Moran, 2002), while horizontal transmission usually occurs through a free-living stage of the symbiont and is often represented as an environmental acquisition of a symbiont (Nyholm and McFall-Ngai, 2004; Heath and Tiffin, 2009). A mixture of the two primary modes of transmission can be termed "leaky vertical transmission", whereby symbionts are both passed vertically via gametes or embryos and also acquired from the surrounding environment (Fig. 2.3; Vrijenhoek, 2010). Each of these modes of transmission provides different life-history advantages and disadvantages (Ewald, 1987).

Under vertical transmission, offspring receive symbionts that are identical to those of the parent, with an advantage of "symbiont assurance" (Vrijenhoek, 2010), in that all offspring will immediately host the symbionts required for growth. However, after a number of host generations, many obligate symbionts display a potential cost of vertical transmission: a reduction in genome size and a consequent loss of metabolic capabilities (Moran, 2002). Moreover, long-distance dispersal of offspring can reveal a second disadvantage of vertical transmission: a particular strain of symbiont may not be optimal in all habitats. Strong natural selection for locally optimal and genetically diverse strains of symbionts can thereby favour hosts that acquire their symbionts from the local environment (Won *et al.*, 2003; Vrijenhoek, 2010). Despite this advantage, horizontal transmission presents its own additional risks. Firstly, offspring might risk dispersing to a habitat where symbionts are absent. Secondly, environmental acquisition can lead to mixed genotypes of symbionts present within a single host, providing an opportunity for the evolution of "cheater" symbionts that exploit the host (Frank, 1996). Thirdly, environmental acquisition implies the existence of pathways for colonization of the host by beneficial symbionts; pathogens can potentially employ these same pathways to infect a host. Thus, host species that exhibit various degrees of leaky vertical transmission might reflect diverse evolutionary solutions that balance the costs and benefits of purely vertical versus horizontal modes of transmission (Vrijenhoek, 2010).

As noted in Section 3.1, obligate vertical transmission is expected to result in highly host-specific lineages of symbionts within each host species.

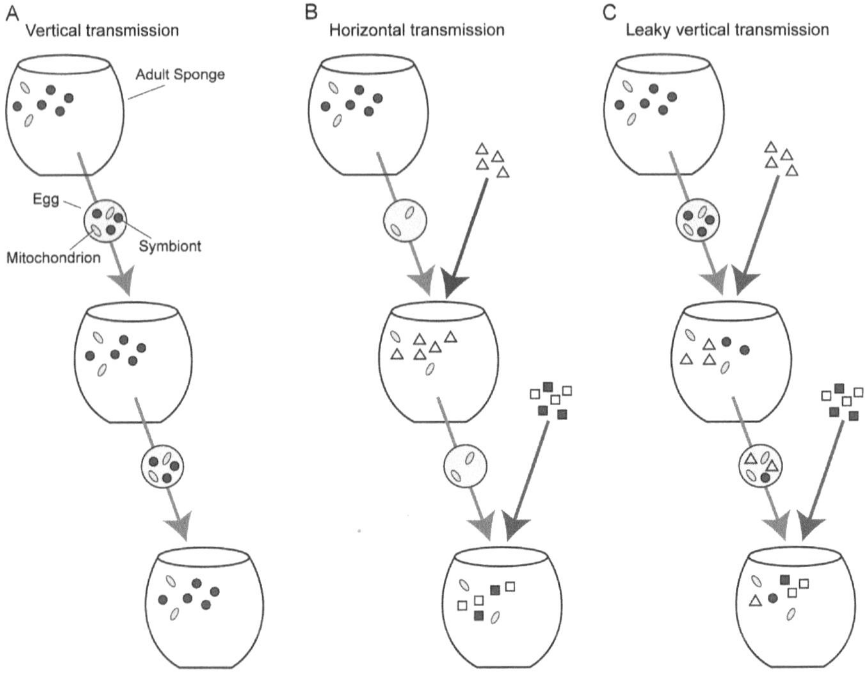

Figure 2.3 Modes of microbial symbiont transmission illustrated for a hypothetical sponge. The white ovals represent mitochondria, while circles, triangles, and squares represent symbiotic microbes. (A) Under obligate vertical transmission, the symbiont is inherited maternally along with the mitochondria, leading to genetic coupling of mitochondrial and symbiont genotypes. (B) Under horizontal transmission, symbionts are acquired anew in each generation, decoupling the cytoplasmic organelles from symbiont genotypes. Furthermore, infections involving environmental bacteria will likely lead to mixed symbiont genotypes within a host and differences among hosts over time and space. (C) Leaky vertical transmission is predominantly vertical with occasional environmental acquisition, or vertical with massive environmental swamping. In either case, the vertical transmission component will create a small lag time in the decoupling of host mitochondrial and symbiont genotypes. The horizontal component, depending how prevalent, can create mixed genotype infections. Figure adapted from Vrijenhoek (2010). (For a colour version of this figure, the reader is referred to the online version of this chapter.)

This expected pattern underlies the search for SSSCs in surveys of sponge microbiomes (Taylor *et al.*, 2007b). In contrast, horizontal transmission and leaky vertical transmission can retain mixed populations of symbionts within each lineage for multiple generations of hosts. When comparing host and symbiont phylogenies, vertical transmission is expected to yield patterns of strict cospeciation, but horizontal transmission can also generate patterns of strict cospeciation if the colonization of hosts is highly regulated or tied to allopatric speciation events (Hafner and Nadler, 1988; Clayton *et al.*, 2003;

Vrijenhoek, 2010). Modes of transmission may also be correlated with genetic variability and physiological capabilities. For example, in Red Sea soft corals, Clade A *Symbiodinium* are vertically transmitted, while Clade C *Symbiodinium* are horizontally transmitted, with Clade A being capable of coping with a wider range of environmental conditions (more generalized), and Clade C being more specialized for particular conditions (Barneah *et al.*, 2004).

4.1. Vertical transmission

Studies of vertical transmission of microbial symbionts are more easily accomplished if (1) the hosts can be observed at various reproductive stages, including adults, gametes, larvae, juveniles; and (2) the symbionts can be easily distinguished using microscopic techniques. For example, Usher *et al.* (2001) visualized the transmission of *S. spongiarum* in the host sponge *Chondrilla australiensis* using electron microscopy. In this species, nurse cells that surround and provision the developing eggs also appear to transmit symbiotic cyanobacteria. Similarly, Ereskovsky *et al.* (2005) used electron microscopy to observe bacteria in the oocytes of the host sponge *Halisarca dujardini*. The bacteria were observed in vacuoles that entered the oocyte through endocytosis. In contrast, Collin *et al.* (2010) observed the transmission of large, filamentous cyanobacteria (*O. spongeliae*) from adult tissue to developing embryos, hypothesizing that the cyanobacterial filaments actively colonize the embryos (Fig. 2.4).

Vertical transmission can also be examined using molecular tools; however, as described above, patterns of strict cospeciation can also result from horizontal transmission. Thus, a molecular approach is strongest when symbiont-derived DNA sequences are used to design probes for fluorescent *in situ* hybridization (FISH). For example, Sharp *et al.* (2007) used a FISH approach to confirm the vertical transmission of three different phylotypes of bacteria by observing cells stained with specific probes in the mesohyl and embryos of the host sponge *Corticium candelabrum*. A similar approach confirmed the vertical transmission of bacteria in the sponge *Tedania ignis* (Fig. 2.5). Only a few studies have used experimental manipulations to document vertical transmission. For example, Schönberg and Loh (2005) transplanted the bioeroding sponge *Cliona orientalis* among corals over a broad geographic range, demonstrating that this sponge acquires its dinoflagellate symbionts through vertical transmission rather than from the corals that it erodes.

4.2. Horizontal transmission

Evidence for horizontal transmission in sponge–microbe symbioses is mostly derived from multiple molecular-based surveys of symbiotic microbial communities. In cases that report phylogenetically similar symbionts

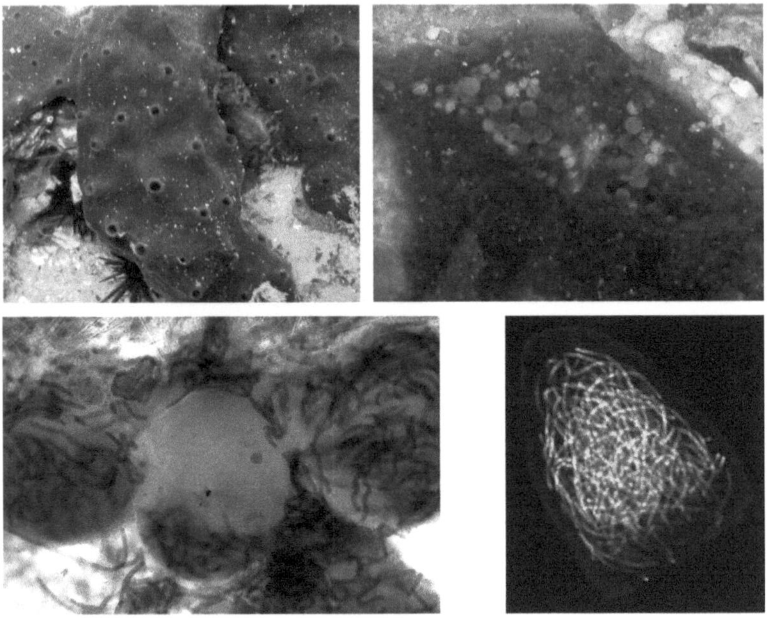

Figure 2.4 Filamentous cyanobacterial symbionts through different life stages of the marine sponge *Xestospongia bocatorensis*. Upper left, the adult sponge *in situ*; upper right, brood chambers on the underside of the sponge contain eggs and developing embryos; lower left, filaments of the symbiotic cyanobacterium *Oscillatoria spongeliae* colonize embryos of *X. bocatorensis* during development; lower right, a free-swimming larva with cyanobacterial filaments visible as chlorophyll autofluoresence. (For a colour version of this figure, the reader is referred to the online version of this chapter.)

(e.g. $<1\%$ divergence in *16S* rRNA gene sequences), horizontal transmission is often invoked as the most likely mechanism to explain these patterns (Enticknap *et al.*, 2006; Erwin and Thacker, 2008a). Similarly, the recent pyrosequencing study of Webster *et al.* (2010a) revealed that several SSSCs proposed by Taylor *et al.* (2007b) include sequences from the ambient environment. These authors argued that these disruptions of SSSCs indicate that these symbionts are horizontally transmitted. Additional evidence for environmental acquisition of microbial symbionts by host sponges has been reported for the Mediterranean sponge, *Petrosia ficiformis*, where microbes are absent in the gametes and larvae, yet occur abundantly in adult sponge tissues (Maldonado and Riesgo, 2009).

In the squid–*Vibrio* and legume–rhizobium symbioses, experiments with cultured symbionts have proved powerful tools for elucidating the mechanistic basis of symbiont acquisition and the ability of hosts to control symbiont uptake (Nyholm and McFall-Ngai, 2004; Heath and Tiffin,

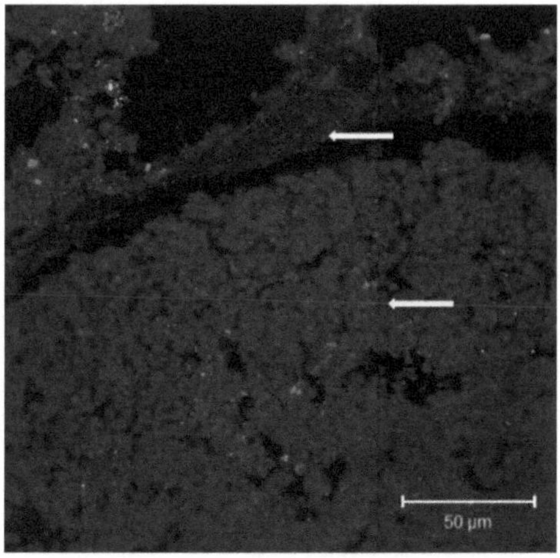

Figure 2.5 Confocal micrograph of microbial symbionts localized to a developing embryo (below) and its surrounding nurse cells (above) in the marine sponge *Tedania ignis*. Fluorescent *in situ* hybridization used a eubacterial probe (magenta) and a negative control probe (green), following methods described by Sharp *et al.* (2007). (For interpretation of the references to colour in this figure legend, the reader is referred to the online version of this chapter.)

2009). Although experiments using labelled sponge-derived symbionts have demonstrated that sponges do have an ability to discriminate between prey and symbiotic bacteria (Wilkinson *et al.*, 1984; Wehrl *et al.*, 2007), additional manipulative experiments are needed to examine the potential for horizontal transmission among sponge hosts, as these could reveal more detail about the pathways by which these symbionts colonize their hosts.

4.3. Leaky vertical transmission

A mixed model of transmission has been supported by several recent studies of sponge–microbe symbioses and may represent the most common evolutionary outcome of these associations. Enticknap *et al.* (2006) cultured an alphaproteobacterium from seven distinct sponge species and found that it was closely related to a previously cultured symbiont (Webster and Hill, 2001) isolated from yet another sponge species. This symbiont, originally termed strain NW001, is closely related to *Pseudovibrio denitrificans* (Webster and Hill, 2001; Enticknap *et al.*, 2006). The presence of closely related phylotypes in diverse sponge hosts suggests this alphaproteobacterium is horizontally transmitted, but it has also been visualized in sponge larvae using FISH. This symbiont may therefore be exploiting both horizontal and vertical modes of transmission.

Similarly, two studies by Schmitt *et al.* (2007, 2008) document a mixture of vertical and horizontal transmission in the Caribbean sponges *Ircinia felix* and *Ectyoplasia ferox*. By combining analyses of *16S* rRNA gene sequences derived from adults, larvae, and juveniles with microscopic observations (transmission electron microscopy and FISH), these studies suggest that adult sponges may vertically transmit a complex microbial community to their offspring, yet simultaneously lose or gain symbionts from the external environment. The mixed model proposed by Schmitt *et al.* (2008) is therefore quite similar to the models of leaky vertical transmission proposed for other host–symbiont interactions (e.g. Vrijenhoek, 2010).

5. Ecological Nature of Symbioses: A Balance of Costs and Benefits

As previous sections of our review have noted, the ecological nature of a symbiosis can influence (and be influenced by) symbiont community diversity, host specificity, and modes of transmission from parents to offspring. Although sponge–microbe symbioses are often assumed to represent coevolved, mutually beneficial interactions, basic ecological theory requires manipulative experiments to determine whether an interaction is a mutualism, with benefit to both partners; a commensalism, with benefit to one partner, but no benefit or harm to the other; or a parasitism, with benefit to one partner and harm to the other (Boucher, 1985; Begon *et al.*, 2005). Without experimental manipulations, we can only conclude that some microbes are found within sponges and that these microbes are capable of certain metabolic pathways; such data provide only circumstantial evidence that diverse microbial associations actually benefit a host sponge (Webster and Blackall, 2009). Indeed, the most recent review by Webster and Taylor (2011) begins by declaring a need for an expansion of the field of sponge microbiology beyond the study of phylogenetic diversity of microbes within sponges. Moreover, evolutionary ecologists emphasize a view of symbiosis as a balance of conflict and cooperation between the host and symbiont (Herre *et al.*, 1999; Bronstein *et al.*, 2003; Gomulkiewicz *et al.*, 2003). This view accentuates mechanisms that can change the outcomes of symbiotic interactions, placing mutualism, commensalism, and parasitism on a continuum that can be strongly influenced by environmental conditions, resource availability, symbiont population size, and interactions with other members of an ecological community (Holland *et al.*, 2002; Gomulkiewicz *et al.*, 2003; Neuhauser and Fargione, 2004). A key parameter in determining these outcomes can be the abundance of symbiont populations within a particular host—not only the total abundance of symbionts but also the relative abundance of individual taxa (Holland *et al.*, 2002; Palmer *et al.*, 2010). In this section, we review our current knowledge of the ecology of sponge–microbe symbioses; the following section describes recent advances in analytical techniques that permit explicit manipulations of these interactions and hypothesis testing based on ecological theory.

5.1. Abundance of microbial symbionts

Microscopic studies by Vacelet and Donadey (1977) first noted that some sponge species host abundant communities of symbiotic microbes (even terming these "bacteriosponges"; Reiswig, 1981), while other sponge species appear to host very few microbial symbionts. Recently, a more formal delineation has divided sponges into two groups: high microbial abundance (HMA) and low microbial abundance (LMA) sponges (Hentschel *et al.*, 2003; Weisz, 2006; Weisz *et al.*, 2008; Schläppy *et al.*, 2010). This dichotomy raises an important series of questions, for example: Why do not all sponges host large microbial populations? Can some sponges better control microbial growth than others? Do some sponges exploit their microbial populations, requiring a high density for optimal growth rates? And finally, is microbial abundance related to the possibility that these symbionts act as mutualists, commensals, or parasites?

Several studies have sought to address these questions using quantitative approaches. Vacelet and Donadey (1977) noted that HMA sponges have a denser mesohyl and longer, narrower canals, which should be correlated with a slower rate of water flow through the sponge. Weisz *et al.* (2008) tested this hypothesis by measuring mesohyl density and water pumping rates for a suite of LMA and HMA sponges, finding higher density and slower pumping rates in HMA than in LMA sponges. These authors conclude that the distinction between LMA and HMA sponges reflect two different life-history strategies, with LMA sponges being capable of pumping larger amounts of water, thereby obtaining a larger amount of nutrition in the form of particulate organic matter (POM). In contrast, the denser mesohyl of HMA sponges might be better suited to hosting abundant microbial communities, with a slower pumping rate allowing more time for symbiotic bacteria to acquire dissolved organic matter (DOM) and inorganic nutrients from circulating seawater (Reiswig, 1974, 1981; Yahel *et al.*, 2003; Weisz *et al.*, 2008). Moreover, since seawater pumped through the mesohyl oxygenates sponge tissues, and sponges that are not actively pumping can become anoxic (Schönberg *et al.*, 2004; Hoffmann *et al.*, 2005a, 2008; Schläppy *et al.*, 2007), temporal and/or spatial variation in oxygenation within a host sponge could act to regulate the availability of anoxic niches within host tissues, impacting the physiology of anaerobic microbes (Schläppy *et al.*, 2010). Indeed, Hoffman *et al.* (2005b) found a correlation between anoxic conditions and sulphate reduction (an anaerobic microbial process) within the sponge *Geodia barrette* (Schläppy *et al.*, 2010). Clearly, additional research is needed to determine which players drive the structural and physiological differences observed between LMA and HMA sponges. Are these differences fixed, or can host sponges alter their body plans and pumping rates to control the abundance and physiology of their symbionts? Alternatively, can microbial symbionts excrete chemical signals to manipulate host morphology and behaviour?

Although some studies have portrayed LMA sponges as devoid of associated microorganisms, LMA species may also receive critical benefits from their microbial symbionts. For example, Schläppy *et al.* (2010) measured higher rates of nitrification and denitrification in an LMA sponge, *Dysidea avara*, than in an HMA sponge, *Chondroisa reniformis*. *D. avara* had previously been categorized as having virtually no microbial symbionts, based on transmission electron microscope (TEM) observations (Turon *et al.*, 1997), whereas observations based on fluorescence microscopy (FISH) revealed a microbial abundance comparable to that of marine sediments (Schläppy *et al.*, 2010). These results suggest that earlier categorizations of HMA and LMA sponges need to be re-evaluated using FISH techniques. Moreover, these data demonstrate that interactions between hosts and their microbial communities may differ significantly among sponges designated as HMA or LMA species. Despite a tendency to automatically label HMA sponges as examples of mutualisms based on microscopic observations, high chlorophyll *a* concentrations, or high respiration rates, these measures of abundance do not necessarily indicate the magnitude of metabolic processes and do not provide evidence of mutualistic or parasitic interactions; instead, experimental measurements and manipulations of host and symbiont metabolism are needed.

5.2. Mutualisms: Benefits to host sponges

In a mutualistic interaction, both partners gain benefits that outweigh the costs of the interaction. In sponge–microbe mutualisms, hypothesized benefits to the host sponge include fixed carbon and/or nitrogen, while the host might provide symbiotic microbes with essential nutrients and shelter (Wilkinson and Fay, 1979; Sarà *et al.*, 1998). While some sponges may gain this nutrition through the phagocytosis and digestion of their microbial symbionts (Rutzler, 1988; Maldonado and Young, 1998), other translocation mechanisms remain unknown (Wilkinson, 1979). Other benefits from cyanobacterial symbionts include the production of secondary metabolites (Unson *et al.*, 1994; Flowers *et al.*, 1998; Flatt *et al.*, 2005; Schmidt, 2008) that can serve a variety of ecological functions, including predator and competitor deterrence (Pawlik *et al.*, 1995; Thacker *et al.*, 1998; Engel and Pawlik, 2000) and resistance to malignant microbial infections (Garson, 2001; Thakur *et al.*, 2003).

Measurements of the relative benefits and costs of symbiotic associations can be best accomplished by experimentally manipulating these interactions. Our laboratory has focused on cyanobacteria–sponge interactions because shading experiments allow us to manipulate rates of photosynthesis by symbiotic cyanobacteria, altering the benefit gained by host sponges. As described above, studies by Wilkinson (1983, 1987) assessed benefits and costs by measuring photosynthesis to respiration (P:R) ratios. These studies generated several hypotheses concerning the nature of phototrophic sponges, including the suggestion that phototrophic symbioses were more prevalent in the Indo-Pacific region than in the Caribbean, where species were rarely demonstrated to be phototrophic. Subsequent P:R measurements at natural irradiances have revealed that, in fact, several Caribbean species are phototrophic (Thacker *et al.*, 2007; Erwin and Thacker, 2008b).

Although high P:R measurements suggest that a sponge can benefit from symbiont photosynthesis, measurements of growth in control and shaded conditions provide a better assessment of whether symbiont photosynthesis impacts host fitness. Not all host sponges respond in the same way to shading. Shading experiments revealed that species such as *Lamellodysidea chlorea* and *Chondrilla nucula* are extremely dependent on symbiont photosynthesis (Arillo *et al.*, 1993; Thacker, 2005); since these two species do not survive experimental shading, they likely form obligate mutualisms with their photosynthetic symbionts. Other taxa, such as *X. muta*, *Neopetrosia exigua*, and *N. subtriangularis* sometimes show no impact of shading, suggesting that their interaction could be a commensalism (Gómez *et al.*, 2002; Thacker, 2005; Erwin and Thacker, 2008b; Freeman and Thacker, 2011). *A. aerophoba* and *A. fulva* display reduced growth under reduced light availability but still maintain positive growth rates, suggesting a facultative mutualism (Wilkinson and Vacelet, 1979; Erwin and Thacker, 2008b) and a substantial contribution of cyanobacterial symbionts to host nutrition. As an alternative explanation, these latter five host sponges might not only receive supplemental nutrition from their symbionts but also vary in the plasticity of their nutritional modes. For example, the Mediterranean sponge *P. ficiformis* activates heterotrophic metabolic pathways in response to the loss of cyanobacterial symbionts (Arillo *et al.*, 1993). Similarly, many coral species are able to increase their heterotrophic feeding rates and replenish nutrient stores during bleaching, when symbiont populations are reduced or absent (Grottoli *et al.*, 2006). Investigations involving direct observations of filtration rates, filtration efficiency, and heterotrophic versus autotrophic nutrient assimilation will be needed to determine whether sponges possess similar flexibility in switching between different modes of nutrition. Notably, different phylotypes of *Synechococcus* and *Oscillatoria* dominate the photosymbiont communities of these diverse host sponges, such that these different experimental outcomes might be correlated with the presence of different symbionts. We currently lack a way to transfer symbionts among host sponges, but future manipulations with cultured symbionts would greatly strengthen our understanding of the mechanisms underlying nutritional symbioses.

5.3. Parasitism: Costs to host sponges and sponge diseases

Earlier sections of our review suggest that approximately one-third of coral reef sponges host cyanobacterial symbionts. An even larger proportion of sponges are known to host other types of bacteria, raising a question: if photosynthetic symbionts provide an important source of supplemental energy and carbon to host sponges, why do not all sponges host them? Hosting diverse communities of abundant microbial symbionts may involve physiological and ecological costs that are not yet well described for sponges, including (1) costs resulting from competition between symbiont and host cells for limited nutrients and (2) costs due to host control of symbiont cell division and population size (Smith, 1991). In addition, although sponges may offer protection from grazing by generalist herbivores to their symbionts, they may also attract specialist predators. For example, the opisthobranch mollusc *Tylodina perversa* prefers to consume tissues of the sponge *A. aerophoba* that contain high concentrations of cyanobacteria (Becerro *et al.*, 2003). Likewise, gastropterid molluscs associated with *Dysidea granulosa* seem to forage on this sponge's symbiotic cyanobacteria (Becerro *et al.*, 2006). Despite these potential costs, individual species of sponges hosting photosymbionts are often found to be more abundant than sympatric species that do not host photosymbionts (Erwin and Thacker, 2007). Quantitative modelling of sponge–microbe symbioses is needed to better understand the roles these costs may play in the ecology and evolution of these associations.

Microbial symbionts may also be the causative agents of some sponge diseases. Diseases of coral reef organisms have been reported more frequently in recent years (Harvell *et al.*, 1999; Rosenberg and Loya, 2004; Weil, 2004), and this increase has been attributed to environmental factors such as anthropogenic pollution and nutrient enrichment (Green and Bruckner, 2000; Sutherland *et al.*, 2004; Kaczmarsky *et al.*, 2005). Similarly, sponge diseases have also been reported with increasing frequency; these diseases have been the subject of several recent articles (Webster *et al.*, 2002; Cervino *et al.*, 2006; Cowart *et al.*, 2006; Olson *et al.*, 2006; Gochfeld *et al.*, 2007, 2012; Webster, 2007; Wulff, 2007; Luter *et al.*, 2010a,b; Maldonado *et al.*, 2010), but it remains unclear whether this increased disease prevalence is a consequence of increased monitoring of sponge communities or of increased degradation of marine ecosystems.

Webster *et al.* (2002) reported the isolation of a pathogenic bacterium from the sponge *R. odorabile*; this study remains the only identification of a causative agent of a sponge disease. Mukherjee *et al.* (2009) documented how this pathogen causes harm to the host: the pathogen produces a collagenase enzyme that degrades the skeletal fibres of *Rhophaloeides*. Interestingly, this pathogen has not been identified in other species of sponges, suggesting that it may be species specific (Webster and Taylor, 2011).

Olson *et al.* (2006) documented a disease termed "*Aplysina* red band syndrome" (ARBS), which affects approximately 15% of rope sponges in the *Aplysina cauliformis*–*A. fulva* species complex on reefs in the Bahamas, as well as smaller proportions of *Aplysina* spp. in Belize and Panama. This disease is characterized by a red band, composed of filamentous cyanobacteria, that surrounds a necrotic lesion (Fig. 2.6). As the band moves down the length of the sponge, the lesion expands; this progression weakens the sponge skeleton, which often breaks at the site of the lesion (Olson *et al.*, 2006). ARBS alters secondary metabolite production and microbial community structure within host sponges, but nutrient enrichment does not exacerbate these effects (Gochfeld *et al.*, 2012).

5.4. Balancing mutualism and parasitism

Since each partner of a symbiosis can respond to specific costs and benefits, mutualisms may represent a trade-off between conflict and cooperation (Herre *et al.*, 1999). In this view of symbiosis, changing environmental conditions are hypothesized to switch the outcome of symbioses between

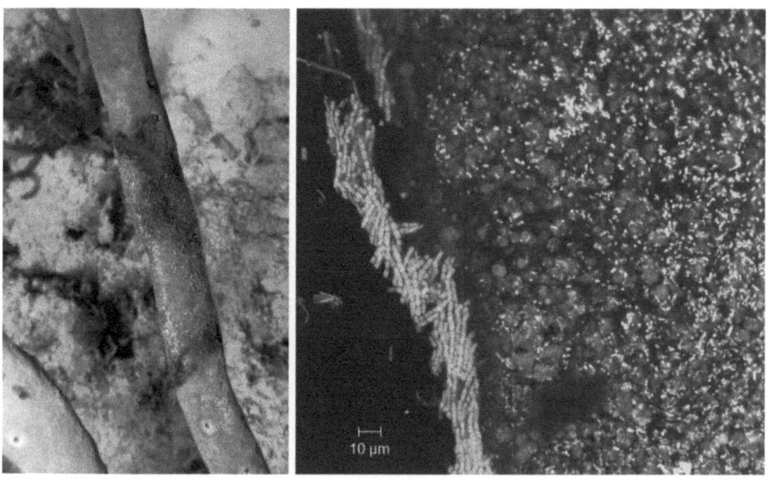

Figure 2.6 *Aplysina* red band syndrome (ARBS). Left: an ARBS lesion spreading along a branch of the Caribbean rope sponge, *Aplysina cauliformis* (Photo by Dr. D. Gochfeld). Right: confocal image of a cross-section through the leading edge of the red band. Bright areas of this image are from chlorophyll autofluoresence. The filamentous cyanobacteria on the outer (left) surface of the sponge are putative agents of ARBS, while the unicellular cyanobacteria inside the sponge are *Synechococcus spongiarum*. (For interpretation of the references to colour in this figure legend, the reader is referred to the online version of this chapter.)

net positive and net negative interactions. Thus, local environmental conditions may mediate trade-offs between microbial symbionts and marine sponges, with environmental changes disrupting cooperation between mutualists and creating a more competitive or parasitic interaction. For example, under shaded conditions, cyanobacterial symbionts are often absent or found at reduced abundance in potential sponge hosts (Rutzler, 1988; Sarà *et al.*, 1998; Gómez *et al.*, 2002), most likely because of the lack of available light for photosynthesis. However, it is not known whether these symbionts can disperse away from a shaded environment or whether they are more readily consumed by their hosts under shaded conditions. Rutzler (1988) described cyanobacterial symbionts overgrowing their host sponges, with cyanobacteria multiplying faster than the host sponge could phagocytose or expel them. Increased water temperature and increased nutrient availability might be able to stimulate rapid symbiont growth that cannot be controlled by the host sponge (Rutzler, 1988). With a high potential for both parasitic and mutualistic interactions, sponges and their microbial symbionts provide ideal model systems for examining the trade-offs between the costs and benefits of symbiosis.

Manipulations of symbiotic microbial communities have revealed differences in the stability of these communities among experimental treatments and host species. For example, T-RFLP analyses showed that the microbial community within *A. cauliformis* was not altered by nutrient enrichment but was significantly different in the presence of disease (Gochfeld *et al.*, 2012). Lemoine *et al.* (2007) held *Halichondria bowerbanki* in three different temperatures and observed no apparent changes in the density or diversity of symbionts using electron microscopy. However, DGGE analyses demonstrated gains and losses of particular symbionts under each temperature treatment. Similarly, Thoms *et al.* (2003) transplanted *Aplysina cavernicola* from deep to shallow sites, observing no changes by electron microscopy, but distinct sets of permanent versus variable DGGE bands. Gerce *et al.* (2009) cultivated *A. aerophoba* under a variety of conditions for 6 months but found a high degree of stability in the associated microbial community in all treatments, despite a loss of biomass and morphological degeneration during cultivation. Likewise, Friedrich *et al.* (2001) also cultivated *A. aerophoba* in aquaria but found that its microbial community could be disturbed with the addition of antibiotics. In contrast, Webster *et al.* (2010b) documented substantial changes in the microbial community of *R. odorabile* following 12 months of cultivation in large mesocosms, but not following cultivation *in situ* or in small aquaria.

The Caribbean giant barrel sponge *X. muta* provides an interesting case study for examining the response of a host sponge and its photosymbionts to environmental stresses, since this sponge displays cyclic bleaching (from which the sponge can recover over time) as well as fatal bleaching (Cowart *et al.*, 2006). López-Legentil *et al.* (2008) measured sponge chlorophyll *a* content as an index of photosymbiont abundance and heat-shock protein gene (*hsp70*) expression as an index of the host's stress response in sponges undergoing cyclic and fatal bleaching in the field, as well as in response to laboratory manipulations of temperature and salinity. Although chlorophyll *a* concentrations were lower in both cyclic and fatally bleached tissues, *hsp70* expression increased only in fatally bleached tissue. Further, although higher temperatures increased *hsp70* expression, chlorophyll *a* concentrations were not significantly impacted. Across all manipulations, *hsp70* expression and photosymbiont abundance were not correlated in this host, indicating that symbiont communities may not always provide an indicator of host responses to environmental stresses.

Stability of microbial community structure does not necessarily imply stability of community function. Although Freeman and Thacker (2011) found that nutritional transfer from microbial symbionts was stably maintained under shading for *A. cauliformis* and *N. subtriangularis*, nutritional transfer was not stable in *A. fulva*. Thus, the degree of stability of microbial community structure and function can vary among host taxa, especially in response to experimental manipulations. There is a clear need for additional studies that manipulate sponge–microbe interactions and measure changes in the functional processes (e.g. carbon/nitrogen metabolism, secondary metabolite production) of the symbiont community, as well as proxies of holobiont fitness, such as growth and reproduction. Metagenomic approaches to sponge–microbe symbioses (Thomas *et al.*, 2010; Liu *et al.*, 2011; Siegl *et al.*, 2011) are expected to provide greater insight into the mechanisms underlying these processes, including changes in gene expression and gene regulation, but such research will be most informative when coupled with experimental studies.

6. Novel Experimental Approaches

The manipulations of sponge–microbe interactions described above demonstrate that symbiont communities can change in response to experimental manipulations (Thoms *et al.*, 2003; Lemoine *et al.*, 2007; Gochfeld *et al.*, 2012) and that these changes can affect holobiont fitness (Thacker, 2005; Erwin and Thacker, 2008b). However, these experiments do not determine the mechanisms underlying symbiotic interactions. For example, if nutritional symbioses are assumed to be based upon a transfer of carbon or nitrogen, one must demonstrate that these transfers occur. Other fundamental questions concerning these interactions include: What specific compounds are transferred? Does transfer vary across host species? What taxa of microbial symbionts are active participants in nutrient transfer? Seminal work by Wilkinson (1979) and Borowitzka *et al.* (1988) clearly demonstrated the power of using radioisotopic ^{14}C to trace carbon fixation pathways in sponge–cyanobacteria symbioses. More recently, less hazardous analyses using the stable isotopes of carbon and nitrogen have allowed investigators to follow these nutrients through marine food webs and to trace the sources of nutrition for marine sponges (Lajtha and Michener, 1994; McCarthy *et al.*, 2007; Weisz *et al.*, 2007, 2010; Freeman and Thacker, 2011).

6.1. Stable isotope analyses

In stable isotope analyses, the abundance of the heavier stable isotope for carbon (^{13}C) and nitrogen (^{15}N) in experimental samples is compared to that of well-defined international standards and is expressed using the delta (δ) or permil notation (‰):

$$\delta X = \left(\frac{R_{\text{sample}}}{[R_{\text{standard}} - 1]} \right) \times 1000$$

where X is the heavy isotope (^{13}C or ^{15}N) and $R = {}^{13}C{:}{}^{12}C$ or $^{15}N{:}{}^{14}N$ (Lajtha and Michener, 1994; Freeman and Thacker, 2011). In food web analyses, $\delta^{13}C$ and $\delta^{15}N$ values can be used to determine the ultimate carbon source of a food web and the trophic level of various consumers within the system, respectively. This capability is due to the phenomenon of trophic enrichment, where the $\delta^{13}C$ values of a consumer's tissues are equal to or slightly enriched (by approximately 0.5–1.0‰) relative to those of a primary producer at the base of the food web (Michener and Schell, 1994). Likewise, the $\delta^{15}N$ values of a consumer's tissues are enriched between 2‰ and 5‰ (average of 3.4‰; Michener and Schell, 1994; Freeman, 2001) relative to its immediate diet. Thus, $\delta^{13}C$ values remain constant or increase by 0.5–1.0‰ at each level of a food web, while $\delta^{15}N$ values increase by about 3.4‰ with each trophic step. Although little is known about how sponges fractionate the carbon and nitrogen obtained by filter feeding on POM or derived from their microbial symbionts, the well-established enrichment values listed above are assumed to apply to marine sponges (Weisz *et al.*, 2007; Freeman and Thacker, 2011).

While stable isotope analyses might allow us to assess whether or not sponge hosts obtain carbon and/or nitrogen from POM or from their microbial symbionts, these analyses must be performed in a manner that correctly partitions the possible sources of nutrition for the host. Thus, to adequately quantify the proportion of carbon and nitrogen that the host receives from each source, one must isolate the microbial cells (source 1) from the sponge cells (sample) and analyse these separately, along with POM (source 2). By separating each of these two possible inputs, one can determine the relative contribution of each source to the host sponge using a two-end member stable isotope mixing model (Fry, 2006). Although the utility of separating host and symbiont cells prior to stable isotope analysis is well established in the coral–zooxanthellae symbiosis literature (Muscatine *et al.*, 1989; Muscatine and Kaplan, 1994; Tanaka *et al.*, 2006), it has only recently been performed in marine sponge–microbe interactions (Freeman and Thacker, 2011). In marine sponges, such separations are crucial when delineating between symbiont versus heterotrophic nutrient inputs because HMA sponges can contain a wide range of bacterial abundance, from 10^8 to 10^{10} bacterial cells per gram of sponge tissue, while LMA sponges contain 10^5 to 10^6 bacterial cells per gram (Hentschel *et al.*, 2006). Thus, if cell separations are not performed, differences among host species in the isotope values of "bulk" tissue (containing both microbial and sponge cells) can simply reflect differences in microbial biomass among hosts, thereby providing minimal data on the relative nutrient contribution of symbionts to hosts.

In the first comprehensive investigations of sponge–microbe symbioses using stable isotopes, Weisz (2006) and Weisz *et al.* (2007) used $\delta^{13}C$ and $\delta^{15}N$ values from bulk sponge tissue of 17 HMA and 9 LMA sponges from the Florida Keys to determine whether HMA sponges obtain nutrition from their abundant microbial communities. These authors concluded that enriched $\delta^{15}N$ values in LMA sponges reflect nutrient inputs from heterotrophic filter feeding, while depleted $\delta^{15}N$ values in most, but not all, HMA sponges reflect internal microbial transformations of nitrogen that might be transferred to the sponge host. For carbon, reduced variability in the $\delta^{13}C$ values of HMA sponges across sites was hypothesized to be due to symbiotic microbial communities within these sponges, while higher variability in the $\delta^{13}C$ values of LMA sponges was attributed to distinct sources of POM among sites. Despite using bulk sponge tissue, this study provided compelling evidence that stable isotope analysis can differentiate between symbiont- and heterotrophically derived nutrients in sponges. In large survey-based studies, bulk tissue analyses can play an important role by delineating sponges into distinct groups that can be further studied at a finer scale using experimental manipulations (Weisz, 2006; Weisz *et al.*, 2007; Freeman and Thacker, 2011).

Freeman and Thacker (2011) coupled manipulative shading experiments with $\delta^{13}C$ and $\delta^{15}N$ isotope analyses of isolated sponge and microbial cell fractions to examine nutritional symbioses in four common Caribbean sponge species. Comparisons of the $\delta^{13}C$ and $\delta^{15}N$ values of sponge cells with those of microbial cells and POM revealed that each of these four species forms a unique nutritional interaction with its symbionts. The sponge *Niphates erecta* is an LMA sponge that receives both carbon and nitrogen by filter feeding on POM. The sponges *A. cauliformis*, *A. fulva*, and *N. subtriangularis* are all HMA sponges, but *A. cauliformis* receives carbon and nitrogen from its symbionts, while *A. fulva* receives only nitrogen and *N. subtriangularis* receives only carbon from its symbionts (Freeman and Thacker, 2011). In addition, since two possible sources (POM vs.

microbial) of nutrients were isolated, a two-end member mixing model (Fry, 2006) could be used to determine the percentage of symbiont- versus heterotrophically derived carbon assimilated by each host species (Freeman and Thacker, 2011). This study provides the first direct evidence that host sponges are indeed obtaining nutrients from their microbial symbiont populations. Since the sponges *A. cauliformis* and *N. subtriangularis* are capable of maintaining the interaction with their symbionts, even under shaded conditions, while the response of *A. fulva* to shading was decoupled from that of its symbionts, it seems that the two former sponge species may be engaged in a more obligate mutualism with their symbionts, while *A. fulva* is engaged in a more facultative mutualism, as previously demonstrated by Erwin and Thacker (2008b). Moreover, the observed genetic diversity of *S. spongiarum* symbionts among these host sponges (Erwin and Thacker, 2008a) suggests that these different outcomes result from genetic differences among the various clades of *S. spongiarum*. Further experimentation is needed to determine whether these correlations reflect differences in gene expression by hosts, symbionts, or both.

6.2. Enriched stable isotope tracers

Studies of coral–zooxanthellae and sponge–zooxanthellae interactions have begun to use compounds that are enriched in heavy isotopes, such as $NaH^{13}CO_3$, to trace nutrient flow through a symbiotic system (Fry, 2006; Tanaka *et al.*, 2006; Weisz *et al.*, 2010). These compounds are ideal for tracing nutrient flow in these systems because typically only one member of the symbiosis is capable of metabolizing a particular tracer. For example, bacterial symbionts are capable of fixing $NaH^{13}CO_3$ into organic carbon, but the host sponge cannot assimilate large quantities of this compound without its partner. Likewise, since substrates containing enriched nitrogen ($^{15}NO_3$, $^{15}NH_4$, and $^{15}N_2$) can generally be assimilated only by microbial symbionts, host cells can only become enriched by nutrient transfer from symbiont to host, for example, through the transfer of enriched amino acids or sugars. By performing "pulse-chase" experiments, it is possible to "pulse" a sponge or coral holobiont by exposing it to an enriched tracer compound that is dissolved in seawater and then to follow the transfer of enriched compounds from symbionts to the host over a subsequent time interval, termed the "chase" period (Tanaka *et al.*, 2006; Weisz *et al.*, 2010).

Weisz *et al.* (2010) combined enriched stable isotope tracers with reciprocal transplant experiments to determine whether *Cliona* sponges obtain nutrients from their zooxanthellae symbionts. These authors found that the pinacoderm of *Cliona varians* was highly enriched in ^{13}C following a pulse with $NaH^{13}CO_3$. While this outer region of the sponge became less enriched over the course of a 1-week chase period, the internal choanosome became increasingly enriched. Although this experiment provides compelling evidence for a transfer of symbiont-derived nutrition from the pinacoderm (which contains the majority of symbiotic zooxanthellae in this sponge species) to the interior of the sponge, the conclusions that can be drawn from these data are limited by the lack of separation of host and symbiont tissue prior to isotope analyses (e.g. as in Tanaka *et al.*, 2006). To better trace the flow of symbiont-derived nutrients into host sponge tissue, future studies should determine whether isolated symbiont cells become less enriched as the host cell fraction becomes more enriched over the course of the chase period.

While pulse-chase experiments provide valuable information on the transfer of symbiont-derived carbon and nitrogen over longer time intervals, a more rapid assessment of carbon and nitrogen transfer can be performed by using filtered seawater laced with enriched compounds ($NaH^{13}CO_3$ and $^{15}NO_3$) in short-term (6 h) incubations in light and dark bottles, as described above for measurements of P:R ratios. We performed a pilot test of this methodology by combining ^{13}C enrichment with light and dark bottle incubations of the Caribbean sponges *A. fulva* and *A. cauliformis* (Fig. 2.7). In this experiment, three individual sponges of each species were divided into four pieces. For each individual, an initial piece was used for natural abundance ^{13}C measurements prior to enrichment and the other three pieces were assigned to a treatment bottle that was clear (light), amber (dark), or clear with two layers of shade cloth wrapped around it (shade). Each bottle was filled with filtered (0.70 μm mesh) seawater laced with $NaH^{13}CO_3$; sponges were incubated for 6 h. Subsequently, bacterial and host cell fractions were separated from bulk tissue as described by Freeman and Thacker (2011). Enrichment values are expressed as $\%^{13}C$ found in these samples. While there was no enrichment in the dark samples relative to the initial samples, the shade and light incubations suggest that the assimilation of $NaH^{13}CO_3$ by photosymbionts and the resulting transfer of enriched compounds to the host differs under varying irradiances (Fig. 2.7). While this experiment provides direct evidence for variation in carbon transfer under different irradiances, nitrogen transfer from symbionts to hosts appears to be much more complex (Weisz *et al.*, 2010; Freeman and Thacker, 2011). Experiments that manipulate a variety of compounds enriched in ^{15}N (e.g. $^{15}NO_3$, $^{15}NH_4$) could allow us to better delineate the nitrogen cycle within marine sponges and determine whether the sponge host acquires the products of microbially mediated nitrogen transformations.

6.3. Identifying compounds translocated between symbionts and their hosts

Although the pulse-chase experiments and light–dark bottle incubations described above provide persuasive evidence that enriched compounds are transferred from microbial symbionts to the host sponge, the identity of the

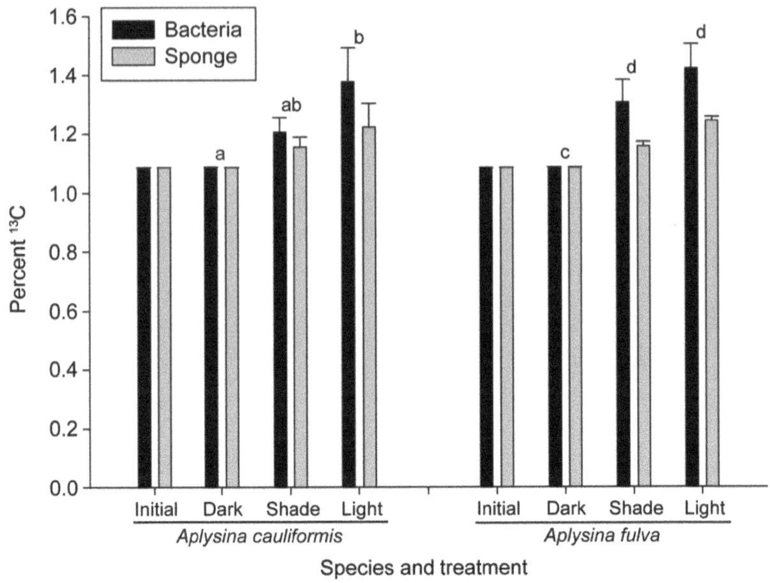

Figure 2.7 Percent ^{13}C in sponge and bacterial cell fractions from initial collections (provided for reference) and 6-h incubations in dark, shade, or light treatments. Bars represent means ± standard errors from $n = 3$ individual sponges per treatment; different letters above bars indicate significantly different means across treatments (dark, shade, and light). Analyses of variance were conducted separately for each species, blocked by individual sponges, and considered treatments and fractions (bacterial cells vs. sponge cells) as effects. For *Aplysina cauliformis*, only a treatment effect was significant ($F_{2,8} = 8.885$, $P = 0.009$), with holobionts held in full light incorporating significantly more ^{13}C than sponges held in darkness. For *A. fulva*, significant effects were observed for fractions ($F_{1,9} = 12.734$, $P = 0.006$) and treatments ($F_{2,9} = 19.816$, $P = 0.001$), with holobionts in shade or light showing greater ^{13}C incorporation than those in darkness, and bacteria cells in shade or light showing greater ^{13}C incorporation than sponge cells. These data suggest that the symbionts of *A. cauliformis* translocate carbon to host cells at a faster rate than those of *A. fulva*.

compounds exchanged remains unknown. An initial attempt to identify compounds that are translocated from cyanobacterial symbionts to host sponges was carried out over 30 years ago. Using $NaHCO_3$ labelled with the ^{14}C radioisotope, Wilkinson (1979) found that some species of Red Sea sponges and their associated cyanobacterial symbionts were capable of fixing carbon in lighted conditions. In addition, using two-dimensional fingerprint co-chromatography, he reported that both isolated cyanobacteria and whole sponge tissue had similar patterns of compounds that were labelled with ^{14}C. In particular, organic acids, alcohols, amino acids, and organic phosphates were all enriched in ^{14}C (Wilkinson, 1979). What Wilkinson found most interesting, however, was that the supernatant above isolated cyanobacterial symbionts contained glycerol and low molecular weight organic phosphates that were enriched in ^{14}C. Since these two products appear to be secreted by isolated cyanobacteria, he argued that these compounds are the products of symbiont carbon fixation that are translocated to the host sponge (Wilkinson, 1979). A later study by Borowitzka *et al.* (1988), also using $NaH^{14}CO_3$, investigated carbon fixation by the filamentous cyanobacterium *O. spongeliae*, hosted by *Lamellodysidea herbacea*. This study confirmed that both the intact sponge–cyanobacteria association and isolated cyanobacteria are photosynthetic. In addition, these authors demonstrated that the low molecular weight products of symbiont carbon fixation are rapidly converted into high molecular weight compounds, such as polynucleotides, polysaccharides, and proteins. Although these seminal papers provide a possible mechanism for nutrient transfer to the sponge host, this area of research has remained largely unexplored to date.

The recent development of methods such as coupled gas chromatography and isotope ratio mass spectrometry (GC-IRMS) allows measurements of the isotope ratios (at natural abundance and after incubation with enriched tracers) of specific compounds in biological samples (Godin *et al.*, 2008). By combining the use of enriched stable isotope tracers with compound-specific isotope analyses, it is now possible for us to build upon these initial investigations of compound translocation and to identify specific compounds. For instance, incubation of marine sponges in $NaH^{13}CO_3$ followed by GC-IRMS analyses of isolated sponge and symbiont cell fractions might allow us to determine which host amino acids are enriched in ^{13}C and, thus, contain the products of photosymbiont carbon metabolism. Since only symbionts are capable of metabolizing this tracer molecule, enriched amino acids within host cells represent direct and strong evidence for symbiont-derived nutrient translocation to the host. This method can also be applied to ^{15}N-enriched tracer compounds to determine which host amino acids contain the products of various microbially mediated nitrogen transformations. These methods can also be coupled with histological techniques to determine the location of tracer compounds within host tissues through MALDI imaging mass spectrometry (Reyzer *et al.*, 2010).

6.4. Identifying specific microbial players

Novel methods allow us not only to identify the specific compounds exchanged between hosts and symbionts but also to identify which microbial symbionts are active participants, in particular metabolic pathways Stable isotope probing (SIP) could prove to be an ideal method for determining which microbes within a complex mixture are responsible for specific metabolic processes (Radajewski *et al.*, 2003; Dumont and Murrell, 2005). DNA-based SIP builds on the idea originally proposed by Meselson and Stahl (1958) that DNA labelled with ^{15}N is heavier than DNA labelled with ^{14}N and can thus be separated using a centrifuge. For

DNA-SIP, researchers can isolate DNA from bacteria that have been incubated on substrates that contain compounds that are highly enriched in ^{13}C. The isolated DNA is then fractionated by density-gradient centrifugation to obtain a heavy ^{13}C-DNA and a light ^{12}C-DNA band. The heavy ^{13}C band contains the genomes of microbes from the sample that could actively acquire and incorporate the labelled substrate (i.e. their DNA contains more ^{13}C than organisms in the ^{12}C band). This ^{13}C-DNA can then be used as a template for *16S* rRNA PCR using group-specific primer sets; subsequent DNA sequencing allows researchers to identify which microbes have assimilated this substrate. Specific *16S* rRNA FISH probes can then be designed for these microbes to visualize their location within the host (Sharp *et al.*, 2007).

Although the application of SIP to organisms that form symbioses with microbes is a relatively new idea and incubation times have yet to be determined (Dumont and Murrell, 2005), we are conducting preliminary short-term experiments to assess which of the vast array of microbes living within a sponge host are capable of acquiring and fixing inorganic carbon. Likewise, incubations with compounds enriched in ^{15}N can be used to better understand which microbes are responsible for certain nitrogen transformations that contribute to the sponge nitrogen cycle (Taylor *et al.*, 2007b; Mohamed *et al.*, 2008; Fiore *et al.*, 2010). Applications of this technique to sponges in wet-lab incubations or *in situ* experiments have the potential to clearly resolve which members of the resident microbial community are responsible for particular metabolic processes within the host and to reveal which members of the microbial community are transients that have limited interactions with the host.

7. A Conceptual Model of Sponge–Symbiont Interactions

The presence of abundant and diverse communities of microbes within a wide variety of sponges provides an ideal set of experimental systems for testing hypotheses about the formation and maintenance of symbiosis. Sponge hosts can be studied along phylogenetic, geographic, and environmental gradients, all of which allow sponge–symbiont interactions to be examined at multiple scales. In this section, we use a conceptual model to summarize our views of how sponges and their symbionts might interact (Fig. 2.8). Our model groups the players in these interactions as a sponge host, its beneficial symbionts, and its harmful symbionts. The double-headed arrow connecting the two groups of symbionts indicates that the benefits and costs of hosting a particular genotype of symbiont can change, such that only the ratio of benefits and costs can be used to define a

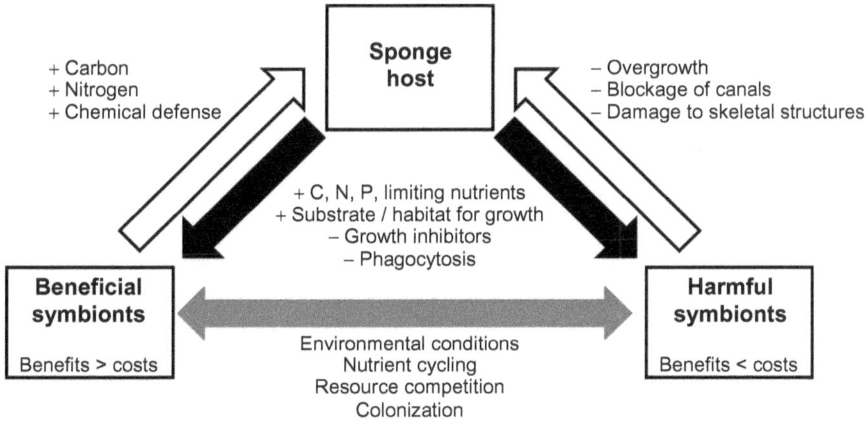

Figure 2.8 A conceptual model of sponge–microbe symbioses. In this view, a variety of factors can alter the benefits and costs of symbiosis for the host sponge, such that individual microbial taxa could be beneficial or harmful, depending on local environmental conditions, nutrient cycling within the host, resource competition, and colonization by other potential symbionts. Beneficial symbionts might supply carbon, nitrogen, or chemical defences to the host, while harmful symbionts might overgrow the host, block the aquiferous system, or cause damage to skeletal structures. Sponge hosts could stimulate symbiont growth by providing carbon, nitrogen, phosphorus, or other limiting nutrients, as well as substrates or critical microhabitat conditions. Finally, sponges might control symbiont populations by releasing growth inhibitors or through phagocytosis and expulsion.

symbiont as "beneficial" or "harmful". Factors that could change these benefits and costs include local environmental conditions, nutrient cycling among different symbiont genotypes within the host, resource competition among symbionts, and the colonization of a host by novel symbionts.

We represent the benefits of hosting microbial symbionts as a gain of carbon, nitrogen, or chemical defence, but, as we have mentioned earlier, many other types of benefits are possible. Some of the potential costs of hosting symbionts include overgrowth of the sponge by microbes (Rutzler, 1988), as well as blockage of the aquiferous system and damage to skeletal structures (Webster *et al.*, 2002). Host sponges are not necessarily passive participants supplying a substrate for growth but could actively supply or regulate symbiont access to carbon, nitrogen, phosphorous, or other limiting nutrients. Moreover, anoxic niches within a host sponge could provide critical habitat for anaerobic symbionts (Schläppy *et al.*, 2010). Host sponges might be able to control microbial growth and select for the presence of specific strains by producing growth inhibitors or other selective agents that manipulate the community structure of symbionts. Sponges can also phagocytose and expel symbionts, but these two mechanisms might be insufficient to counter microbial growth rates (Rutzler, 1988).

Since sponge–microbe symbioses appear to display a predominate pattern of leaky vertical transmission, these communities are highly likely to be colonized by "cheater" species or genotypes (Sachs and Simms, 2006). Such microbes might acquire nutrients from the host or other symbionts, using them for growth and reproduction without returning any benefits to the host or to other members of the symbiotic community. Cheaters are frequently reported from symbioses, for example, in fig tree–fig wasp (Jandér and Herre, 2010), legume–*Rhizobia* (Simms *et al.*, 2006), and ant–*Acacia* (Palmer *et al.*, 2010) symbioses. The presence of cheating symbionts could explain why most sponges do not demonstrate a tightly coevolved, obligate symbiosis with their microbial communities and why even closely related species display different interactions with their symbionts (Freeman and Thacker, 2011). However, no studies of sponge–microbe symbioses have explicitly included experiments to provide evidence of cheating symbionts. Such studies need to consider a variety of questions, including (1) What are the costs of hosting these symbionts to the sponge? (2) Can a sponge regulate its symbiotic microbial communities? (3) Can a sponge differentiate between cheaters and true mutualists? (4) Does hosting genetically distinct symbionts benefit the sponge host? These questions also require us to consider the lifetime fitness of a host and whether different symbiont communities can be favoured during ontogeny. For example, recent research on ant–*Acacia* symbioses reveals that multi-species mutualisms that include cheaters can be favoured if hosts enhance their lifetime fitness by optimizing survivorship and reproduction in different stages of their life history (Palmer *et al.*, 2010).

Although several recent studies of sponge microbiology have provided qualitative models of symbiont transmission and potential benefits of symbioses, quantitative models are needed to provide a stronger theoretical basis for future experimental work. Many categories of ecological models, including adjusted Lotka–Volterra models (Neuhauser and Fargione, 2004), biological market models (Cowden and Peterson, 2009), and functional response models (Holland *et al.*, 2002), can be applied readily to these systems. However, these models require us to better understand how we quantify the potential benefits and costs for each member of the symbiosis (Bronstein, 2001). For sponge–microbe associations, this process is complicated by the fact that many of these interactions likely represent complex multiple-partner mutualisms that also include cheaters. Multiple partners could differ in their benefit to cost ratios, implying that it might be ecologically cheaper for the host to trade resources with some potential symbionts than others. These features of sponge–symbiont interactions suggest that biological market models could provide a high degree of explanatory power (Cowden and Peterson, 2009).

Finally, there is a clear need for the field of sponge microbiology to integrate across multiple levels of biological organization and to bridge traditionally distinct fields in future investigations. For example, T-RFLP and clone library-based methods reveal that the common mangrove sponge *T. ignis* hosts a diverse bacterial assemblage (Yang *et al.*, 2011), while FISH techniques indicate that bacteria are vertically transmitted from parent to offspring (Fig. 2.5). The presence of symbiotic, heterotrophic bacteria could potentially explain previous observations of DOM assimilation by the lecithotrophic larvae of *T. ignis* (Jaeckle, 1995), while production of defensive compounds by these bacteria could potentially explain the observed unpalatability of these larvae (Lindquist and Hay, 1996). Even for this relatively well-studied host species, multiple experimental manipulations are still needed to fully document the interactions between the symbiotic microbial community and the host. Moreover, each one of the diverse host–symbiont interactions described in our review poses similar sets of questions and hypotheses in need of experimental work.

8. Conclusions

Although a large number of recent studies and reviews have documented that sponges can host microbes from nearly all bacterial and archaeal lineages, we still understand little about the evolutionary, ecological, and physiological processes that control the abundance of microbes within sponges and the specificity of these associations. Amazingly, we still have little idea of how sponges maintain the integrity of their body plans while hosting a high density and diversity of seawater-derived bacteria in the mesohyl. Future studies of body plan development could manipulate the presence of microbial symbionts to better assess how cellular-level interactions influence these associations and host physiology. In addition, studies of sponge–microbe associations should use stronger, quantitative hypothesis testing to determine whether microbes found within a host sponge truly reflect coevolved symbioses rather than transient interactions dictated by local environments.

Recent studies have revealed evidence of both vertical and horizontal transmission of microbial symbionts in sponges. These data suggest that a "leaky vertical" or mixed mode of transmission may be most common for marine sponges, with some frequency of horizontal transmission always available; however, this frequency likely varies among host species, with some hosts exerting greater control over the symbiont community than others. Future experiments should be designed to explicitly target these different modes of symbiont transmission, as different taxa of symbionts may be more likely to be transmitted by one mode or the other. Reciprocal transplants of sponges among habitats as well as investigations of multiple life-history stages will accelerate such research. In addition, quantitative ecological and evolutionary models will greatly enhance our understanding of the impacts of different modes of transmission on the outcomes of these interactions.

The studies reviewed here demonstrate that not only do some sponges obtain carbon and nitrogen from their microbial symbionts, but also that there is likely to be a high degree of variability among host–symbiont interactions in the identity and amount of nutrients transferred. Current metagenomic sequencing projects promise to reveal additional metabolic capabilities of symbiotic microbes, allowing us to understand the genetic basis of nutritional exchanges as well as other potential benefits and costs to the host sponge. Although such projects are highly likely to reveal how host sponges might influence their symbiont populations, metagenomic studies alone will not be able to address these questions. Our review has emphasized the need for additional experimental manipulations coupled with advanced laboratory techniques to determine whether potential microbial symbionts are active within host sponges and whether symbionts of interest act as mutualists, commensals, or parasites. Cultured symbionts are needed for rigorous experimentation, and the development of novel methods to culture and transplant sponge-associated microbes should be a high priority for our research community.

Many recent studies of sponge–microbe symbioses parallel studies of coral–zooxanthellae symbioses (Erwin and Thacker, 2008a; Hill *et al.*, 2011); future work should expand upon these parallels and integrate knowledge across both systems. For example, in both associations, symbionts originally described as single species, *S. spongiarum* in sponges and *Symbiodinium microadriaticum* Freudenthal, 1962 in sponges and corals, are now thought to represent genetically diverse species complexes with a tremendous range of range of host specificity (Rowan and Powers, 1991; Rowan, 1998; LaJeunesse, 2001; Hill *et al.*, 2011). *Symbiodinium* clades display a wide variety of physiological capabilities and host-derived benefits (Coles and Brown, 2003; Rowan, 2004), so it will not be surprising to find similar levels of physiological variation in sponge–microbe interactions. Corals might be able to acclimate to changing environmental conditions by expelling current symbiont populations (i.e. bleaching) and establishing new assemblages that are better adapted to ambient conditions (Buddemeier and Fautin, 1993; Baker, 2003; Sotka and Thacker, 2005), but we have little understanding of a similar process in marine sponges (Lemoine *et al.*, 2007). Such knowledge will be critical to predicting the responses of sponges to global climate change.

The evolution of sponge–microbe symbioses can be considered an adaptation that expands the biochemical repertoire of sponges, adding metabolic processes critical for survival in a tremendous variety of habitats, from deep-sea to shallow coral reefs. We expect that future studies of sponge–microbe symbioses will continue to document the amazing diversity of these complex associations and to reveal the importance of these symbionts to the ecology and evolution of sponges, as well as their impacts on nutrient cycling and other ecosystem functions.

ACKNOWLEDGEMENTS

We thank the organizing committee of the Eighth World Sponge Conference (M. J. Uriz, M. Becerro, M. Maldonado, and X. Turon) for the opportunity to write this review. D. Baker, M. Becerro, D. Gochfeld, M. Fogel, M. Hill, A. Martin, J. Olson, J. Pawlik, and R. Vrijenhoek stimulated discussions that improved our review. Additional comments by M. Becerro and two anonymous reviewers also improved the chapter. The confocal images shown here would not be possible without the assistance of M. Haygood and K. Sharp; additional figures were provided by D. Gochfeld, A. Martin, and R. Vrijenhoek. Stable isotope analyses were conducted by D. Baker and M. Fogel. Financial support was provided by the U.S. National Science Foundation under Grants 0726944 and 0829986 to R. W. T.

REFERENCES

Amann, R., Ludwig, W., and Schleifer, K. H. (1995). Phylogenetic identification and in situ detection of individual microbial cells without cultivation. *Microbiological Reviews* **59,** 143–169.

Anderson, S. A., Northcote, P. T., and Page, M. J. (2010). Spatial and temporal variability of the bacterial community in different chemotypes of the New Zealand marine sponge *Mycale hentscheli. FEMS Microbiology Ecology* **72,** 328–342.

Arillo, A., Bavestrello, G., Burlando, B., and Sarà, M. (1993). Metabolic integration between symbiotic cyanobacteria and sponges: A possible mechanism. *Marine Biology* **117,** 159–162.

Baker, A. C. (2003). Flexibility and specificity in coral-algal symbiosis: Diversity, ecology, and biogeography of *Symbiodinium. Annual Review of Ecology, Evolution, and Systematics* **34,** 661–689.

Barneah, O., Weis, V. M., Perez, S., and Benayahu, Y. (2004). Diversity of dinoflagellate symbionts in Red Sea soft corals: Mode of symbiont acquisition matters. *Marine Ecology Progress Series* **275,** 89–95.

Becerro, M. A. (2008). Quantitative trends in sponge ecology research. *Marine Ecology* **29,** 167–177.

Becerro, M. A., Turon, X., Uriz, M. J., and Templado, J. (2003). Can a sponge feeder be a herbivore? *Tylodina perversa* (Gastropoda) feeding on *Aplysina aerophoba* (Demospongiae). *Biological Journal of the Linnean Society* **78,** 429–438.

Becerro, M. A., Starmer, J. A., and Paul, V. J. (2006). Chemical defenses of cryptic and aposematic gastropterid mollusks feeding on their host sponge *Dysidea granulosa. Journal of Chemical Ecology* **32,** 1491–1500.

Begon, M., Townsend, C. R., and Harper, J. L. (2005). Ecology: From Individuals to Ecosystems. Blackwell Publishing, London.

Berthold, R. J., Borowitzka, M. A., and Mackay, M. A. (1982). The ultrastructure of *Oscillatoria spongeliae*, the blue-green algal endosymbiont of the sponge *Dysidea herbacea. Phycologia* **21,** 327–335.

Bertrand, J.-C., and Vacelet, J. (1971). L'association entre éponges cornées et bactéries. *Comptes Rendus Hebdomadaires des Séances de l'Académie des Sciences Paris* **273,** 638–641.

Borowitzka, M. A., Hinde, R., and Pironet, F. (1988). Carbon fixation by the sponge *Dysidea herbacea* and its endosymbiont *Oscillatoria spongeliae*. In "Proceedings of the 6th International Coral Reef Symposium." Vol. 3, pp. 151–156.

Boucher, D. H. (1985). The Biology of Mutualism: Ecology and Evolution. Oxford University Press, New York.

Boyer, S. L., Flechtner, V. R., and Johansen, J. R. (2001). Is the 16S-23S rRNA internal transcribed spacer region a good tool for use in molecular systematics and population genetics? A case study in cyanobacteria. *Molecular Biology and Evolution* **18,** 1057–1069.

Bronstein, J. L. (2001). The costs of mutualism. *American Zoologist* **41,** 825–839.

Bronstein, J. L., Wilson, W. G., and Morris, W. F. (2003). Ecological dynamics of mutualist/antagonist communities. *American Naturalist* **162,** S24–S39.

Brück, W. M., Brück, T. B., Self, W. T., Reed, J. K., Nitecki, S. S., and McCarthy, P. J. (2010). Comparison of the anaerobic microbiota of deep-water *Geodia* spp. and sandy sediments in the Straits of Florida. *The ISME Journal* **4,** 686–699.

Bruns, T. D., Bidartondo, M. I., and Taylor, D. L. (2002). Host specificity in ectomycorrhizal communities: What do the exceptions tell us? *Integrative and Comparative Biology* **42,** 352–359.

Buchner, P. (1965). Endosymbiosis of Animals with Plant Microorganisms. Wiley, New York.

Buddemeier, R. W., and Fautin, D. G. (1993). Coral bleaching as an adaptive mechanism: A testable hypothesis. *BioScience* **43,** 320–326.

Carballo, J. L., and Ávila, E. (2004). Population dynamics of a mutualistic interaction between the sponge *Haliclona caerulea* and the red alga *Jania adherens*. *Marine Ecology Progress Series* **279,** 93–104.

Cervino, J. M., Winiarski-Cervino, K., Polson, S. W., Goreau, T., and Smith, G. W. (2006). Identification of bacteria associated with a disease affecting the marine sponge *Ianthella basta* in New Britain, Papua New Guinea. *Marine Ecology Progress Series* **324,** 139–150.

Clark, M. A., Moran, N. A., Baumann, P., and Wernegreen, J. J. (2000). Cospeciation between bacterial endosymbionts (*Buchnera*) and a recent radiation of aphids (*Uroleucon*) and pitfalls of testing for phylogenetic congruence. *Evolution* **54,** 517–525.

Clayton, D. H., Bush, S. E., Goates, B. M., and Johnson, K. P. (2003). Host defense reinforces host-parasite cospeciation. *Proceedings of the National Academy of Sciences of the United States of America* **100,** 15694–15699.

Clement, B. G., Kehl, L. E., DeBord, K. L., and Kitts, C. L. (1998). Terminal restriction fragment patterns (TRFPs), a rapid, PCR-based method for the comparison of complex bacterial communities. *Journal of Microbiological Methods* **31,** 135–142.

Coles, S. L., and Brown, B. E. (2003). Coral bleaching—Capacity for acclimatization and adaptation. *Advances in Marine Biology* **46,** 183–223.

Collin, R., Mobley, A. S., Lopez, L. B., Leys, S. P., Diaz, M. C., and Thacker, R. W. (2010). Phototactic responses of larvae from the marine sponge *Neopetrosia proxima* and *Xestospongia bocatorensis* (Haplosclerida: Petrosiidae). *Invertebrate Biology* **129,** 121–128.

Cowart, J. D., Henkel, T. P., McMurray, S. E., and Pawlik, J. R. (2006). Sponge orange band (SOB): A pathogenic-like condition of the giant barrel sponge, *Xestospongia muta*. *Coral Reefs* **25,** 513.

Cowden, C. C., and Peterson, C. J. (2009). A multi-mutualist simulation: Applying biological market models to diverse mycorrhizal communities. *Ecological Modelling* **220,** 1522–1533.

Custódio, M. R., Lôbo-Hajdu, G., Hajdu, E., and Muricy, G. (2007). Porifera Research: Biodiversity, Innovation, and Sustainability. Série Livros, Vol. 28. Museu Nacional, Rio de Janeiro.

Diaz, M. C., and Rützler, K. (0308). Sponges: An essential component of Caribbean coral reefs. *Bulletin of Marine Science* **69,** 535–546.

Diaz, M. C., Thacker, R. W., Rützler, K., and Piantoni, C. (2007). *Haliclona (Soestella) walentinae* (Chalinidae, Haplosclerida) and *Xestospongia bocatorensis* (Petrosiidae, Haplosclerida), two new sponge species from Caribbean Panamá with filamentous cyanobacterial symbionts. In "Porifera Research: Biodiversity, Innovation, and Sustainability" (M. R. Custódio, G. Lôbo-Hajdu, E. Hajdu and G. Muricy, eds), Série Livros, Vol. 28, pp. 31–36. Museu Nacional, Rio de Janeiro.

Dorazio, R. M., Royle, J. A., Söderström, B., and Glimskär, A. (2006). Estimating species richness and accumulation by modeling species occurrence and detectability. *Ecology* **87,** 842–854.

Dosse, G. (1939). Bakterien und Pilzbefunde sowie pathologische und Fäulnisvorgänge in Meeres- und Süsswasserschwämmen. *Zeitschrift für Parasitenkunde* **11,** 331–356.

Drake, H. L., and Horn, M. A. (2007). As the worm turns: The earthworm gut as a transient habitat for soil microbial biomes. *Annual Review of Microbiology* **61,** 169–189.

Dumont, M. G., and Murrell, J. C. (2005). Stable isotope probing-linking microbial identity to function. *Nature Reviews* **3,** 499–504.

Engel, S., and Pawlik, J. R. (2000). Allelopathic activities of sponge extracts. *Marine Ecology Progress Series* **207,** 273–281.

Enticknap, J. J., Kelly, M., Peraud, O., and Hill, R. T. (2006). Characterization of a culturable alphaproteobacterial symbiont common to many marine sponges and evidence for vertical transmission via sponge larvae. *Applied and Environmental Microbiology* **72,** 3724–3732.

Ereskovsky, A. V., Gonobobleva, E., and Vishnyakov, A. (2005). Morphological evidence for vertical transmission of symbiotic bacteria in the viviparous sponge *Halisarca dujardini* Johnston (Porifera, Demospongiae, Halisarcida). *Marine Biology* **146,** 869–875.

Erwin, P. M., and Thacker, R. W. (2007). Incidence and identity of photosynthetic symbionts in Caribbean coral reef sponge assemblages. *Journal of the Marine Biological Association of the United Kingdom* **87,** 1683–1692.

Erwin, P. M., and Thacker, R. W. (2008a). Cryptic diversity of the symbiotic cyanobacterium *Synechococcus spongiarum* among sponge hosts. *Molecular Ecology* **17,** 2937–2947.

Erwin, P. M., and Thacker, R. W. (2008b). Phototrophic nutrition and symbiont diversity of two Caribbean sponge-cyanobacteria symbioses. *Marine Ecology Progress Series* **362,** 139–147.

Erwin, P. M., López-Legentil, S., Gonzalez-Pech, R., and Turon, X. (2011a). A specific mix of generalists: Bacterial symbionts in Mediterranean *Ircinia* spp. *FEMS Microbiology Ecology* **79,** (3), 619–637.

Erwin, P. M., Olson, J. B., and Thacker, R. W. (2011b). Phylogenetic diversity, host-specificity and community profiling of sponge-associated bacteria in the northern Gulf of Mexico. *PLoS One* **6,** e26806.

Ewald, P. W. (1987). Transmission modes and evolution of the parasitism-mutualism continuum. *Annals of the New York Academy of Sciences* **503,** 295–306.

Fiore, C. L., Jarett, J. K., Olson, N. D., and Lesser, M. P. (2010). Nitrogen fixation and nitrogen transformations in marine symbioses. *Trends in Microbiology* **18,** 455–463.

Flatt, P. M., Gautschi, J. T., Thacker, R. W., Musafija-Girt, M., Crews, P., and Gerwick, W. H. (2005). Identification of the cellular site of polychlorinated peptide biosynthesis in the marine sponge *Dysidea (Lamellodysidea) herbacea* and symbiotic cyanobacterium *Oscillatoria spongeliae* by CARD-FISH analysis. *Marine Biology* **147,** 761–774.

Flowers, A. E., Garson, M. J., Webb, R. I., Dumdei, E. J., and Charan, R. D. (1998). Cellular origin of chlorinated diketopiperazines in the dictyoceratid sponge *Dysidea herbacea* (Keller). *Cell and Tissue Research* **292,** 597–607.

Frank, S. A. (1996). Host-symbiont conflict over mixing of symbiotic lineages. *Proceedings of the Royal Society of London B* **263,** 339–344.

Freeman, K. H. (2001). Isotopic biogeochemistry of marine organic carbon. In "Stable Isotope Geochemistry" (J. W. Valley and D. R. Cole, eds), pp. 579–597. Mineralogical Society of America, Washington, DC.

Freeman, C. J., and Thacker, R. W. (2011). Complex interactions between marine sponges and their symbiotic microbial communities. *Limnology and Oceanography* **56,** (5), 1577–1586.

Friedrich, A. B., Merkert, H., Fendert, T., Hacker, J., Proksch, P., and Hentschel, U. (1999). Microbial diversity in the marine sponge *Aplysina cavernicola* (formerly *Verongia cavernicola*) analyzed by fluorescence *in situ* hybridization. *Marine Biology* **134,** 461–470.

Friedrich, A. B., Fisher, I., Proksch, P., Hacker, J., and Hentschel, U. (2001). Temporal variation of the microbial community associated with the Mediterranean sponge *Aplysina aerophoba. FEMS Microbiology Ecology* **38,** 105–113.

Fry, B. (2006). Stable isotope ecology. Springer, New York.

Garson, M. J. (2001). Ecological perspectives on marine natural product biosynthesis. In "Marine Chemical Ecology" (J. B. McClintock and B. J. Baker, eds), pp. 71–114. CRC Press, Boca Raton.

Gerce, B., Schwartz, T., Voigt, M., Rühle, S., Kirchen, S., Putz, A., Proksch, P., Obst, U., Syldatk, C., and Hausmann, R. (2009). Morphological, bacterial, and secondary metabolite changes of *Aplysina aerophoba* upon long-term maintenance under artificial conditions. *Microbial Ecology* **58,** 865–878.

Gloor, G. B., Hummelen, R., Macklaim, J. M., Dickson, R. J., Fernandes, A. D., MacPhee, R., and Reid, G. (2010). Microbiome profiling by illumina sequencing of combinatorial sequence-tagged PCR products. *PLoS One* **5,** e15406.

Gochfeld, D. J., Schlöder, C., and Thacker, R. W. (2007). Sponge community structure and disease prevalence on coral reefs in Bocas del Toro, Panama. In "Porifera Research: Biodiversity, Innovation, and Sustainability" (M. R. Custódio, G. Lôbo-Hajdu, E. Hajdu and G. Muricy, eds), Série Livros, Vol. 28, pp. 335–343. Museu Nacional, Rio de Janeiro.

Gochfeld, D. J., Easson, C. G., Freeman, C. J., Thacker, R. W. and Olson, J. B. (2012). Disease and nutrient enrichment as potential stressors on the Caribbean sponge Aplysina cauliformis and its bacterial symbionts. *Marine Ecology Progress Series.* In Press.

Godin, J. P., Breuillé, D., Obled, C., Papet, I., Schierbeek, H., Hopfgartner, G., and Fay, L.-B. (2008). Liquid and gas chromatography coupled to isotope ratio mass spectrometry for the determination of ^{13}C-valine isotopic ratios in complex biological samples. *Journal of Mass Spectrometry* **43,** 1334–1343.

Gómez, R., Erpenbeck, D., Van Dijk, T., Richelle-Maurer, E., Devijver, C., Braekman, J. C., Woldringh, C., and Van Soest, R. W. M. (2002). Identity of cyanobacterial symbiont of *Xestospongia muta. Bollettino dei Musei e degli Istituti Biologici dell' Università di Genova* **66–67,** 82–83.

Gomulkiewicz, R., Nuismer, S. L., and Thompson, J. N. (2003). Coevolution in variable mutualisms. *American Naturalist* **162,** S80–S93.

Green, E. P., and Bruckner, A. W. (2000). The significance of coral disease epizootiology for coral reef conservation. *Biological Conservation* **96,** 347–361.

Grottoli, A. G., Rodrigues, L. J., and Palardy, J. E. (2006). Heterotrophic plasticity and resilience in bleached corals. *Nature* **440,** 1186–1189.

Hafner, M. S., and Nadler, S. A. (1988). Phylogenetic trees support the coevolution of parasites and their hosts. *Nature* **332,** 258–259.

Harvell, C. D., Kim, K., Burkholder, J. M., Colwell, R. R., Epstein, P. R., Grimes, D. J., Hofman, E. E., Lipp, E. K., Osterhaus, A. D. M. E., Overstreet, R. M., Porter, J. W., Smith, G. W., and Vasta, G. R. (1999). Emerging marine diseases: Climate links and anthropogenic factors. *Science* **285,** 1505–15010.

Head, I. M., Saunders, J. R., and Pickup, R. W. (1998). Microbial evolution, diversity and ecology: A decade of ribosomal RNA analysis of uncultivated microorganisms. *Microbial Ecology* **35,** 1–21.

Heath, K. D., and Tiffin, P. (2009). Stabilizing mechanisms in a legume-rhizobium mutualism. *Evolution* **63,** 652–662.

Hentschel, U., Hopke, J., Horn, M., Friedrich, A. B., Wagner, M., Hacker, J., and Moore, B. S. (2002). Molecular evidence for a uniform microbial community in sponges from different oceans. *Applied and Environmental Microbiology* **68,** 4431–4440.

Hentschel, U., Fieseler, L., Wehrl, M., Gernert, C., Steinert, M., Hacker, J., and Horn, M. (2003). Microbial diversity of marine sponges. In "Molecular Marine Biology of Sponges" (W. E. G. Muller, ed.), pp. 59–88. Springer-Verlag, Berlin.

Hentschel, U., Usher, K. M., and Taylor, M. W. (2006). Marine sponges as microbial fermenters. *FEMS Microbiology Ecology* **55,** 167–177.

Herre, E. A., Knowlton, N., Mueller, U. G., and Rehner, S. A. (1999). The evolution of mutualisms: Exploring the paths between conflict and cooperation. *Trends in Ecology and Evolution* **14,** 49–53.

Hill, M., Hill, A., Lopez, N., and Harriott, O. (2006). Sponge-specific bacterial symbionts in the Caribbean sponge, *Chondrilla nucula* (Demospongiae, Chondrosida). *Marine Biology* **148,** 1221–1230.

Hill, M., Allenby, A., Ramsby, B., Schönberg, C., and Hill, A. (2011). *Symbiodinium* diversity among host clionaid sponge from Caribbean and Pacific reefs: Evidence of heteroplasmy and putative host-specific symbiont lineages. *Molecular Phylogenetics and Evolution* **59,** 81–88.

Hinde, R., Pironet, F., and Borowitzka, M. A. (1994). Isolation of *Oscillatoria spongeliae,* the filamentous cyanobacterial symbiont of the marine sponge *Dysidea herbacea. Marine Biology* **119,** 99–104.

Hoffmann, F., Larsen, O., Rapp, H. T., and Osinga, R. (2005a). Oxygen dynamics in choanosomal sponge explants. *Marine Biology Research* **1,** 160–163.

Hoffmann, F., Larsen, O., Thiel, V., Rapp, H. T., Pape, T., Michaelis, W., and Reitner, J. (2005b). An anaerobic world of sponges. *Geomicrobiology Journal* **22,** 1–10.

Hoffmann, F., Røy, H., Bayer, K., Hentschel, U., Pfannkuchen, M., Brümmer, F., and de Beer, D. (2008). Oxygen dynamics and transport in the Mediterranean sponge *Aplysina aerophoba. Marine Biology* **153,** 1257–1264.

Holland, J. N., DeAngelis, D. L., and Bronstein, J. L. (2002). Population dynamics and mutualism: Functional responses of benefits and costs. *American Naturalist* **159,** 231–244.

Hosokawa, T., Kikuchi, Y., Nikoh, N., Shimada, M., and Fukatsu, T. (2006). Strict host–symbiont cospeciation and reductive genome evolution in insect gut bacteria. *PLoS Biology* **4,** e337.

Hunt, D. E., David, L. A., Gevers, D., Preheim, S. P., Alm, E. J., and Polz, M. F. (2008). Resource partitioning and sympatric differentiation among closely related bacterioplankton. *Science* **320,** 1081–1085.

Isaacs, L. T., Kan, J., Nguyen, L., Videau, P., Anderson, M. A., Wright, T. L., and Hill, R. T. (2009). Comparison of the bacterial communities of wild and captive sponge *Clathria prolifera* from the Chesapeake Bay. *Marine Biotechnology* **11,** 758–770.

Iteman, I., Rippka, R., Tandeau de Marsac, N., and Herdman, M. (2000). Comparison of conserved structural and regulatory domains within divergent 16S rRNA-23S rRNA spacer sequences of cyanobacteria. *Microbiology* **146,** 1275–1286.

Jaeckle, W. B. (1995). Transport and metabolism of alanine and palmitic acid by field-collected larvae of *Tedania ignis* (Porifera, Demospongiae): Estimated consequences of limited label translocation. *Biological Bulletin* **189,** 159–167.

Jandér, K. C., and Herre, E. A. (2010). Host sanctions and pollinator cheating in the fig tree-fig wasp mutualism. *Proceedings of the Royal Society of London B* **277,** 1481–1488.

Kaczmarsky, L. T., Draud, M., and Williams, E. H. (2005). Is there a relationship between proximity to sewage effluent and the prevalence of coral disease? *Caribbean Journal of Science* **41,** 124–137.

Knowlton, N., and Rohwer, F. (2003). Multispecies microbial mutualisms on coral reefs: The host as a habitat. *American Naturalist* **162,** S51–S62.

Kraft, N. J. B., Cornwell, W. K., Webb, C. O., and Ackerly, D. D. (2007). Trait evolution, community assembly, and the phylogenetic structure of ecological communities. *American Naturalist* **170,** 271–283.

LaJeunesse, T. C. (2001). Investigating the biodiversity, ecology, and phylogeny of endosymbiotic dinoflagellates in the genus *Symbiodinium* using the internal transcribed spacer region: In search of a "species" level marker. *Journal of Phycology* **37,** 866–880.

Lajtha, K., and Michener, R. H. (1994). Stable Isotopes in Ecology and Environmental Science. Blackwell Scientific Publications, Oxford.

Law, R. (1985). Evolution in a mutualistic environment. In "The Biology of Mutualism: Ecology and Evolution" (D. H. Boucher, ed.), pp. 145–170. Oxford University Press, New York.

Lee, Y. K., Lee, J., and Lee, H. K. (2001). Microbial symbiosis in marine sponges. *Journal of Microbiology* **39,** 254–264.

Lee, O. O., Wang, Y., Yang, J., Lafi, F. F., Al-Suwailem, A., and Qian, P.-Y. (2011). Pyrosequencing reveals highly diverse and species-specific microbial communities in sponges from the Red Sea. *ISME Journal* **5,** 650–664.

Lemloh, M., Fromont, J., Brummer, F., and Usher, K. M. (2009). Diversity and abundance of photosynthetic sponges in temperate Western Australia. *BMC Ecology* **9,** 1–13.

Lemoine, N., Buell, N., Hill, A., and Hill, M. (2007). Assessing the utility of sponge microbial symbiont communities as models to study global climate change: A case study with *Halichondria bowerbanki*. In "Porifera Research: Biodiversity, Innovation, and Sustainability" (M. R. Custódio, G. Lôbo-Hajdu, E. Hajdu and G. Muricy, eds), Série Livros, Vol. 28, pp. 419–425. Museu Nacional, Rio de Janeiro.

Lindquist, N., and Hay, M. E. (1996). Palatability and chemical defense of marine invertebrate larvae. *Ecological Monographs* **66,** 431–450.

Liu, M. Y., Kjelleberg, S., and Thomas, T. (2011). Functional genomic analysis of an uncultured δ-proteobacterium in the sponge *Cymbastela concentrica*. *The ISME Journal* **5,** 427–435.

López-Legentil, S., Song, B., McMurray, S. E., and Pawlik, J. R. (2008). Bleaching and stress in coral reef ecosystems: *hsp70* expression by the giant barrel sponge *Xestospongia muta*. *Molecular Ecology* **17,** 1840–1849.

López-Legentil, S., Erwin, P. M., Pawlik, J. R., and Song, B. (2010). Effects of sponge bleaching on ammonia-oxidizing *Archaea*: Distribution and relative expression of ammonia monooxygenase genes associated with the barrel sponge *Xestospongia muta*. *Microbial Ecology* **60,** 561–571.

Luter, H. M., Whalan, S., and Webster, N. S. (2010a). Prevalence of tissue necrosis and brown spot lesions in a common marine sponge. *Marine and Freshwater Research* **61,** 481–484.

Luter, H. M., Whalan, S., and Webster, N. S. (2010b). Exploring the role of microorganisms in the disease-like syndrome affecting the sponge *Ianthella basta*. *Applied and Environmental Microbiology* **76,** 5736–5744.

Madri, P. P., Hermel, M., and Claus, G. (1971). The microbial flora of the sponge *Microciona prolifera* Verrill and its ecological implications. *Botanica Marina* **14,** 1–5.

Maldonado, M., and Riesgo, A. (2009). Gametogenesis, embryogenesis, and larval features of the oviparous sponge *Petrosia ficiformis* (Haplosclerida, Demospongiae). *Marine Biology* **156,** 2181–2197.

Maldonado, M., and Young, C. M. (1998). Limits on the bathymetric distribution of keratose sponges: A field test in deep water. *Marine Ecology Progress Series* **174,** 123–139.

Maldonado, M., Sánchez-Tocino, L., and Navarro, C. (2010). Recurrent disease outbreaks in corneous demosponges of the genus *Ircinia*: Epidemic incidence and defense mechanisms. *Marine Biology* **157,** 1577–1590.

Margulis, L., and Fester, R. (1991). Symbiosis as a Source of Evolutionary Innovation. MIT Press, Boston.

Martin, A. P. (2002). Phylogenetic approaches for describing and comparing the diversity of microbial communities. *Applied and Environmental Microbiology* **68,** 3673–3682.

McCarthy, M. D., Benner, R., Lee, C., and Fogel, M. L. (2007). Amino acid nitrogen isotopic fractionation patterns as indicators of heterotrophy in plankton, particulate, and dissolved organic matter. *Geochimica et Cosmochimica Acta* **71,** 4727–4744.

Meselson, M., and Stahl, F. W. (1958). The replication of DNA in *Escherichia coli*. *Proceedings of the National Academy of Sciences of the United States of America* **44,** 671–683.

Meyer, B., and Kuever, J. (2008). Phylogenetic diversity and spatial distribution of the microbial community associated with the Caribbean deep-water sponge *Polymastia* cf. *corticata* by 16S rRNA, *aprA*, and *amoA* gene analysis. *Microbial Ecology* **56,** 306–321.

Michener, R. H., and Schell, D. M. (1994). Stable isotope ratios as tracers in marine aquatic food webs. In "Stable Isotopes in Ecology and Environmental Science" (K. Lajtha and R. H. Michener, eds), pp. 138–157. Blackwell Scientific Publications, Oxford.

Mira, A., and Moran, N. A. (2002). Estimating population size and transmission bottlenecks in maternally transmitted endosymbiotic bacteria. *Microbial Ecology* **44,** 137–143.

Mohamed, N. M., Colman, A. S., Tal, Y., and Hill, R. T. (2008). Diversity and expression of nitrogen fixation genes in bacterial symbionts of marine sponges. *Environmental Microbiology* **10,** 2910–2921.

Moore, L. R., Rocap, G., and Chisholm, S. W. (1998). Physiology and molecular phylogeny of coexisting *Prochlorococcus* ecotypes. *Nature* **393,** 464–467.

Moran, N. A. (2002). Microbial minimalism: Genome reduction in bacterial pathogens. *Cell* **108,** 583–586.

Mukherjee, J., Webster, N. S., and Llewellyn, L. E. (2009). Purification and characterization of a collagenolytic enzyme from a pathogen of the Great Barrier Reef sponge, *Rhopaloeides odorabile*. *PLoS One* **4,** e7177.

Muscatine, L., and Kaplan, I. R. (1994). Resource partitioning by reef corals as determined from stable isotope composition II. δ^{15}N of zooxanthellae and animal tissue versus depth. *Pacific Science* **48,** 304–312.

Muscatine, L., Porter, J. W., and Kaplan, I. R. (1989). Resource partitioning by reef corals as determined from stable isotope composition. I. δ^{13}C of zooxanthellae and animal tissue vs. depth. *Marine Biology* **100,** 185–193.

Neuhauser, C., and Fargione, J. E. (2004). A mutualism-parasitism continuum model and its application to plant-mycorrhizal interactions. *Ecological Modelling* **177,** 337–352.

Nishijima, M., Lindsay, D. J., Hata, J., Nakamura, A., Kasai, H., Ise, Y., Fisher, C. R., Fujiwara, Y., Kawato, M., and Maruyama, T. (2010). Association of thioautotrophic bacteria with deep-sea sponges. *Marine Biotechnology* **12,** 253–260.

Nocker, A., Burr, M., and Camper, A. K. (2007). Genotypic microbial community profiling: A critical technical review. *Microbial Ecology* **54,** 276–289.

Nyholm, S. V., and McFall-Ngai, M. J. (2004). The winnowing: Establishing the squid-*Vibrio* symbiosis. *Nature* **2,** 632–642.

Oh, P. L., Benson, A. K., Peterson, D. A., Patil, P. B., Moriyama, E. N., Roos, S., and Walter, J. (2010). Diversification of the gut symbiont *Lactobacillus reuteri* as a result of host-driven evolution. *The ISME Journal* **4,** 377–387.

Olson, J. B., and Kellogg, C. A. (2010). Microbial ecology of corals, sponges and algae in mesophotic coral environments. *FEMS Microbiology Ecology* **73,** 17–30.

Olson, J. B., and McCarthy, P. J. (2005). Associated bacterial communities of two deep-water sponges. *Aquatic Microbial Ecology* **39,** 47–55.

Olson, J. B., Gochfeld, D. J., and Slattery, M. (2006). *Aplysina* red band syndrome: A new threat to Caribbean sponges. *Diseases of Aquatic Organisms* **71,** 163–168.

Pace, N. R. (1997). A molecular view of microbial diversity and the biosphere. *Science* **276,** 734–740.

Pace, N. R., Stahl, D. A., Lane, D. J., and Olsen, G. J. (1986). The analysis of natural microbial populations by ribosomal RNA sequences. *Advances in Microbial Ecology* **9,** 1–55.

Page, R. D. M. (2003). Tangled Trees: Phylogeny, Cospeciation and Evolution. Chicago University Press, Chicago.

Palmer, T. M., Doak, D. F., Stanton, M. L., Bronstein, J. L., Kiers, E. T., Young, T. P., Green, J. R., and Pringle, R. M. (2010). Synergy of multiple partners, including free-loaders, increases host fitness in a multispecies mutualism. *Proceedings of the National Academy of Sciences of the United States of America* **107,** 17234–17239.

Panagiotis, T., Kemsley, E. K., Ridgway, K. P., Toscano, M. J., Humphrey, T. J., and Narbad, A. (2010). PCR-denaturing gradient gel electrophoresis of complex microbial communities: A two-step approach to address the effect of gel-to-gel variation and allow valid comparisons across a large dataset. *Microbial Ecology* **59,** 776–786.

Pawlik, J. R., Chanas, B., Toonen, R. J., and Fenical, W. (1995). Defenses of Caribbean sponges against predatory reef fish. I. Chemical deterrency. *Marine Ecology Progress Series* **127,** 183–194.

Preston, C. M., Wu, K. Y., Molinski, T. F., and DeLong, E. F. (1996). A psychrophilic crenarchaeon inhabits a marine sponge: *Cenarchaeum symbiosum* genus nov., sp. nov. *Proceedings of the National Academy of Sciences of the United States of America* **93,** 6241–6246.

Radajewski, S., McDonald, I. R., and Murrell, J. C. (2003). Stable-isotope probing of nucleic acids: A window to the function of uncultured microorganisms. *Current Opinion in Biotechnology* **14,** 296–302.

Reeder, J., and Knight, R. (2009). The 'rare biosphere': A reality check. *Nature Methods* **6,** 636–637.

Regoli, F., Cerrano, C., Chierici, E., Bompadre, S., and Bavestrello, G. (2000). Susceptibility to oxidative stress of the Mediterranean demosponge *Petrosia ficiformis*: Role of endosymbionts and solar irradiance. *Marine Biology* **137,** 453–461.

Reiswig, H. M. (1974). Water transport, respiration and energetics of three tropical marine sponges. *Journal of Experimental Marine Biology and Ecology* **14,** 231–249.

Reiswig, H. M. (1981). Partial carbon and energy budgets of the bacteriosponge *Verongia fistularis* (Porifera: Demospongiae) in Barbados. *Pubblicazioni della Stazione Zoologica di Napoli I: Marine Ecology* **2,** 273–293.

Reyzer, M. L., Chaurand, P., Angel, P. M., and Caprioli, R. M. (2010). Direct molecular analysis of whole-body animal tissue sections by MALDI imaging mass spectrometry. In "Mass Spectrometry Imaging" (S. S. Rubakhin and J. V. Sweedler, eds), pp. 285–301. Springer, New York.

Ridley, C. P., Bergquist, P. R., Harper, M. K., Faulkner, D. J., Hooper, J. N. A., and Haygood, M. G. (2005). Speciation and biosynthetic variation in four dictyoceratid sponges and their cyanobacterial symbiont, *Oscillatoria spongeliae*. *Chemistry and Biology* **12,** 397–406.

Roberts, D. E., Cummins, S. P., Davis, A. R., and Pangway, C. (1999). Evidence for symbiotic algae in sponges from temperate coastal reefs in New South Wales, Australia. *Memoirs of the Queensland Museum* **44,** 493–497.

Rocap, G., Distel, D. L., Waterbury, J. B., and Chisholm, S. W. (2002). Resolution of *Prochlorococcus and Synechococcus* ecotypes by using 16S-23S ribosomal DNA internal transcribed spacer sequences. *Applied and Environmental Microbiology* **68,** 1180–1191.

Rocap, G., Larimer, F. W., Lamerdin, J., Malfatti, S., Chain, P., Ahlgren, N. A., Arellano, A., Coleman, M., Hauser, L., Hess, W. R., Johnson, Z. I., Land, M., Lindell, D., Post, A. F., Regala, W., Shah, M., Shaw, S. L., Steglich, C., Sullivan, M. B., Ting, C. S., Tolonen, A., Webb, E. A., Zinser, E. R., and Chisholm, S. W. (2003). Genome divergence in two *Prochlorococcus* ecotypes reflects oceanic niche differentiation. *Nature* **424,** 1042–1047.

Ronquist, F. (1997). Phylogenetic approaches in coevolution and biogeography. *Zoologica Scripta* **26,** 313–322.

Ronquist, F. (2004). Bayesian inference of character evolution. *Trends in Ecology and Evolution* **19,** 475–481.

Rosenberg, E., and Loya, Y. (2004). Coral Health and Disease. Springer, Berlin.

Rowan, R. (1998). Diversity and ecology of zooxanthellae on coral reefs. *Journal of Phycology* **34,** 407–417.

Rowan, R. (2004). Thermal adaptation in reef coral symbionts. *Nature* **430,** 742.

Rowan, R., and Powers, D. A. (1991). A molecular genetic classification of zooxanthellae and the evolution of animal-algal symbioses. *Science* **251,** 1348–1351.

Rutzler, K. (1988). Mangrove sponge disease induced by cyanobacterial symbionts: Failure of a primitive immune system? *Diseases of Aquatic Organisms* **5,** 143–149.

Rutzler, K. (1990). Associations between Caribbean sponges and photosynthetic organisms. In "New Perspectives in Sponge Biology" (K. Rützler, ed.), pp. 455–466. Smithsonian Institution Press, Washington, DC.

Sachs, J. L., and Simms, E. L. (2006). Pathways to mutualism breakdown. *Trends in Ecology and Evolution* **21,** 585–592.

Santavy, D. L., and Colwell, R. R. (1990). Comparison of bacterial communities associated with the Caribbean sclerosponge *Ceratoporella nicholsoni* and ambient seawater. *Marine Ecology Progress Series* **67,** 73–82.

Santos, S. R., Shearer, T. L., Hannes, A. R., and Coffroth, M. A. (2004). Fine-scale diversity and specificity in the most prevalent lineage of symbiotic dinoflagellates (*Symbiodinium*, Dinophyceae) of the Caribbean. *Molecular Ecology* **13,** 459–469.

Sarà, M., and Vacelet, J. (1973). Ecologie des Demosponges. In "Traite de Zoologie 3(1), Spongiaires" (P.-P. Grasse, ed.), pp. 462–576. Masson, Paris.

Sarà, M., Bavestrello, G., Cattaneo-Vietti, R., and Cerrano, C. (1998). Endosymbiosis in sponges: Relevance for epigenesis and evolution. *Symbiosis* **25,** 57–70.

Schläppy, M.-L., Hoffmann, F., Røy, H., Wijffels, R. H., Mendola, D., Sidri, M., and de Beer, D. (2007). Oxygen dynamics and flow patterns of *Dysidea avara* (Porifera: Demospongiae). *Journal of the Marine Biological Association of the United Kingdom* **86,** 1677–1682.

Schläppy, M.-L., Schöttner, S. I., Lavik, G., Kuypers, M. M. M., de Beer, D., and Hoffmann, F. (2010). Evidence of nitrification and denitrification in high and low microbial abundance sponges. *Marine Biology* **157,** 593–602.

Schmidt, E. W. (2008). Trading molecules and tracking targets in symbiotic interactions. *Nature Chemical Biology* **4,** 466–473.

Schmitt, S., Weisz, J. B., Lindquist, N., and Hentschel, U. (2007). Vertical transmission of a phylogenetically complex microbial consortium in the viviparous sponge *Ircinia felix*. *Applied and Environmental Microbiology* **73,** 2067–2078.

Schmitt, S., Angermeier, H., Schiller, R., Lindquist, N., and Hentschel, U. (2008). Molecular microbial diversity survey of sponge reproductive stages and mechanistic insights into vertical transmission of microbial symbionts. *Applied and Environmental Microbiology* **74,** 7694–7708.

Schmitt, S., Tsai, P., Bell, J., Fromont, J., Ilan, M., Lindquist, N., Perez, T., Rodrigo, A., Schupp, P. J., Vacelet, J., Webster, N., Hentschel, U., and Taylor, M. W. (2011). Assessing the complex sponge microbiota: Core, variable, and species-specific bacterial communities in marine sponges. *ISME Journal* **6,** (3), 564–57610.1038/ismej.2011.116.

Schönberg, C. H. L., and Loh, W. K. W. (2005). Molecular identity of the unique symbiotic dinoflagellates found in the bioeroding demosponge *Cliona orientalis. Marine Ecology Progress Series* **299,** 157–166.

Schönberg, C. H. L., Hoffmann, F., and Gatti, S. (2004). Using microsensors to measure sponge physiology. *Bollettino dei Musei e degli Istituti Biologici dell'Università di Genova* **68,** 593–604.

Sharp, K. H., Eam, B., Faulkner, D. J., and Haygood, M. G. (2007). Vertical transmission of diverse microbes in the tropical sponge *Corticium* sp. *Applied and Environmental Microbiology* **73,** 622–629.

Siegl, A., Kamke, J., Hochmuth, T., Piel, J., Richter, M., Liang, C., Dandekar, T., and Hentschel, U. (2011). Single-cell genomics reveals the lifestyle of *Poribacteria*, a candidate phylum symbiotically associated with marine sponges. *The ISME Journal* **5,** 61–70.

Simms, E. L., Taylor, D. L., Povich, J., Shefferson, R. P., Sachs, J. L., Urbina, M., and Tausczik, Y. (2006). An empirical test of partner choice mechanisms in a wild legume-rhizobium interaction. *Proceedings of the Royal Society of London B* **273,** 77–81.

Smith, D. C. (1991). Why do so few animals form endosymbiotic associations with photosynthetic microbes? *Philosophical Transactions of the Royal Society B* **333,** 225–230.

Sogin, M. L., Morrison, H. G., Huber, J. A., Welch, D. M., Huse, S. M., Neal, P. R., Arrieta, J. M., and Herndl, G. J. (2006). Microbial diversity in the deep sea and the underexplored 'rare biosphere'. *Proceedings of the National Academy of Sciences of the United States of America* **103,** 12115–12120.

Sotka, E. E., and Thacker, R. W. (2005). Do some corals like it hot? *Trends in Ecology and Evolution* **20,** 59–62.

Stanley, G. D. (2006). Photosymbiosis and the evolution of modern coral reefs. *Science* **312,** 857–858.

Steindler, L., Beer, S., and Ilan, M. (2002). Photosymbiosis in intertidal and subtidal tropical sponges. *Symbiosis* **33,** 1–11.

Steindler, L., Huchon, D., Avni, A., and Ilan, M. (2005). 16S rRNA phylogeny of sponge-associated cyanobacteria. *Applied and Environmental Microbiology* **71,** 4127–4131.

Stewart, F. J., and Cavanaugh, C. M. (2007). Intragenomic variation and evolution of the internal transcribed spacer of the rRNA operon in bacteria. *Journal of Molecular Evolution* **65,** 44–67.

Sun, W., Dai, S., Wang, G., Xie, L., Jiang, S., and Li, X. (2010). Phylogenetic diversity of bacteria associated with the marine sponge *Agelas robusta* from South China Sea. *Acta Oceanologica Sinica* **29,** 65–73.

Sutherland, K. P., Porter, J. W., and Torres, C. (2004). Disease and immunity in Caribbean and Indo-Pacific zooxanthellate corals. *Marine Ecology Progress Series* **266,** 273–302.

Tanaka, Y., Miyajima, T., Koike, I., Hayashibara, Y., and Ogawa, H. (2006). Translocation and conservation of organic nitrogen within the coral-zooxanthella symbiotic system of *Acropora pulchra*, as demonstrated by dual isotope-labeling techniques. *Journal of Experimental Marine Biology and Ecology* **336,** 110–119.

Taylor, M. W., Hill, R. T., Piel, J., Thacker, R. W., and Hentschel, U. (2007a). Soaking it up: The complex lives of marine sponges and their microbial associates. *The ISME Journal* **1,** 187–190.

Taylor, M. W., Radax, R., Steger, D., and Wagner, M. (2007b). Sponge-associated microorganisms: Evolution, ecology, and biotechnological potential. *Microbiology and Molecular Biology Reviews* **71,** 295–347.

Taylor, M. W., Thacker, R. W., and Hentschel, U. (2007c). Evolutionary insights from sponges. *Science* **316,** 1854–1855.

Thacker, R. W. (2005). Impacts of shading on sponge-cyanobacteria symbioses: A comparison between host-specific and generalist associations. *Integrative and Comparative Biology* **45,** 369–376.

Thacker, R. W., and Starnes, S. (2003). Host specificity of the symbiotic cyanobacterium *Oscillatoria spongeliae* in marine sponges, *Dysidea* spp. *Marine Biology* **142,** 643–648.

Thacker, R. W., Becerro, M. A., Lumbang, W. A., and Paul, V. J. (1998). Allelopathic interactions between sponges on a tropical reef. *Ecology* **79,** 1740–1750.

Thacker, R. W., Diaz, M. C., Rützler, K., Erwin, P. M., Kimble, S. J. A., Pierce, M. J., and Dillard, S. L. (2007). Phylogenetic relationships among the filamentous cyanobacterial symbionts of Caribbean sponges and a comparison of photosynthetic production between sponges hosting filamentous and unicellular cyanobacteria. In "Porifera Research: Biodiversity, Innovation, and Sustainability" (E. Hajdu and G. Muricy, eds), pp. 621–626. Museu Nacional, Rio de Janeiro.

Thakur, N. L., Hentschel, U., Krasko, A., Pabel, C. T., Anil, A. C., and Müller, W. E. G. (2003). Antibacterial activity of the sponge *Suberites domuncula* and its primmorphs: Potential basis for epibacterial chemical defense. *Aquatic Microbial Ecology* **31,** 77–83.

Thomas, T., Rusch, D., DeMaere, M. Z., Yung, P. Y., Lewis, M., Halpern, A., Heidelberg, K. B., Egan, S., Steinberg, P. D., and Kjelleberg, S. (2010). Functional genomic signatures of sponge bacteria reveal unique and shared features of symbiosis. *The ISME Journal* **4,** 1557–1567.

Thompson, J. N. (2005). The Geographic Mosaic of Coevolution. University of Chicago Press, Chicago.

Thoms, C., Horn, M., Wagner, M., Hentschel, U., and Proksch, P. (2003). Monitoring microbial diversity and natural product profiles of the sponge *Aplysina cavernicola* following transplantation. *Marine Biology* **142,** 685–692.

Turon, X., Galera, J., and Uriz, M. J. (1997). Clearance rates and aquiferous systems in two sponges with contrasting life-history strategies. *Journal of Experimental Zoology* **278,** 22–36.

Unson, M. D., and Faulkner, D. J. (1992). Cyanobacterial symbiont biosynthesis of chlorinated metabolites from *Dysidea herbacea* (Porifera). *Experientia* **49,** 349–353.

Unson, M. D., Holland, N. D., and Faulkner, D. J. (1994). A brominated secondary metabolite synthesized by the cyanobacterial symbiont of a marine sponge and accumulation of the crystalline metabolite in the sponge tissue. *Marine Biology* **119,** 1–11.

Usher, K. M. (2008). The ecology and phylogeny of cyanobacterial symbionts in sponges. *Marine Ecology* **29,** 178–192.

Usher, K. M., Kuo, J., Fromont, J., and Sutton, D. C. (2001). Vertical transmission of cyanobacterial symbionts in the marine sponge *Chondrilla australiensis* (Demospongiae). *Hydrobiologia* **461,** 15–23.

Usher, K. M., Fromont, J., Sutton, D. C., and Toze, S. (2004). The biogeography and phylogeny of unicellular cyanobacterial symbionts in sponges from Australia and the Mediterranean. *Microbial Ecology* **48,** 167–177.

Usher, K. M., Kuo, J., Fromont, J., Toze, S., and Sutton, D. C. (2006). Comparative morphology of five species of symbiotic and nonsymbiotic coccoid cyanobacteria. *European Journal of Phycology* **41,** 179–188.

Usher, K. M., Bergman, B., and Raven, J. A. (2007). Exploring cyanobacterial mutualisms. *Annual Reviews of Ecology, Evolution and Systematics* **38,** 255–273.

Vacelet, J. (1975). Étude en microscopie électronique de l'association entre bactéries etspongiaires du genre *Verongia* (Dictyoceratida). *Journal de Microscopie et de Biologie Cellulaire* **23,** 271–288.

Vacelet, J., and Donadey, C. (1977). Electron microscope study of the association between some sponges and bacteria. *Journal of Experimental Marine Ecology* **30,** 301–314.
Vrijenhoek, R. C. (2010). Genetics and evolution of deep-sea chemosynthetic bacteria and their invertebrate hosts. In "The Vent and Seep Biota, Topics in Geobiology 33" (S. Kiel, ed.), pp. 15–50. Springer, Berlin.
Webster, N. S. (2007). Sponge disease: A global threat? *Environmental Microbiology* **9,** 1363–1375.
Webster, N. S., and Blackall, L. L. (2009). What do we really know about sponge-microbial symbioses? *The ISME Journal* **3,** 1–3.
Webster, N. S., and Hill, R. T. (2001). The culturable microbial community of the Great Barrier Reef sponge *Rhopaloeides odorabile* is dominated by an alpha Proteobacterium. *Marine Biology* **138,** 843–851.
Webster, N. S., and Taylor, M. W. (2011). Marine sponges and their microbial symbionts: Love and other relationships. *Environmental Microbiology* **14,** (2), 335–346.
Webster, N. S., Wilson, K. J., Blackall, L. L., and Hill, R. T. (2001). Phylogenetic diversity of bacteria associated with the marine sponge *Rhopaloeides odorabile*. *Applied and Environmental Micrbiology* **67,** 434–444.
Webster, N. S., Negri, A. P., Webb, R. I., and Hill, R. T. (2002). A spongin-boring alpha proteobacterium is the etiological agent of disease in the Great Barrier Reef sponge, *Rhopaloeides odorabile*. *Marine Ecology Progress Series* **232,** 305–309.
Webster, N. S., Negri, A. P., Munro, M. M. H. G., and Battershill, C. N. (2004). Diverse microbial communities inhabit Antarctic sponges. *Environmental Microbiology* **6,** 288–300.
Webster, N. S., Taylor, M. W., Behnam, F., Lücker, S., Rattei, T., Whalan, S., Horn, M., and Wagner, M. (2010a). Deep sequencing reveals exceptional diversity and modes of transmission for bacterial sponge symbionts. *Environmental Microbiology* **12,** 2070–2082.
Webster, N. S., Cobb, R. E., Soo, R., Anthony, S. L., Battershill, C. N., Whalan, S., and Evans-Illidge, E. (2010b). Bacterial community dynamics in the marine sponge *Rhopaloeides odorabile* under in situ and ex situ cultivation. *Marine Biotechnology* **13,** 296–304.
Wehrl, M., Steinert, M., and Hentschel, U. (2007). Bacterial uptake by the marine sponge *Aplysina aerophoba*. *Microbial Ecology* **53,** 355–365.
Weil, E. (2004). Coral reef diseases in the wider Caribbean. In "Coral Health and Disease" (E. Rosenberg and Y. Loya, eds), pp. 35–68. Springer-Verlag, New York.
Weisz, J. B. (2006). *Measuring impacts of associated microbial communities on Caribbean reef sponges: Searching for symbiosis*. Doctoral Dissertation, University of North Carolina at Chapel Hill.
Weisz, J. B., Hentschel, U., Lindquist, N., and Martens, C. S. (2007). Linking abundance and diversity of sponge-associated microbial communities to metabolic differences in host sponges. *Marine Biology* **152,** 475–483.
Weisz, J. B., Lindquist, N., and Martens, C. S. (2008). Do associated microbial abundances impact marine demosponge pumping rates and tissue densities? *Oecologia* **155,** 367–376.
Weisz, J. B., Massaro, A. J., Ramsby, B. D., and Hill, M. S. (2010). Zooxanthellar symbionts shape host sponge trophic status through translocation of carbon. *Biological Bulletin* **219,** 189–197.
Wilkinson, C. R. (1978a). Microbial associations in sponges. I. Ecology, physiology and microbial populations of coral reef sponges. *Marine Biology* **49,** 161–167.
Wilkinson, C. R. (1978b). Microbial associations in sponges. II. Numerical analysis of sponge and water bacterial communities. *Marine Biology* **49,** 169–176.
Wilkinson, C. R. (1979). Nutrient translocation from symbiotic cyanobacteria to coral reef sponges. In "Biologie des Spongiaires" (C. Levi and N. Boury-Esnault, eds), pp. 373–380. Colloques Internationaux du Centre National de la Recherche Scientifique, Paris.
Wilkinson, C. R. (1983). Net primary productivity in coral reef sponges. *Science* **219,** 410–412.

Wilkinson, C. R. (1987). Interocean differences in size and nutrition of coral reef sponge populations. *Science* **236,** 1654–1657.

Wilkinson, C. R., and Fay, P. (1979). Nitrogen fixation in coral reef sponges with symbiotic cyanobacteria. *Nature* **279,** 527–529.

Wilkinson, C. R., and Vacelet, J. (1979). Transplantation of marine sponges to different conditions of light and current. *Journal of Experimental Marine Biology and Ecology* **37,** 91–104.

Wilkinson, C. R., Nowak, M., Austin, B., and Colwell, R. R. (1981). Specificity of bacterial symbionts in Mediterranean and Great Barrier Reef Sponges. *Microbial Ecology* **7,** 13–21.

Wilkinson, C. R., Garrone, R., and Vacelet, J. (1984). Marine sponges discriminate between food bacteria and bacterial symbionts: Electron microscope radioautography and in situ evidence. *Proceedings of the Royal Society of London, Series B* **220,** 519–528.

Won, Y.-J., Hallam, S. J., O'Mullan, G. D., Pan, I. L., Buck, K. R., and Vrijenhoek, R. C. (2003). Environmental acquisition of thiotrophic endosymbionts by deep-sea mussels of the genus *Bathymodiolus*. *Applied and Environmental Microbiology* **69,** 6785–6792.

Wulff, J. L. (2007). Disease prevalence and population density over time in three common Caribbean coral reef sponge species. *Journal of the Marine Biological Association of the United Kingdom* **87,** 1715–1720.

Yahel, G., Sharp, J. H., Marie, D., Häse, C., and Genin, A. (2003). *In situ* feeding and element removal in the symbiont-bearing sponge *Theonella swinhoei*: Bulk DOC is the major source for carbon. *Limnology and Oceanography* **48,** 141–149.

Yang, J., Sun, J., Lee, O. O., Wong, Y. H., and Qian, P. Y. (2011). Phylogenetic diversity and community structure of sponge-associated bacteria from mangroves of the Caribbean Sea. *Aquatic Microbial Ecology* **62,** 231–240.

Young, J. P. W., Mutch, L. A., Ashford, D. A., Zézé, A., and Mutch, K. E. (2003). The molecular evolution of host specificity in the rhizobium-legume symbiosis. In "Genes in the Environment" (R. Hails, H. C. J. Godfray and J. Beringer, eds), pp. 245–257. Blackwell Science, Oxford.

CHAPTER THREE

Nutrient Fluxes Through Sponges: Biology, Budgets, and Ecological Implications

Manuel Maldonado*,[1], Marta Ribes† and Fleur C. van Duyl‡

Contents

* Department of Marine Ecology, Centro de Estudios Avanzados de Blanes (CEAB-CSIC), Blanes, Girona, Spain
† Institut de Ciències del Mar (ICM-CSIC), Passeig Marítim de la Barceloneta, Barcelona, Spain
‡ Department of Biological Oceanography, Royal Netherlands Institute for Sea Research, P.O. Box 59, Texel, The Netherlands
[1]Corresponding author: Email: maldonado@ceab.csic.es

Advances in Marine Biology, Volume 62
ISSN 0065-2881, DOI: 10.1016/B978-0-12-394283-8.00003-5

Abstract

Marine sponges are able to process a variety of carbon (C), nitrogen (N), phosphorous (P), and silicon (Si) dissolved compounds, in addition to the particulate C, N, and P obtained through regular feeding. While Si fluxes through sponges are exclusively related to the elaboration of their skeleton of biogenic silica, C, N, and P fluxes derive from a complex combination of metabolic processes that include feeding, respiration, egestion, excretion, as well as hosting of large microbial populations within the sponge body. Because of the remarkable abundance of sponges in many benthic marine communities, they have the potential to impact the availability of the compounds they take up and release, affecting the benthic–pelagic coupling and cycling rates of chemical elements that are crucial to determine growth of bacterioplankton and primary producers at the ecosystem level. Unfortunately, our knowledge and understanding of the magnitude of the sponge-meditated nutrient fluxes and their ecological implications depends much on the compound type (i.e. C, N, P, or Si). Herein, we review the available knowledge on the subject with emphasis on recent developments.

Key Words: carbon balance; carbon metabolism; DOC uptake; nitrogen balance; nitrogen metabolism; phosphorous balance; POM uptake; silicate; silicon balance; silicon metabolism

1. Introduction

Sponges are ubiquitous marine organisms, occurring with moderate to high abundance on continental shelves, slopes, abyssal plains, and even hadal bottoms. Because of their ubiquity and abundance, sponges are now understood as an ecologically important benthic group that plays a variety of functional roles in marine ecosystems: (1) They have long been recognized to influence the structure of benthic communities through substrate occupation. Sponges participate in processes of both spatial exclusion of competitors (Rützler, 1970; Sarà, 1970; Thacker *et al.*, 1998; and references therein) and cooperative spatial interactions, being relevant habitat builders that add complexity to the communities by favouring increased abundance and diversity of other organisms (Dayton *et al.*, 1974; McClintock *et al.*, 2005; Wulff, 2008). (2) In carbonated ecosystems, such as coral reefs, sponges are thought to influence the carbonate framework through processes of both cementation (Wulff, 1984, 2001) and bioerosion (Diaz and Rützler, 2001). (3) The suspension-feeding activity of sponges (mostly on bacterioplankton and microphytoplankton) has also been vindicated as a significant trophic link between the benthos and the overlying water column, being suspected to impact the benthic–pelagic coupling of particulate carbon fluxes at a diversity of scales (e.g. Reiswig, 1971b; Pile, 1997; Gili and Coma, 1998; Ribes *et al.*, 1999b). (4) Yet, sponges have a further functional role that remains relatively understudied despite its multiple ecological and biogeochemical derivations. During the past decades, the notion has emerged that sponges may influence the availability of dissolved nutrients, such as dissolved carbon (Yahel *et al.*, 2003; De Goeij *et al.*, 2008b), various nitrogen compounds (Corredor *et al.*, 1988; Bayer *et al.*, 2007; Jiménez and Ribes, 2007), and silicate (Reincke and Barthel, 1997; Scheffers *et al.*, 2004; Maldonado *et al.*, 2005, 2010a). These dissolved nutrients exert a major impact on primary production and their use by the phytoplankton is, in turn, responsible for establishing interconnections of much ecological, environmental, and biogeochemical relevance between C, N, P, and Si cycles. Therefore, it is urgent to identify the level at which sponges affect the availability and cycling of those chemical elements. Here, we review the available information to date with the objective of providing a better global understanding of the role of sponges as sources, sinks, and cyclers of a variety of relevant C, N, P, and Si dissolved and particulate compounds, usually referred to as "nutrients".

1.1. Measurement of nutrient fluxes

Nutrient fluxes in sponges are commonly evaluated through estimates of uptake rates and efflux rates in individual organisms, then extrapolated to the population or community level whenever possible; however, comparison among methods is not always straightforward (Riisgård, 2001; Yahel *et al.*, 2005). Different methodologies can be used, but not all of them are suitable for certain organisms. Based on Yahel *et al.* (2005), methods can be classified as indirect or direct techniques.

1.1.1. Indirect techniques

They have been widely used in research of nutrition and physiology of filter feeders (Wafar *et al.*, 1990; Díaz and Ward, 1997; Ribes *et al.*, 2000; Pile *et al.*, 2003; Jiménez and Ribes, 2007). For these techniques, animals are placed in a vessel with little or no renewal of water. Changes in compound concentration (cells, coloured beads, dissolved radioactive label, excretion products, etc.) in the surrounding water over time are attributed to the animal's activity and are used to quantify production and/or removal rates (Riisgård, 2001). The comparison with a control situation (same vessel without organisms) is used to correct for changes in compound concentration not due to the organisms' activity. Indirect methods cannot differentiate pumping rates and retention efficiency, so in feeding studies the term "clearance rate" is used, that is, volume of water cleared of particles (assuming an efficiency in the removal of 100%) per time unit (Riisgård, 2001). Incubation in closed vessels is one of the most popular indirect methods used in the study of diverse benthic suspension feeders, such as bivalves (Roditi and Fisher, 1999; Petersen *et al.*, 2004), sponges (Díaz and Ward, 1997; Ribes *et al.*, 1999b; Jiménez and Ribes, 2007; Bayer *et al.*, 2008; De Goeij *et al.*, 2008a; Van Duyl *et al.*, 2008), corals (Ribes *et al.*, 2000), and ascidians (Ribes *et al.*, 2000).

There are several problems related to the use of closed systems for the study of feeding and excretion of suspension feeders. In particular, decline in oxygen and/or food concentration and accumulation of excretion compounds in the incubation water are reported to alter-feeding behaviour and pumping (Díaz and Ward, 1997; Yahel *et al.*, 2005; Hadas *et al.*, 2008). Other biases may be induced by re-filtration of the incubation water, which can cause underestimates of removal rates if sampling times are not well adjusted to the pumping rates of the studied animal (Ribes *et al.*, 2000; Yahel *et al.*, 2005).

In flow-through methods (also an indirect technique), animals are placed in a vessel with a continuous water flow, and their activity measured by determining differences in compound concentration between inflow and outflow water. The flow-through method would overcome some of the problems associated with incubations (i.e. stagnation), as water is renewed during the experiments. Some of the shortcomings of this methodology are that all water flowing through the chamber has to be in contact with the filter-feeder and that re-filtration has to be avoided. Also, choosing the best flow rate in the chamber is not trivial: flow rate has to be low enough to detect the organism signal in the outflow but high enough to avoid re-filtration. Additionally, flow rate can have an effect over the regular pumping behaviour (Yahel *et al.*, 2005; Hadas *et al.*, 2008), with unusual flows either stimulating or inhibiting pumping.

1.1.2. Direct techniques

They imply sampling water immediately upon completing a single run through the organism's filtration system. Changes in compound concentration between the inflow and the outflow, combined with organisms' pumping rates (measured with Dye Front Speed method, flowmeters, dye-video records, etc.) result in ingestion/uptake or production rates (Reiswig, 1971a,b, 1974; Savarese *et al.*, 1997; Yahel *et al.*, 2005; Weisz *et al.*, 2008). Problems associated to indirect methodology, such as stagnation or flux in the vessel, are avoided with the application of direct methodology. As concentrations are measured directly from inflows and outflows, and not from ambient water, direct techniques facilitate "*in situ*" approaches both in the field (Reiswig, 1971a; Pile *et al.*, 1996; Yahel *et al.*, 2003; Southwell *et al.*, 2008b) or in tanks with high renovation rates (Hadas *et al.*, 2006; Yahel *et al.*, 2006). Also with the application of direct methodology, retention efficiency and pumping rate are clearly discriminated, allowing more detailed studies of nutrition, including food selectivity (Pile *et al.*, 1996;

Yahel *et al.*, 2006). Despite the advantages of the direct techniques (Riisgård, 2001, 2004; Yahel *et al.*, 2005), they are not free of limitations, as the organisms must have a well-defined excurrent aperture to allow reliable direct sampling without contacting the animal tissue. Also, sampling suction rate has to be much lower than organism excurrent flow rate to avoid contamination of the outgoing-water sample with ambient water (Yahel *et al.*, 2005). These requirements make the In-Ex method unsuitable for organisms with extremely low pumping rates, those with an excurrent aperture < 1 mm, and those lacking well-defined outflows.

Different methodologies have been applied indiscriminately to sponges and their comparisons that can result in conflicting estimates of uptake and/or release rates and consequently in large differences when extrapolated as community-scale fluxes. Although some efforts have been done to evaluate the effect of different methods for some particular organisms, such as, for instance, mussels (Riisgård, 2001; Yahel *et al.*, 2005), few methodological comparisons have been performed for sponges (but see Southwell *et al.*, 2008b).

1.2. Nutrient availability: Particulate and dissolved compounds

It has been documented that marine sponges are able to remove and return a variety of both particulate and dissolved "nutrients" and related compounds, that is, those known or suspected to interfere with primary production and bacterioplankton growth.

1.2.1. Particulate nutrients

Particulate organic matter (POM) can be generally divided into live particulate organic matter (LPOM) composed of various planktonic cells, and non-living particulate organic matter (hereafter "detritus"), which is a structurally and chemically diverse fraction deriving from several sources (Ribes *et al.*, 2003; Wilson *et al.*, 2003; Hadas *et al.*, 2009). Detritus significance as a food source for marine sponges is poorly documented, but in some cases it can contribute up to 50% of particulate organic carbon (POC; Hadas *et al.*, 2009). Sponges can retain LPOM from viruses (< 0.2 μm) to larger cells as phytoplankton, but, due to the features of their filtration system, they are particularly efficient in retaining 0.2–2 μm picoplankton cells (Pile *et al.*, 1996; Ribes *et al.*, 1999b; Hadas *et al.*, 2006; Yahel *et al.*, 2006).

Regular particle capture can take place at three functional sites inside sponges. Large particles (> 50 μm) that do not enter the ostia are taken up at the surface by epithelial pinacocytes. Smaller particles (< 50 μm) that enter the ostia are taken up by pinacocytes lining the canal walls. The smallest particles (< 5 μm) are often engulfed by the choanocytes in the choanochambers. After capture by pinacocytes and choanocytes, food particles in incipient digestion stage are passed to the mesohyl cells by transcytosis. Once inside the mesohyl, they are phagocytosed mostly by archaeocytes, which complete digestion and deliver the resulting assimilable compounds to other cells. The general perception from the earliest studies on sponge feeding is that particle uptake by sponges is highly efficient at the 1–2 μm size range but largely unselective (Reiswig, 1971a; Pile *et al.*, 1996; Ribes *et al.*, 1999b). Nevertheless, more recent studies have shown that contribution of different planktonic groups may differ among sponge species due to variation in choanocyte numbers, feeding methods (Leys and Eerkes-Medrano, 2006), and complexity of aquiferous systems (Weisz *et al.*, 2008).

Table 3.1 shows retention efficiency of various picoplankton cells by several sponges. Some species, such as *Callyspongia* sp. from southwestern Australia, exhibited consistently negative selectivity for LDNA bacteria (i.e. smaller and less active heterotrophic bacteria) and positive selectivity for *Synechococcus* spp.; HDNA bacteria (i.e. bigger and more active heterotrophic bacteria) experienced neutral or positive selection (Hanson *et al.*, 2009). In the Mediterranean demosponge *Spongia officinalis,* it has also been found a positive selection on *Synechococcus* spp. and picoeukaryotes (Topçu *et al.*, 2010). Additional experimental evidence of selective feeding on particular functional categories of external bacteria and bacterial symbionts has been reported in the Mediterranean demosponge *Aplysina aerophoba* (Wehrl *et al.*, 2007). Contrary to expectations, the hexactinellid *Aphrocallistes vastus* was found to select against *Synechococcus* spp. during July (Yahel *et al.*, 2006; 2007). This negative selection was intriguing, as *Synechococcus* spp. and *Prochlorococcus* spp. appear to be the most preferred food type by most marine demosponges examined to date (Pile *et al.*, 1996; Pile, 1999; Ribes *et al.*, 1999b; Van Duyl *et al.*, 2002; Yahel *et al.*, 2003; 2005). More recently, Maldonado *et al.* (2010b), through a combination of laboratory experiments and transmission electron microscopy study, assessed the potential of the marine sponge *Hymeniacidon perlevis* to feed on three pathogenic microbes common in coastal waters: two bacteria (*Escherichia coli* and *Vibrio anguillarum*) and the marine yeast *Rhodotorula* sp. All three microbes were ingested by the sponge, but selectively, at different rates and using different cellular mechanisms. Such differences in the ingestion and digestion pathways led to large differences in the effectiveness of the sponge to remove a particular microbial type from the ambient water.

1.2.2. Dissolved nutrients and sponge-associated microbes

Many studies have shown that the ability of sponges to remove and/or release large quantities of organic and inorganic dissolved compounds may be related to photoautotrophy and chemoautotrophy processes mediated by sponge-associated microbial communities.

Table 3.1 Retention efficiency (%) of heterotophic bacteria (Hb), *Synechococcus* spp. (Syn), *Proclorococcus* spp. (Pro), nanoeucaryotes (Nk), and pikoeucaryotes (Pk) by sponge species from different locations

Sponge	Location	Bacterial type	Retention efficiency (%)					Source
			Hb	Syn	Pk	Nk	Pro	
Spongia officinalis	Mediterranean		48	77–89	77–87	41–64	–	Topçu *et al.* (2010)
Callyspongia sp.	NW Australia	LDNA	40	86	–	–	–	Hanson *et al.* (2009)
		HDNA	85					
Negombata magnifica	Red Sea		83–90	98–99	65–90	>55	–	Hadas *et al.* (2009)
Theonella swinhoei	Red Sea		85–95		20–40		–	Yahel *et al.* (2003)
Rhabdocalyptus dawsoni	NE Pacific	LDNA	92	80–86	84–91	100	NP	Yahel *et al.* (2006, 2007)
		HDNA	89–97					
Aphrocallistes vastus	NE Pacific	LDNA	84	68	82	80	NP	Yahel *et al.* (2006, 2007)
		HDNA	82					
Sericolophus hawaiicus	Central Pacific		47–54			–	–	Pile and Young (2006)
Dysidea avara	Mediterranean		98	99	91	–	87	Jiménez (2011)
Chondrosia reniformis	Mediterranean		98	96	92	–	93	Jiménez (2011)
Agelas oroides	Mediterranean		79	94	94	–	93	Jiménez (2011)
Neopetrosia problematica	NE Pacific		80	90	98	–	NP	Jiménez (2011)
Haliclona mollis	NE Pacific		88	91	98	–	NP	Jiménez (2011)
Haliclona sp.	NE Pacific		83	98	99	–	NP	Jiménez (2011)
Neopetrosia vanilla	NE Pacific		77	82	100	–	NP	Jiménez (2011)
Sycon coactum	NE Pacific		81	79	98	–	NP	Jiménez (2011)
Thethya leysae	NE Pacific		91	90	91	–	NP	Jiménez (2011)
Halocliona sp.	Red Sea		68	93	92	–	93	Jiménez (2011)
Diacarnus sp.	Red Sea		65	100	98	–	100	Jiménez (2011)
Theonella swinhoei	Red Sea		90	100	98	–	100	Jiménez (2011)
Niphates rowi	Red Sea		69	98	95	–	98	Jiménez (2011)

LDNA, low DNA heterotrophic bacteria; HDNA, high DNA heterotrophic bacteria; NP, not present; –, data not available.

Most marine sponges establish a persistent association with microorganisms (archaea, bacteria, cyanobacteria, yeasts, dinoflagellates, diatoms, etc.), being in many cases a symbiosis, as understood by Moya *et al.* (2008): long-term association between two or more organisms of different species that is integrated at the behavioural, metabolic, or genetic level. Microbial associates often occur intercellularly in the sponge mesohyl (Fig. 3.1), although small populations of "microrganisms" may also occur intracellularly

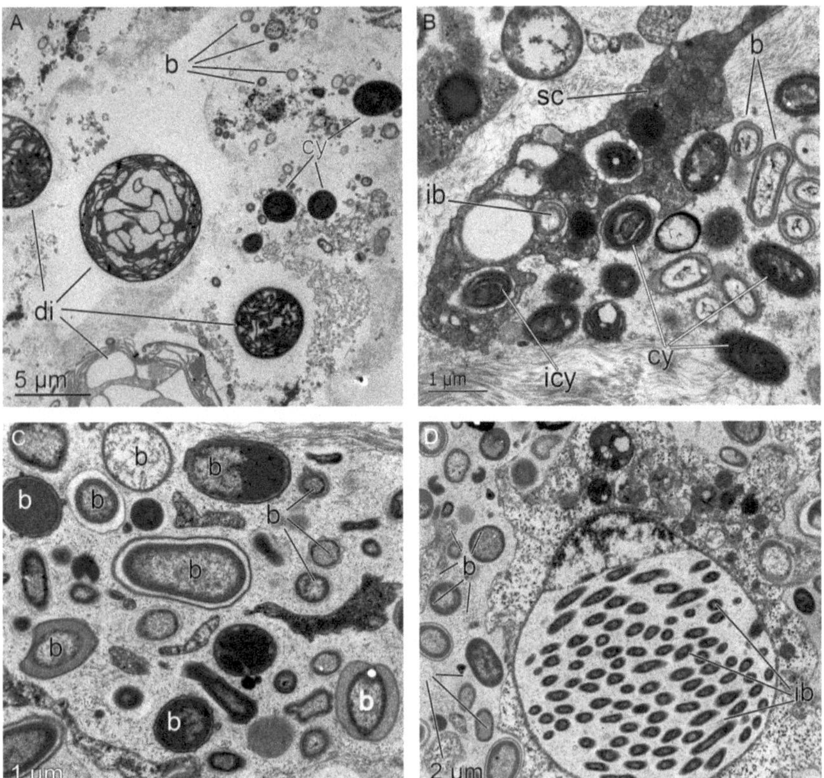

Figure 3.1 Transmission electron microscopy showing microbial communities in the tissues of marine sponges. (A) Detail of the mesohyl of the demosponge *Ircinia variabilis*, showing intercellularly associated dinoflagellates (di), cyanobacteria (cy), and bacteria (b). (B) Detail of the mesohyl of the demosponge *Chondrilla nucula,* showing abundance of intercellular bacteria (b) and cyanobacteria (cy) in the vicinity of a sponge cell (sc), which has incorporated intracellularly some cyanobacteria (icy) and bacteria (ib). (C) Abundance and diversity of intercellular bacteria (b) in the mesohyl of the demosponge *Aplysina cavernicola.* (D) Micrograph of *A. cavernicola* showing abundance and diversity of intercellular bacteria (b) in the mesohyl, but also a sponge cell, the nucleus of which contains abundant intranuclear bacteria (ib), recalling a virus infection.

(Fig. 3.1D). The acquisition of these microbes can be by either environmental transmission, that is, recruiting the microbes from the surrounding water or vertical transmission, that is, transference of the paternal microbes to the progeny via their incorporation in gametes and early embryo stages (reviewed in Maldonado, 2007). Whether one or both acquisition pathways operate for a given sponge species will largely determine the diversity and specific composition of its associated microbial community, as well as the level at which these microbial associates resemble the microbial community in the surrounding ambient seawater. It appears that a given sponge species may contain a mixture of generalist and specialist microorganisms (Taylor *et al.*, 2004).

Based on the density of hosted microbes, sponges have been categorized as either "high-microbial abundance" (HMA; former bacteriosponges) or "low-microbial abundance" (LMA) sponges. HMA sponges are characterized by a microbial population exceeding in 2–4 orders of magnitude, the average microbial density in seawater. In contrasts, LMA sponges host microbial densities of 10^5–10^6 microbes g^{-1} sponge wet weight, which fall in the range of natural bacterial concentration in seawater (Vacelet and Donadey, 1977; Wilkinson, 1978; Reiswig, 1981; 10^5–10^6 microbes g^{-1} sponge wet weight; Hentschel *et al.*, 2006) and also reflect approximately the phylogenetic diversity of the natural microbial assemblage (Schmitt *et al.*, 2007). HMA and LMA are considered as two basic life strategies resulting from evolutionary processes with contrasting characteristics. LMA species have well-irrigated tissues, high pumping rates, and low-diversity microbial communities. Contrarily, HMA species are characterized by a dense tissue, low pumping rates, and high concentration and diversity of microbial associates that are different from the ones occurring in the surrounding water column (e.g. Siegl *et al.*, 2008; Weisz *et al.*, 2008).

Microbial symbionts are hypothesized to contribute to the health and nutrition of sponges in different ways, including production of defending antibiotics, acquisition of limiting nutrients, and processing of metabolic waste (Hoffmann *et al.*, 2005; Taylor *et al.*, 2007; Siegl *et al.*, 2008). Although consumption and remineralization of POM by sponges may have an important role in the ecosystems where they are abundant, the association of sponges with microbial communities cause their functional role to be more complex than mere heterotrophy. While a wide array of phylogenetically and physiologically diverse microbes associated with sponges has been described, their exact role in nutrient cycling remains poorly known. Most of our current understanding about the microbial diversity associated with marine sponges is based on 16S rRNA gene library construction, functional genes surveys, and metagenomics, which are used to attempt unravelling the metabolic routes integrated in those of the sponge metabolism. These metabolic pathways include a variety of processes with alternative energy (photo- or chemotrophic) and carbon (hetero- or autotrophic) sources under different oxygen conditions (Vacelet *et al.*, 1995, 1996; Hoffmann *et al.*, 2005, 2009; Hentschel *et al.*, 2006; Taylor *et al.*, 2007).

2. Use of Carbon by Sponges

Sponges are remarkable suspension feeders that remove substantial amounts of organic matter from the passing water. Wherever they are abundant, they are suspected to play an important role in organic matter processing and benthic–pelagic coupling. Here, we review the knowledge on carbon fluxes through sponges and highlight recent developments, anticipating that to quantify realistically their impact on the carbon flux at the ecosystem level, we still need further insight in the carbon metabolism of different functional groups of sponges.

2.1. Carbon metabolism: The carbon balance

The carbon balance of sponges has challenged sponge scientists for more than 40 years. Compiling carbon balances of sponges confront us with the fact that the basic physiology of sponges is still not fully understood. Of essential aspects, such as (1) the ingestion of total organic carbon, (2) its assimilation and excretion, and (3) the role of sponge-associated microbes in the sponge growth, we know surprisingly little more than when Henry Reiswig reviewed the literature on the subject and presented his pioneering energy budgets of several tropical sponges some 30 years ago (Reiswig, 1974, 1981). He conducted the first *in situ* studies on particulate feeding by marine sponges, on water transport through the aquiferous canals, and on respiration and energetics of sponges, shortly followed by Frost's work (1976, 1978) on fresh water sponges. Since then, main findings for improving our understanding of carbon metabolism of sponges comprise:

1. The low energy costs of pumping and filtration in sponges compared with other benthic invertebrates and the large effect of temperature on filtration rate (Riisgård *et al.*, 1993).
2. The high energy cost of net growth in sponges compared with other benthic suspension feeders (Thomassen and Riisgård, 1995).
3. A better understanding of the particulate diet and selective filtration of particles by sponges (Pile *et al.*, 1996, 1997; Turon *et al.*, 1997; Ribes *et al.*, 1999b; e.g. Maldonado *et al.*, 2010c).
4. Demonstration of bulk dissolved organic matter (DOM) uptake by sponges *in situ* (Yahel *et al.*, 2003; De Goeij *et al.*, 2008b).
5. Demonstration of direct consumption of DOM by both sponge cells and sponge-associated prokaryotes (De Goeij *et al.*, 2008a).

6. Demonstration that HMA sponges pump per unit volume 52–94% less water through their aquiferous system than LMA sponges (Weisz *et al.*, 2008).
7. Realization that sponges may experience rapid cell turnover and choanocyte shedding (De Goeij *et al.*, 2009).

The discovery of bulk uptake of DOM by tropical sponges is an important breakthrough in carbon balance studies (Yahel *et al.*, 2003; De Goeij *et al.*, 2008a). DOM is a large source of organic matter in the aquatic environments and consumption may solve imbalances between carbon consumption and respiration by sponges. Another challenge in sponge carbon metabolism is improving our understanding of energy expenditure on net production by HMA and LMA sponges (Thomassen and Riisgård, 1995; Weisz *et al.*, 2008; De Goeij *et al.*, 2009).

This review focuses on carbon ingestion, growth, respiration, and waste disposal of sponges. Literature is screened for ingestion and respiration data of different sponge species, also in relation to their microbial abundance. The hypothesis that will be tested is whether mismatches between ingestion and respiration in studies that measured POC and/or plankton uptake and did not measure dissolved organic carbon (DOC) uptake, are on average larger in HMA than LMA sponges. In addition, the carbon balance of several sponge species are discussed in the light of recent findings.

2.1.1. Carbon balance equation

Reiswig (1971a, 1974, 1981) was the pioneer putting up carbon balances based on *in situ* measurements for different sponge species. It took decennia to confirm or refute with new measurements some of his interpretations and suggestions (Thomassen and Riisgård, 1995; De Goeij *et al.*, 2008b; Hadas *et al.*, 2008; 2009; Koopmans *et al.*, 2010). The carbon balance in its most simple form is described as:

$$I = P + R + E.$$

It comprises ingestion (I) of carbon, its use for biomass production (P) and respiration (R), and disposal of waste (E), which includes excretion as well as egestion. The P and R terms together represent the assimilation (A). Ingestion is defined as the incorporation of carbon in the sponge and, therefore, not the total amount of organic matter entering the sponge aquiferous system with the water inflow. After ingestion, the carbon is used for biomass production and generation of energy. Biomass production is defined as growth in volume, size, or weight, including gonadal development (net synthesis of organic compounds). Respiration is the metabolic process through which organic matter is oxidized with release of energy. The oxidation of organic matter for energy coincides with the release of carbon dioxide under aerobic conditions in the sponge (=respiration). The net growth efficiency (NGE) is determined as $P/(P+R)$. Since P in relation to R is only sporadically measured in sponges, as far as we know, I/R is used as a measure of metabolic efficiency. Besides carbon loss through respiration, there is organic matter excretion and egestion by the sponge. Waste organic carbon is disposed of and consists of detrital material (DOC as well as POC), comprising digested as well as indigested material. Determination of the ingestion of undefined fractions of organic matter (such as detrital POC and DOC) is biased by the excretion/egestion of organic matter by the sponge. What is measured in most studies is ingestion minus "excretion + egestion" ($I-E$) and it is referred to as the net ingestion rate. However, the E term does not usually cover the full excretion/egestion, because part of the waste disposal by sponges is via faecal pellets, which may settle on the bottom and become unavailable for suspension feeders (Witte *et al.*, 1997).

2.1.2. Carbon ingestion

Organic matter enters the sponge via diffusion, epithelial pinocytosis, and phagocytosis, all facilitated by water pumping at transport rates ranging from 0.002 to 0.84 $cm^3\ cm^{-3}$ sponge s^{-1} (Simpson, 1984). Particle uptake is constrained by the maximum diameter of sponge ostia, generally 50 μm. The main retention and ingestion sites are the choanocyte chambers, although pinacocytes are also able to engulf and pinocyte particles. Sponges have traditionally been regarded as little selective in their particle uptake and therefore incorporating organic particles, regardless of their nutritional value (e.g. Reiswig, 1971a; Wolfrath and Barthel, 1989; Turon *et al.*, 1997). Nevertheless, there is growing evidence that at least some sponges perform a size-independent selection of their food (Frost, 1976, 1980; Van de Vyver *et al.*, 1990; Yahel *et al.*, 2006; Maldonado *et al.*, 2010c).

Carbon capturing rates are thought to be dependent on a wide variety of factors that differ between sponge species, such as body size and morphology (Vogel, 1977, 1978), feeding strategy, such as filtering, carnivorism, methanoptrophy, and others (Vacelet *et al.*, 1996; Vacelet and Duport, 2004; Leys and Eerkes-Medrano, 2006), complexity of the aquiferous system (Reiswig, 1971a; Vacelet and Donadey, 1977; Turon *et al.*, 1997; Weisz *et al.*, 2008), etc. Ingestion of organic carbon also depends on its availability in the ambient water, water transport rate through the sponge, particle-size retention efficiency (e.g. heterotrophic bacterioplankton, cyanobacteria), life-cycle stage, environmental conditions, etc. (Gerrodette and Flechsig, 1979; Riisgård *et al.*, 1993; Ribes *et al.*, 1999a).

Sponges ingest plankton or living particulate organic carbon (LPOC) with preference for the nano- and picoplankton. The highest particle retention efficiencies are found for particles between 0.1 and 10 μm in size. Retention efficiency can rise to 100% for plankton size classes ranging in diameter from 0.2 to 6 μm (e.g., Thomassen and Riisgård, 1995; Pile, 1997; Ribes *et al.*,

1999b, see also Table 3.1). Plankton ingestion rates are part of the POC removal rates of sponges listed in Table 3.2, with plankton ingestion rates (LPOC) varying from 0.04 to 1.17 μmol C cm^{-3} h^{-1}.

Detrital particulate organic carbon (POC_{det}) is also retained by sponges and accounts for the fraction of POC minus LPOC (Reiswig, 1971b; Witte *et al.*, 1997; Yahel *et al.*, 2003; Hadas *et al.*, 2009). POC_{det} ingestion is not always easy to assess. Retention efficiencies of detrital POC are biased by faecal production of the sponge (Reiswig, 1971b; Stuart and Klumpp, 1984; Kowalke, 2000). Re-filtration of suspended waste products of the sponge may particularly occur in closed systems. Based on *in situ* measurements, Hadas *et al.* (2009) ascribe 1/3 of the ingested POC by the reef sponge *Negombata magnifica* to detritus, implying that detritus can form an important part of the diet. Sponges are regarded as opportunistic feeders within their specific particle size spectrum (Coma *et al.*, 2001). Ingestion of detritus is, therefore, dependent on the size of the detritus particle and the preferred size spectrum of the sponge. Total POC ingestion ($LPOC + POC_{det}$) range from 15 to 232 μmol C g DW^{-1} h^{-1} and from 0.252 to 1.4 μmol C cm^{-3} h^{-1} for the sponges in Table 3.2. The POC ingestion rates do not differ between temperate and tropical sponges. Low POM concentrations in oligotrophic waters appear to be compensated by sponges showing either higher filtration rates or increased consumption of DOM.

DOC has long been considered as a food source for sponges. The composition of natural DOM is basically unknown and comprises minute particles, as well as truly dissolved DOM. The DOC size fraction in ambient water is operationally defined as the organic carbon passing through a GF/F (nominal pore size 0.7 μm) or a 0.2-μm pore size filter. Therefore, small particles, such as viruses (<0.2 μm diameter) and 0.1 μm beads, fall in the size fraction of natural DOM. These small size fractions have been reported to be retained by sponges (Hadas *et al.*, 2006; Leys and Eerkes-Medrano, 2006). There are also some indications that colloids and aggregates in the DOM fraction may be consumed by sponges (Reiswig, 1990; Alber and Valiela, 1995), but there are no uptake rates available to date. The capacity for retention and uptake of truly DOM, such as glucose, amino-acids palmitic acid by sponges, or fluorescently labelled DOM in laboratory experiments, has been known for quite some time (Stephens and Schinske, 1961; Schmidt, 1970; Weissenfels, 1976; Wilkinson and Garrone, 1979; Jaeckle, 1995), but experimental evidence for bulk uptake of DOC by sponges was missing until recently. On the tropical sponge *Theonella swinhoei*, Yahel *et al.* (2003) provided the first quantitative assessment of *in situ* retention of ambient bulk DOC. The DOC uptake by *T. swinhoei* is on average 1.56 μmol C cm^{-3} h^{-1}, equalling the highest recorded plankton uptake rates by sponges (Table 3.2). Similarly, De Goeij *et al.* (2008b) reported DOC removal from ambient water at even higher rates by the encrusting coral cavity sponges *Halisarca caerulea*, *Mycale*

Table 3.2 Overview of metabolic rates of different sponge species measured under *in situ* environmental conditions

Sponge	Host type	T (°C)	DOC	POC	WTR	CR_{POC}	I_{DOC}	I_{POC}	R	I_{POC}/R	$I_{POC+DOC}/R$	FTE	Source
Per g DW													
Haliclona oculata	LMA	19	–	40.83		5.67	0	231.53	95.54	2.45		0.06	Koopmans *et al.* (2010)
Halichondria panicea	LMA	14	–	16.67[a]	1.7	1.7	–	28.33	26.75[b]	1.06		0.06	Thomassen and Riisgård (1995)
Dysidea avara	LMA	17	–	32.25	0.06	0.65	0	20.96	9.25	2.28		0.07	Ribes *et al.* (1999b), Coma *et al.* (2002), Turon *et al.* (1997)
Negombata magnifica	LMA	22.5	–	4		4.2	–	16.8	14.88	1.13		0.28	Hadas *et al.* (2008a,b)
Mycale laxissima	LMA	29	–	5.33	12.29	12.29	–	65.55	25.26	2.59		0.49	Reiswig (1973, 1974)
Tectitethya crypta	LMA	30	–	7.08	5.35	3.72	–	26.35	6.64	3.97		0.56	Reiswig (1973, 1974)
Verongula reiswigi	HMA	29	–	5.33	3.05	2.77	–	14.77	27.39	0.54		0.1	Reiswig (1973, 1974)
Aplysina fistularis	HMA	27	–	4.47	3.15		–	3.87	33.64	0.11			Reiswig (1981)
Spongilla lacustris				0.075				30.31					Frost (1978), Karchenko and Lyashenko (1986)
Per cm^3													
Theonella swinhoei	HMA				0.16	0.156	1.56	0.252	1.404	0.18	1.29	0.11	Yahel *et al.* (2003)
Halisarca caerulea	L/HMA	27	127	5.3		0.248[c]	17.1	1.4	2.7	0.52	6.85		De Goeij *et al.* (2008a)
Aplysina fistularis	HMA	27	–		0.44		–	0.545	4.74	0.12		0.09	Reiswig (1981)
Aplysina aerophoba	HMA	18	–	–			–	0.003	1.25	0.0026			Wehrl *et al.* (2007): 20 fg C/bacterium; 1 cm^3 sp = 1.0469 g WW, from Reiswig (1981)
Per g AFDW													
Mycale acerata	HMA	1.8				0.18			0.00393			45.8	Kowalke (2000)
Isodictya kerguelensis	–	1				0.22			0.00156			140.9	Kowalke (2000)

T, temperature in degree Celsius; DOC, dissolved organic carbon in μmol C l^{-1}; POC, particulate organic carbon in μmol C l^{-1}; I, ingestion of DOC or POC in μmol C h^{-1}; R, respiration in μmol O_2 h^{-1}; WTR, specific water transport rate in l h^{-1}; CR, clearance rate in l h^{-1}; I/R, ingestion in μmol C. μmol O_2^{-1}; FTE, filtration efficiency (convection requirement) in L μmol O_2^{-1}; –, not measured.

[a] Bacterial ingestion rate.

[b] Maintenance respiration.

[c] Chl a plus bacterium-based POC.

microsigmatosa, and *Merlia normani,* which ranged from 11 to 21 μmol C cm^{-3} h^{-1}. Interestingly, more than 90% of the diet of these sponges consists of DOM.

Sponges in clear shallow water often have photosynthetic properties and harbour zooxanthellae and cyanobacteria (Rützler, 1990; Schönberg *et al.*, 2005; Thacker, 2005; Erwin and Thacker, 2007; Usher, 2008). Sponges may benefit by their photosynthetic associates and carbon metabolism therefore varies between sponges with and without photosynthetic symbionts (Thacker, 2005; Weisz *et al.*, 2010). In addition, there are sponges harbouring chemolithoautotrophic prokaryotes fixing CO_2 (Vacelet *et al.*, 1995; Van Duyl *et al.*, 2008; Hoffmann *et al.*, 2009; Nishijima *et al.*, 2010; Siegl *et al.*, 2011). So there are various pathways potentially contributing to the total carbon acquisition, but organic carbon is generally the most important food source for sponges. Therefore, the emphasis of this review will be on net heterotrophic feeding of sponges.

2.1.3. Assimilation (respiration and production)

Respiration by sponges was reviewed by Osinga *et al.* (1999). *In situ* oxygen consumptions of sponges, as obtained from bulk measurements range from 0.21 to 24.6 μmol O_2 cm^{-3} sponge h^{-1} (Osinga *et al.*, 1999 and references therein). Oxygen consumption is markedly species specific, but it varies with temperature, pumping activity, and with physiological processes, such as growth and reproductive stage (Coma *et al.*, 1995, 2002). Consistent differences in oxygen consumption are not apparent between temperate and tropical sponges (Osinga *et al.*, 1999; Coma *et al.*, 2002; De Goeij *et al.*, 2008b), while polar and boreal sponges tend to show lower respiration rates on average (Witte and Graf, 1996; Kowalke, 2000; Gatti *et al.*, 2002; Hoffmann *et al.*, 2005). Higher transport- and size-specific respiration rates in HMA than in LMA sponges are expected and, accordingly, Reiswig (1974, 1981) reported higher oxygen demands by bacteriosponges (HMA sponges: *Aplysina fistularis*, *Verongula reiswigi*) than those by LMA sponges (*Tectitethya crypta* and *Mycale laxissima*, previously reported as *Tethya cripta and Mycale* sp. by Reiswig in 1974). The mesohyl in HMA sponges is packed with microbes, coinciding with a denser mesohyl than in LMA sponges (Vacelet and Donadey, 1977; Weisz *et al.*, 2008). This may be a reason for enhanced respiration rates in HMA sponges, because bacteria often have a specific respiration (lower NGE or growth yield, less than 10–25%) higher than that of eukaryotic organisms (Del Giorgio and Cole, 2000). In addition, HMA sponges, characterized by more complex aquiferous systems, have a comparatively larger gas exchange surface (Vacelet and Donadey, 1977; Boury-Esnault *et al.*, 1990). Whether respiration is higher on average in HMA than LMA sponges remains to be elucidated.

Respiration comprises both the respiration for growth/reproduction and the respiration for basal metabolic maintenance. Energy costs of growth and maintenance have been determined for few sponges. The maintenance costs of sponges appear to be high in relation to total respiration, when compared to other benthic invertebrates (e.g. Coma *et al.*, 2002). This is not primarily due to the pumping activity, because studies demonstrate that the energy costs of water pumping do not restrict benthic suspension-feeding animal distribution (Jørgensen *et al.*, 1990; Petersen and Riisgård, 1992; Riisgård *et al.*, 1993). There is apparently always enough potential energy available in natural waters to fuel pumping activity. Pumping, as a part of the maintenance costs, accounts on average for 0.85% in *Halichondria panicea* and 25% in *N. magnifica* of total respiration (Thomassen and Riisgård, 1995; Hadas *et al.*, 2008). Hadas *et al.* (2008) report high maintenance costs, including water pumping activity requirements for *N. magnifica*, accounting for 74% of total respiration. During field experiments with *H. panicea* at 14 °C and ambient POC concentration of 1.7 mg C l^{-1}, maintenance respiration (26 μmol O_2 g DW^{-1} h^{-1}) virtually equals ingestion (24.08 μmol POC g DW^{-1} h^{-1}), leaving no organic carbon for net growth (Thomassen and Riisgård, 1995). Moreover the respiratory costs associated with net growth may be high in sponges. Thomassen and Riisgård (1995) related the specific respiration to the specific growth rate of the sponge and found a linear relation to conclude that the respiratory cost associated with the net synthesis of new tissue in *H. panicea* was 139%. This means that 239 units carbon are required to net produce 100 units of biomass. This is surprisingly high in comparison with other invertebrates (12–40%; e.g. Vahl, 1984; Nielsen *et al.*, 1995; Petersen *et al.*, 1995; Clausen and Riisgård, 1996). Sponges apparently have a higher energy demand for growth than other invertebrates (Thomassen and Riisgård, 1995). These studies indicate that rapid growth in size is probably not a key strategy in sponges. These organisms appear to invest relatively more energy in maintenance than in net body growth compared with other invertebrates.

Surprisingly few studies address sponge biomass production in combination with respiration (but see Reiswig, 1974, 1981; Thomassen and Riisgård, 1995; Koopmans *et al.*, 2010). Growth of sponges is usually measured as increase in volume, weight, or area. Sponges are generally considered as slow growing (e.g. Pansini and Pronzato, 1990). Volume increases of less than 5–60% y^{-1} have been recorded *in situ* for temperate and tropical sponges (Reiswig, 1973; Hoppe, 1988: 7.7–19.4 y^{-1}; Koopmans and Wijffels, 2008: 0.3–3% day^{-1}; McMurray *et al.*, 2008: 52% y^{-1}). Growth of cold-water sponges can be particularly slow, with growth rates ranging from 0.003% to 0.07%y^{-1} (Dayton, 1979; Leys and Lauzon, 1998; Van Duyl *et al.*, 2008). Nevertheless, rapid volume increases have been reported for various farmed sponges, such as *Latrunculia wellingtonensis* and *Polymastia croceus*, which grew respectively 960% and 730% of their initial volume on average in only 6 months (Duckworth and Battershill, 2003). Young stages of sponges show high growth rates, doubling their volume in only a few days (e.g. Maldonado and Young, 1999) and small individuals usually grow faster than larger individuals of the same species (Reiswig, 1973; de Caralt *et al.*, 2008). Rapid volume increase has also been noticed in sponge individuals transplanted out of their original microhabitat or habitat (Wilkinson and Vacelet, 1979; Maldonado and Young, 1998). Seasonal variation in growth rate is known, with lower growth rates generally during the cold season (Barthel and Theede, 1986; Ribes *et al.*, 1999b; Garrabou and Zabala, 2001; Koopmans and Wijffels, 2008). Temperature, food availability, and life-cycle stage are also known to influence growth rates. Reshaping, shrinkage, fission, and fusion are also processes affecting growth rates and that vary seasonally with the life cycle of the sponge and the predation pressure (e.g. Garrabou and Zabala, 2001; Tanaka, 2002; de Caralt *et al.*, 2008). Shrinkage rates may equal or nearly equal growth rates leading to major individual reshaping but no substantial net growth. For instance, in *Corticium candelabrum*, maximum monthly shrinkage rate was measured at 0.15 month^{-1} while maximum growth rate was 0.19 month^{-1} (de Caralt *et al.*, 2008). Sponges regenerating damaged body parts show much faster growth than undamaged sponges, demonstrating a large growth potential (Ayling, 1983; Duckworth, 2003; Walters and Pawlik, 2005; Koopmans *et al.*, 2010).

Little is known about differences in growth between HMA and LMA sponges. Microbes, in general, have lower growth efficiencies than higher organisms, usually less than 20%. Therefore, heterotrophic microbes living in association with the sponge may claim a substantial portion of the energy supply to the sponge holobiont, depending of their growth rate (Van Duyl *et al.*, 2008). The carbon they process may not be directly available for the sponge, but sponges may subsequently use the microbially fixed carbon for sponge cell growth. Evidence for such a trophic transfer in sponges is still limited (Wilkinson and Garrone, 1979; Hentschel *et al.*, 2006). Phototrophic and chemoautotrophic microbes associated with the sponge fix dissolved inorganic carbon (DIC) and the photosynthate may come available for the sponge host. Nutrient translocation of photosynthates, such as glycerol and organic phosphate, from cyanobacterial symbionts to sponge hosts have been demonstrated (Wilkinson, 1979). Contributions of phototrophic cyanobacteria to the carbon assimilation account for up to 80% of the carbon budget of the sponge (Wilkinson, 1983; Wilkinson and Cheshire, 1990; Thacker, 2005; Usher, 2008). Cyanosponges are probably faster growing than sponges without photosynthetic symbionts (Wilkinson and Cheshire, 1990). Thacker (2005) found that the specialized filamentous cyanobacterial symbiont *Oscillatoria spongiarum* benefits its sponge host *Lamellodysidea chlorea* with respect to growth. Cyanobacteria associated with sponges may also be commensals (e.g. *Synechococcus spongiarum* in *Xestospongia exigua*) that do not benefit the host sponge (Thacker, 2005; Lopez-Legentil *et al.*, 2008). Recently also trophic transfer of C from sponge-associated zooxanthellae (*Symbiodinium* spp.) to sponge cells has been confirmed for certain sponges (Weisz *et al.*, 2010).

Inorganic carbon fixation by chemolithoautotrophs tend to be low in dark habitats (Van Duyl *et al.*, 2008; Hoffmann *et al.*, 2009). Van Duyl *et al.* (2008) argue that the carbon incorporation rate of DOC and DIC in the dark by sponge-associated microbes of *Higginsia thielei* and *Rossella nodastrella* of cold-water coral reefs represented approximately 10% to the total net carbon production of the sponge. Values of $\delta^{13}C$ and $\delta^{15}N$ measured by Vacelet *et al.* (1996) in a carnivorous demosponge with vertically transmitted methanotrophic symbionts revealed that the sponge derives a substantial portion of its nutrition from its symbionts.

In addition to measuring growth directly, the energy available for growth can be determined from a bioenergetic budget. Physiological responses of sponges can be converted to energy equivalents and used in the balanced energy equation to calculate the production (Widdows and Johnson, 1988), which is then called the scope for growth (SFG). Energy in the ingested food and the absorption efficiency of the energy from the food need to be assessed. A problem with this approach is that there are, as far as we know, no absorption efficiencies available for sponges. Typical non-bivalve absorption efficiencies for benthic suspension feeders are 58.5% (Newell *et al.*, 1982), but whether this is applicable to sponges is unknown. Trussell *et al.* (2006) derive SFG assuming an absorption efficiency of 100%, based on the low energy expenditure on ammonia release by the sponge *Callyspongia vaginalis*. In this review, we focus, however, on the carbon balance and not on the energy balance.

2.1.4. Excretion and egestion

Excretion in sponges occurs through both the outgoing-water flow and the external surface of the body. Special amoebocytes disintegrate in the mesohyl and their granules and vesicles are expelled through the exhalant canals (Simpson, 1984). Likewise, amoeboid cells charged with large membrane-bound inclusions have been seen to squeeze between the epithelial cells until reaching either the lumen of the aquiferous canals or the external surface, where they discharge their contents or even discard themselves out of the sponge body. This putative cell excretion system has been reported from adult sponges and larval stages (Vacelet, 1967; Maldonado, 2009).

Several authors have also reported elimination (i.e. egestion) of pellets of undigested materials and detrital particles (Reiswig, 1971b; Weissenfels, 1976; Stuart and Klumpp, 1984; Witte *et al.*, 1997; Yahel *et al.*, 2003; De Goeij *et al.*, 2009), as well as net DOM efflux (Ribes *et al.*, 1999b). Wolfrath and Barthel (1989) made a detailed study of the production of faecal pellets by *H. panicea*. Feeding experiments with indigestible beads and algal cultures demonstrated production of 15–55 μm in diameter, oval to rounded faecal pellets consisting of densely packed aggregates surrounded by a thin, fragile, membrane-like envelope. Aggregates harboured remnants of up to 560 digested algal cells. In addition to remnants of digested food, faecal pellets can contain residue of digested food as well as indigestible materials. Indigestible material may, however, also be egested as single particles without envelope, which pass more rapidly through the sponge. To date the only estimates of "excretion" rates are from Reiswig (1971b). He reported negative clearance rates of detritus for three tropical sponges ranging from 0.042 to 0.382 mg POC m^{-3}, comprising approximately 1–5% of the ingested particles, though the real level of detrital particle production remained unclear. Sharp and sudden declines in the retention efficiency of 1-μm sized particles in Antarctic sponges and of 2- to 4.5-μm sized particles in temperate and tropical sponges has been interpreted as indirect evidence of detrital particle elimination, probably break-down products of digestion in that size range (Stuart and Klumpp, 1984; Kowalke, 2000). Nevertheless, rates were not derived. The closest to estimating a particulate excretion rate for sponges may be the work by De Goeij *et al.* (2009). They report rapid proliferation of choanocytes in *H. caerulea.* Choanocyte turnover had a periodicity of about 6 h in steady state conditions, with "old" cells being replaced by new cells and aggregates of detrital material (probably remains of shred cells) being expelled to the exhalant canals. Considering the limited growth of *H. caerulea*, the average cell proliferation rate of 13.8 ± 0.9 μmol C cm^{-3} sponge h^{-1} nearly equals cell shedding. Egestion of digested and undigested detrital particulate material from the mesohyl is probably low in this sponge considering the fact that it mainly feeds on DOM (De Goeij *et al.*, 2008b). The rate of choanocyte proliferation may therefore be a reasonable estimate for egestion in the carbon balance of *H. caerulea.*

2.2. Carbon balance in HMA and LMA sponges (and POC vs. DOC)

Imbalances in carbon budgets between ingestion and respiration have been reported for several HMA and LMA sponge species (Reiswig, 1974, 1981; Thomassen and Riisgård, 1995; Hadas *et al.*, 2008, 2009). Discrepancies have been ascribed to potential underestimation of carbon ingestion. For sponges in Table 3.2, the I_{poc}/R index ranges from 0.003 to 3.97. For sponges with an I_{poc}/R smaller than 1, insufficient particulate carbon is most likely assimilated to meet the CO_2 release. In those cases, respiration also exceeds carbon ingestion. DOC is usually invoked as the missing carbon source. Considering recent studies of tropical sponges, in which DOC consumption was evident and substantial in all four sponge species tested (Yahel *et al.*, 2003; De Goeij *et al.*, 2008b), it is likely that other sponge species rely on DOC consumption as well. Microbes are pre-eminently DOM consumers and may mediate DOM consumption in sponges. The general assumption therefore is that HMA sponges are better equipped to feed on DOM than LMA sponges. Of the examined DOC-feeding sponges, *T. swinhoei* is an HMA sponge (Yahel *et al.*, 2003). The DOM-feeding *H. caerulea*—hosting 2.1×10^9 microbes cm^{-3} sponge, according to DAPI counts (De Goeij *et al.*, 2008a)—could be regarded as an HMA sponge, *sensu* Hentschel *et al.* (2003). However, electron microscopy examination by Vacelet and Donadey (1987) found only few bacteria in the mesohyl of this sponge. Absence of mid-chain-branched fatty acids, as well as low % of bacterium-specific fatty acids (<10%), in *H. caerulea* compared to *Chondrilla caribensis* (Van Duyl *et al.*, 2011) would also classify *H. caerulea* as an LMA sponge (Hochmuth *et al.*, 2010). The DOM-feeding sponges *M. microsigmatosa* and *M. normani* have amounts of associated microorganisms comparable to that in *H. caerulea* (De Goeij *et al.*, 2008a), but it remains unclear whether these sponges have to be categorized as LMA or as HMA. Anyway, they contain approximately 10^9 prokaryotes cm^{-3} sponge, which is 2–3 orders of magnitude higher than in ambient seawater (De Goeij *et al.*, 2008a). For several LMA sponges with prokaryote concentrations up to 1 order of magnitude higher than in seawater (i.e. *Dysidea avara, Haliclona oculata),* net DOC uptake could not be established (Ribes *et al.*, 1999b; Koopmans and Wijffels, 2008). Whether the LMA sponges *H. panicea* and *N. magnifica* feed on DOM besides POM remains to be examined. Imbalances have been reported for these sponges, as well as for HMA sponges (Reiswig, 1974; Thomassen and Riisgård, 1995; Hadas *et al.*, 2008, 2009). The emerging view is that the extent of the mismatch between ingestion of POC and respiration tends to be larger in HMA than in LMA sponges (Table 3.2).

To investigate whether DOM is primarily assimilated by sponge-associated microbes or by sponge cells, De Goeij *et al.* (2008b) conducted a labelling study with *H. caerulea*. Labelling of the sponge with diatom derived ^{13}C-DOC demonstrated instantaneous incorporation of ^{13}C in bacterium-specific as well as sponge-related fatty acids of *H. caerulea*, implying uptake of ambient DOC by both sponge cells and sponge-associated microbes (De Goeij *et al.*, 2008b). Although there are doubts about the HMA or LMA status of *H. caerulea*, it is evident that sponge cells take up DOM directly from the ambient water, as do the sponge-associated prokaryotes. Since most microbes in typical HMA sponges dwell in the mesohyl, there is no direct contact between the ambient DOC and the microbes. DOC needs to pass the pinacocyte and/or choanocyte epithelia before being accessible for the microbes. Therefore, it is unlikely that DOM uptake is restricted to sponge-associated microbes and to HMA sponges.

Between-species comparison of sponge metabolic rates in Table 3.2 shows that several sponges show evident imbalances between POC ingestion (I_{poc}) and respiration (R). The I_{poc}/R index is smaller than 1 for 6 of the 12 sponges. This may indicate carbon shortage for maintenance and growth assuming a respiratory quotient of 1 (mol CO_2. mol O_2^{-1}). All species in

Table 3.2 with I_{poc}/R smaller than 1 are clearly HMA sponges, except *H. caerulea*. Sponges with I_{poc}/R just above 1, such as the LMA sponges *N. magnifica* and *H. panicea,* may need additional carbon to account for detrital waste disposal and for net production of biomass in case they grow (Thomassen and Riisgård, 1995; Hadas *et al.*, 2008, 2009). Sponges cannot digest all the ingested carbon (Witte *et al.*, 1997). Exclusion of DOC uptake by *T. swinhoei* and *H. caerulea* from the equation results respectively in I_{poc}/R values of 0.18 and 0.52, showing imbalances comparable with those for *V. reiswigi* and *A. fistularis* (Reiswig, 1974, 1981). Adding the consumed DOC by these sponges, the index exceeds 1 and comes into balanced conditions. The clearance efficiency of total organic matter ($I_{poc+doc}/R$) for *T. swinhoei,* however, remains still quite low compared to *H. caerulea,* being 1.29 versus 6.85. Carbon uptake may still be underestimated because *T. swinhoei* also acquires carbon via its photosynthetic symbionts (Magnino *et al.*, 1999). Likewise, *A. fistularis* and *V. reiswigi.* (Reiswig, 1981) may have also acquired carbon from their associated photosymbionts. Lowering the respiratory quotient (RQ) to 0.7 would alleviate the imbalances found for *N. magnifica* and *H. panicea.* The RQ of a specific sponge is determined by the elemental composition of its food and the physiological state, which may differ between sponge species and growth phases. Sponges mainly feeding on DOC are expected to have higher RQs than sponges feeding on algae or actively growing sponges (e.g. Roy *et al.*, 1999). Lower values have been used for calculations on sponges (e.g. Koopmans *et al.*, 2010), but since we do not know the composition of the diet and physiological state of the sponges, we are maintaining an RQ equal to 1. Summarizing, we can conclude that HMA as well as LMA sponges probably take up DOC. HMA sponges, however, may rely more on DOC as a source of carbon than LMA sponges.

2.3. Between-species carbon balance differences

Studies in which total ingestion (DOC and POC), growth, and respiration have been measured synchronously *in situ* or under comparable conditions in sponges are scarce. The carbon metabolism of seven different sponge species for which at least three variables of the equation $I = P + R + E$ were comparatively quantified is summarized in Table 3.3. None of the studies provide sufficient data to check whether fluxes measured together really meet the balance over a certain time period. Always an unknown flux remained, which was subsequently estimated on basis of the measured fluxes. For the tropical sponges studied by Reiswig (1981), the shortage of carbon ranged from 0.34 μmol C h^{-1} cm^{-3} sponge in *M. laxissima* to 2.9 in *A. fistularis*. In the balance, no account is taken of exuded material that is no longer available for re-filtration. Therefore, the total amount actually required may even be much higher for all sponges, except *H. caerulea*. For

Table 3.3 Carbon balance equations for five tropical and two temperate sponges

Sponge	Host type	$I-E_{partial}$	P	R	$I_{shortage}$	E_{rest}	NGE	%	$R/(I-E_{partial})$	DOC%	Source
Tropical species (μmol C h^{-1} cm^{-3})											
Tectitethya crypta	LMA	0.93	0.03	0.46		0.44	0.06	6	0.49	0.0	Reiswig (1971b, 1973)
Mycale laxissima	LMA	1.45	0.11	1.67	0.34		0.06	6	1.16	18.8	Reiswig (1971b, 1973)
Verongula reiswigi	HMA	0.63	0.02	2.48	1.87		0.01	1	3.94	74.7	Reiswig (1971b, 1973)
Aplysina fistularis	HMA	0.48	0.16	3.22	2.90		0.05	5	6.77	85.9	Reiswig (1981)
Halisarca caerulea	L/HMA	18.50	2.00	2.70		13.8[a]	0.43	43	0.15	92.4	De Goeij *et al.* (2008a, 2009)
Temperate species (μmol C h^{-1} g DW^{-1})											
Haliclona oculata	LMA	231.50	7.60	95.54		128.36	0.07	7	0.41	0.0	Koopmans *et al.* (2010)
Halichondria panicea	LMA	51.64	13.81	37.93	0.10		0.27	27	0.73	0.2	Thomassen and Riisgård (1995)

$I-E_{partial}$, net ingestion of organic carbon; P, net growth in biomass; R, respiration (RQ = 1); E_{rest}, detrital material not available for recirculation; NGE, net growth efficiency (= $P/(P+R)$); shortage, imbalances between Ingestion and assimilation; $R/(I-E_{partial})$, index estimating the amount of the organic matter ingested which was lost in respiration; DOC%, DOC contribution values to total. Grey cells contain calculated values.

[a] Rate of cell shedding.

this sponge, choanocyte shedding (De Goeij *et al.*, 2009) is assumed as a reliable estimate of particle excretion.

Comparing net growth rates measured over long time scales with oxygen consumption measurements yields NGEs of 0.6–6% (derived from data in Reiswig, 1973; 1974; Table 3.3). The growth efficiency of *H. oculata* based on volume increase in time is 7%, assuming an RQ of 1 or 10% for an RQ of 0.75 (Koopmans and Wijffels, 2008; Koopmans *et al.*, 2010). Higher efficiencies of 20–30% have been found at growth rates of 1–4% day^{-1} in laboratory and field experiments (Thomassen and Riisgård, 1995). The NGE of *H. caerulea* is high (43%) when compared to other sponges. It is tempting to assume that the high NGE in *H. caerulea* is due to the efficient and rapid growth rate of choanocytes. De Goeij *et al.* (2009) reported high choanocyte proliferation rates coupled to shedding of old cells in *H. caerulea.* Rapid cell renewal may be a maintenance strategy similar to that in higher animals. Therefore, cell proliferation rate is listed under excretion in Table 3.3, implementing the still undemonstrated assumption that proliferation rate should equal cell shedding rate. Whenever the cell proliferation exceeds shedding, sponge net growth may take place. Whenever sponges invest energy in cell renewal, their net growth may be low. This process probably takes the bulk of the energy delivered to *H. caerulea,* not leaving much energy for net growth in size (De Goeij *et al.*, 2008a). Average production of *H. caerulea* calculated at 2 µmol C cm^{-3} sponge h^{-1} on the C balance equation is probably an overestimate. Further research is needed to establish the role of choanocyte renewal in sponge metabolism and growth process.

2.4. Ecological significance of carbon use

Sponges ingest 29–1970 mg C m^{-2} sponge day^{-1} (Gili and Coma, 1998 and references therein) and 0.04–18.5 µmol C cm^{-3} sponge h^{-1} (De Goeij *et al.*, 2008a and references therein). Wherever sponges are an important component of the benthic community, they withdraw substantial amounts of organic matter from the ambient water, playing a potentially relevant role in the benthic–pelagic coupling of C cycling (Reiswig, 1971b; Lesser, 2006). The pelagic microbial food web represents the main food source for sponges throughout their entire bathymetric and latitudinal range of occurrence (Yahel *et al.*, 2005; Pile and Young, 2006; Bell, 2008). Picoplankton is frequently depleted in water layer overlying coral reefs (Ayukai, 1995; Gast *et al.*, 1998; Yahel *et al.*, 1998; Richter *et al.*, 2001; Van Duyl *et al.*, 2002; Genin *et al.*, 2009; Monismith *et al.*, 2010), an effect largely due to sponges (Richter and Wunsch, 1999; Scheffers *et al.*, 2004). Also DOM depletion in coral reef waters and in coral reef cavity waters has been reported (De Goeij and Van Duyl, 2007; Nelson *et al.*, 2011). The diet of various reef sponges consists for more than 90% of total net carbon uptake of

DOM (Yahel *et al.*, 2003; De Goeij *et al.*, 2008a), accounting for DOC depletion in cavities (De Goeij and Van Duyl, 2007). At least part of DOM removed from the ambient water by DOM-feeding sponges is converted to detrital POM (De Goeij *et al.*, 2008a; 2009). Moreover, more than 50% of the carbon ingested by reef sponges has a clear reef signature with respect to $\delta^{13}C$, $\delta^{15}N$, and fatty acids, suggesting that these sponges obtain the bulk of their food from benthic production (Van Duyl *et al.*, 2011).

Benthic primary producers, such as macroalgae, seagrasses, and corals, produce substantial amounts of DOC and mucus during the daylight period (e.g. Haas *et al.*, 2010; Naumann *et al.*, 2010). This organic matter presents an important source of energy for sponges in the benthic boundary layer and the benthos of shallow-water habitats (e.g. Wild *et al.*, 2004; Behringer and Butler, 2006; Granek *et al.*, 2009; Van Duyl *et al.*, 2011). In Fig. 3.2, a schematic impression of the main net fluxes of carbon in shallow-water habitats is given. Sponges capture plankton, POC_{det}, and DOC, and fuel the detrital pathways by enriching their surrounding benthic environment with waste organic matter. As such, sponges attract commensals and support the

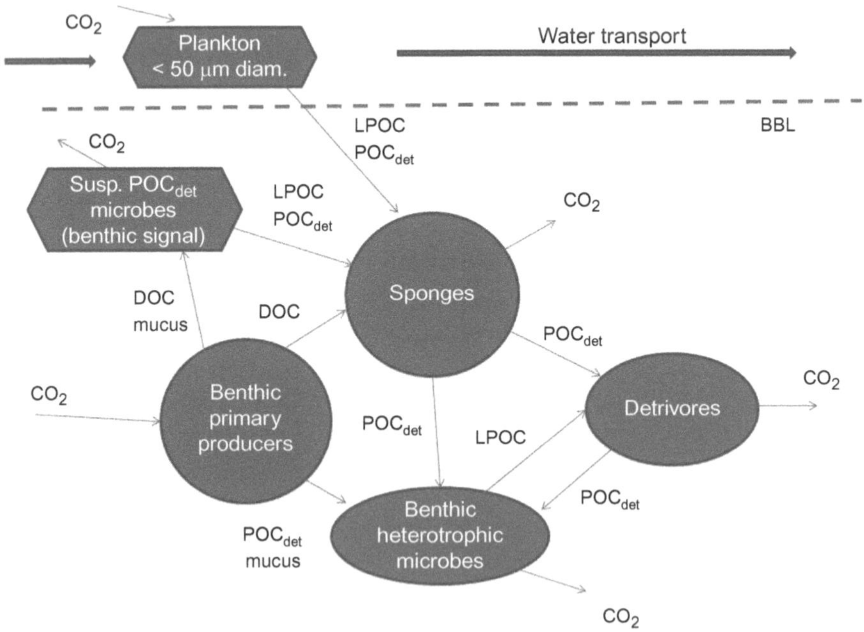

Figure 3.2 Schematic outline of major net fluxes of organic carbon through sponges. Arrows indicate flux of organic carbon to and between state variables representing standing stocks of organic matter and CO_2 pools. Hexagons refer to pelagic state variables, while circles and ellipses refer to benthic state variables. BBL, benthic boundary layer. (For colour version of this figure, the reader is referred to the online version of this chapter.)

increase in biomass of detritivores and benthic microbes (e.g. Queric *et al.*, 2008). Bacterioplankton in the benthic boundary layer may obtain a "benthic signal" indicating that the microbes receive nutrition via benthic release (Nelson *et al.*, 2011; Van Duyl *et al.*, 2011). In aphotic deep waters, sponges rely mainly on bacterioplankton and POC_{det} (e.g. Yahel *et al.*, 2007). It is evident that sponges play a key role in many ecosystems by trapping organic matter from both the pelagic and the benthic compartment, transferring organic matter to the benthic compartment. As a result, sponges enrich benthic communities with organic matter in the form of sponge biomass and by their detrital POM waste disposal.

3. Use of Nitrogen and Phosphorous by Sponges

The nitrogen cycle is highly and thoroughly associated with microbes. Essential and unique steps in the nitrogen cycle are performed by a variety of bacteria, archaea, and eukaryotes (Ward *et al.*, 2007). The nitrogen cycle controls the availability of nitrogenous nutrients, hence markedly influencing the biological productivity in marine systems. Main steps of the nitrogen cycling are: (1) nitrogen fixation, (2) nitrification, (3) denitrification, (4) anaerobic ammonium oxidation (anammox), and (5) remineralizarion. The three last processes are carried out under suboxic conditions. While there are some processes that *per se* do not involve input or loss of nitrogen, as for example nitrification and remineralization, others imply either direct acquisition (nitrogen fixation) or release (denitrification, anammox) of nitrogen. Detection and quantification of all these processes are important for understanding the role of sponges as nitrogen sources or sinks in ecosystems.

In relation to the N cycle, the emerging notion in recent years is that sponges have to be understood as a "consortium" integrating the metabolic reactions of the sponge cells and those of the microbial associates. Rapid pumping rates of sponges indicate that large volumes of water charged with dissolved and particulate nutrients are transported through the body on a nearly constant basis and, somehow, interact with the dense microbial communities in the sponges, offering a high potential for rapid rates of biogeochemical processes. The application of oxygen-sensitive microelectrodes in the tissues of different sponge species showed remarkable oxygen deficiencies in their mesohyl during non-pumping periods (Schläppy *et al.*, 2007; Hoffmann *et al.*, 2009). Consequently both anaerobic and aerobic microbial processes can be expected in sponge tissues. Different steps of the nitrogen cycle have been identified in the same sponge species, suggesting that nitrogen cycling in sponges goes through a complex network of metabolic pathways (Hoffmann *et al.*, 2009). Most studies to date are based on the detection and description of microbial components that could mediate different nitrogen cycling steps (Taylor *et al.*, 2007 and references therein). Few of them included data on resulting fluxes and, overall, no consistent pattern has clearly emerged on the role of marine sponges as sink or source of a particular nitrogen compound. Herein, we review the nitrogen cycle processes noted in marine sponges.

3.1. Nitrogen fluxes in sponges

3.1.1. Remineralization of POM

This process involves the conversion of the particulate organic nitrogen filtered by the sponge to dissolved organic nitrogen and ammonia (also to DOC and phosphate; Fig. 3.3). Nevertheless, if associated microorganisms

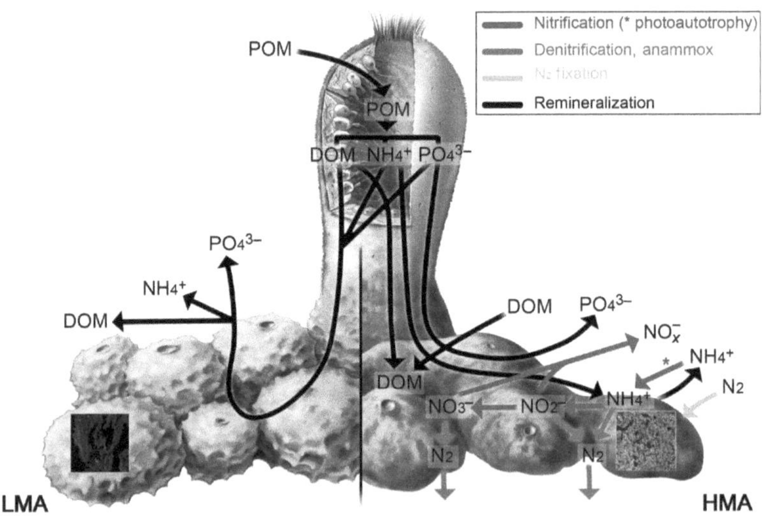

Figure 3.3 Common and specific nutrient fluxes and pathways in marine sponges related to their microbial components abundance. LMA, low-microbial abundance species; HMA, high-microbial abundance species. Remineralization includes the removal of particulate organic matter (POM) by all sponge species. As result of POM metabolism, dissolved organic matter (DOM), ammonia (NH_4^+), and phosphate (PO_4^{3-}) will be generate inside the sponge. Those compounds will be released directly to the water column by LMA species but would be processed through nitrification, photautotrophy, denitrification, and/or anammox in HMA species. External uptake of NH_4^+ and release of NO_x^- will appear in HMA species as well as uptake of dissolved organic compounds. Pictures are confocal microscopy FISH images with universal bacteria probes (bright green area) showing the relative abundance of symbiotic bacteria in each sponge type. (For interpretation of the references to colour in this figure legend, the reader is referred to the online version of this chapter.)

are abundant (HMA sponges), dissolved components can also be uptaken directly from the seawater and transformed by them (Scheffers *et al.*, 2004). Remineralization of POM is a common nutrient step in many sponges, irrespective of the abundance and composition of their associated microbial communities. In this sense, availability of POM was a major determinant of growth rates in three Caribbean species with different microbial abundance (i.e. *C. vaginalis* is LMA, *Agelas conifera* and *A. fistularis* are HMA), pointing out the importance of particulate nutrients in sponge metabolism (Lesser, 2006). Ammonia excretion has also been detected in many sponge species, irrespective of their microbial abundance (Table 3.3; Fig. 3.5), showing that marine sponges can act as a net ammonia source through remineralization of particulate material. The relative importance of different ingested particles as nitrogen source may be related to the reported sponge ability for food selection. Nevertheless, due to differences in food particle size and C:N content, a positive selection over a particular food type does not always imply a net nitrogen source. For instance, whereas cyanobacteria and picoeukaryotes were the major food source in terms of particle abundance for the Mediterranean sponge *S. officinalis*, isotopic analyses revealed that most of the carbon (and probably also nitrogen) retained in terms of biomass came from nanoeukaryotic cells (Topçu *et al.*, 2010). Due to its size, *Synechoccocus* is often the dominant prokaryotic contributor to sponge carbon fluxes (Ribes *et al.*, 2005; Pile and Young, 2006), even though it is often much less abundant than heterotrophic bacteria (Ribes *et al.*, 1999a). The heterotrophic cells, however, are better sources of nitrogen than autotrophic picoplankton due to their lower C:N ratio (Wheeler and Kirchman, 1986; Geider and La Roche, 2002; Bertilsson *et al.*, 2003). Detritus may also contribute to the acquisition of organic nitrogen by sponges, with values ranging from trace levels hard to detect (Yahel *et al.*, 2007) to about 20% (Hadas *et al.*, 2009).

3.1.2. Nitrification

This consists of the oxidation of ammonia to nitrite and nitrate for energy purposes and is mediated by two phylogenetically distinct groups: the ammonia oxidizing bacteria or archaea and the nitrite oxidizing bacteria. Sponge-mediated nitrification (Table 3.4) has been detected in many Caribbean sponges (Corredor *et al.*, 1988; Díaz and Ward, 1997; Southwell *et al.*, 2008a), in several sublittoral Mediterranean sponges (Jiménez and Ribes, 2007; Bayer *et al.*, 2008; Schläppy *et al.*, 2010; Ribes *et al.*, 2012), in *Geodia barretti* from the North Atlantic coast (Hoffmann *et al.*, 2009), and in two species inhabiting a deep-water coral mound (Van Duyl *et al.*, 2008). Also, vertical transmission of ammonium oxidizing archaea has been described in *Luffariella variabilis* and *Rophaloeides odorabile*, two sponge species from the Great Barrier Reef (Steger *et al.*, 2008). Ammonium and nitrite are suspected to accumulate within the sponge body during periods

Table 3.4 Dissolved nitrogen and phosphorous fluxes in marine sponges from different locations

Sponge	Location	Host type	Phototrophic components	NH_4^+	NO_x^-	PO_4^{3-}	DON	Source
μmol L^{-1}								
Agelas conifera	Florida keys	HMA	Low Chl a	**0.15**	−0.57	–	–	Southwell *et al.* (2008a), Weisz (2006)
Aplysina archeri	Florida keys	HMA	High Chl a	−0.3	−0.72	–	–	Southwell *et al.* (2008a), Weisz (2006)
Aplysina lacunosa	Florida keys	HMA	High Chl a	−0.09	−1.14	–	–	Southwell *et al.* (2008a), Weisz (2006)
Ircinia strobilina	Florida keys	HMA	Low Chl a	−0.07	−0.67	–	–	Southwell *et al.* (2008a), Weisz (2006)
Niphates digitalis	Florida keys	LMA	Low Chl a	−0.21	–	–	–	Southwell *et al.* (2008a), Weisz (2006)
Xestospongia muta	Florida keys	HMA	High Chl a	**0.05**	−0.86	–	–	Southwell *et al.* (2008a), Weisz (2006)
Aphrocallistes vastus	North Pacific	LMA	NP	−0.16	0	−0.04	0	Yahel *et al.* (2007)
Rhabdocalyptos dawsoni	North Pacific	LMA	NP	−0.4	−0.13	−0.04	0	Yahel *et al.* (2007)
Dysidea avara	Mediterranean	LMA	DGGE	**NS**	NS	−0.02	−0.48	Ribes *et al.* (2012)
Chondrosia reniformis	Mediterranean	HMA	DGGE	**1.69**	−0.99	−0.02	**NS**	Ribes *et al.* (2012)
Agelas oroides	Mediterranean	HMA	DGGE	**0.57**	−0.99	−0.04	**0.01**	Ribes *et al.* (2012)
μmol g DW^{-1} h^{-1}								
Aplysina cauliformis	Florida keys	HMA	High Chl a	**0.5**	−1.7	–	–	Southwell *et al.* (2008b)
Smenospongia aurea	Florida keys	HMA	Low Chl a	0	−1.8	–	–	Southwell *et al.* (2008b)
Niphates erecta	Florida keys	LMA	Low Chl a	−1.5	0	–	–	Southwell *et al.* (2008b)
Dysidea avara	Mediterranean	LMA	DGGE	−0.65	0	−0.05	–	Jiménez and Ribes (2007)
Agelas oroides	Mediterranean	HMA	DGGE	0	−0.36	−0.06	–	Jiménez and Ribes (2007)

Chondrosia reniformis	Mediterranean	HMA	DGGE	**0.64**	−0.7	0	–	Jiménez and Ribes (2007)
Ircinia oros	Mediterranean	HMA	–	0	−0.23	0	–	Jiménez and Ribes (2007)
Aplysina aerophoba	Mediterranean	HMA	Cyanos	**0.3**	−0.77	−0.03	–	Jiménez and Ribes (2007)
Aplysina aerophoba	Mediterranean	HMA	Cyanos	**0,012** *1	−0.33	–	–	Bayer *et al.* (2008)
				−0,73 *2	−0.23	–	–	Bayer *et al.* (2008)
Chondrilla nucula	Caribbean	HMA	Cyanos	**0.35**	−0.6	–	–	Corredor *et al.* (1988)
Anthosigmella varians	Caribbean	HMA	Zoox	**0.27**	−0.02	–	–	Corredor *et al.* (1988)
Chondrilla nucula	Caribbean	HMA	Cyanos	−0.019	−0.37	–	–	Díaz and Ward (1997)
Pseudaxinella zeai	Caribbean	HMA	Cyanos	−0.04	0	–	–	Díaz and Ward (1997)
Oligoceras violacea	Caribbean	HMA	Cyanos	0.18	−0.57	–	–	Díaz and Ward (1997)
Plakortis halichondroides	Caribbean	HMA	–	−0.1	−0.02	–	–	Díaz and Ward (1997)
Aplysina sp.	W. Australia	–	–	−0.69	0	−0.005	**0.21**	Hatcher (1994)
Iotrochota baculifera	W. Australia	–	–	−0.52	0	−0.004	**0.39**	Hatcher (1994)
Haliclona cymiformis	GBR	–	Rhodophyte	**2.4**	**0.1**			Davy *et al.*, (2002)
μmol cm^{-3} h^{-1}								
Geodia barreti	North Atlantic	HMA	NP	**0.067**	−0.024	–	–	Hoffmann *et al.* (2009)
Dysidea avara	Mediterranean	LMA	DGGE	−0.109	0	−0.009		Jiménez and Ribes (2007)
Agelas oroides	Mediterranean	HMA	DGGE	0	−0.127	−0.02		Jiménez and Ribes (2007)
Chondrosia reniformis	Mediterranean	HMA	DGGE	**0.13**	−0.157	0		Jiménez and Ribes (2007)
Ircinia oros	Mediterranean	HMA	–	0	−0.046	0		Jiménez and Ribes (2007)
Aplysina aerophoba	Mediterranean	HMA	Cyanos	**0.104**	−0.24	−0.008		Jiménez and Ribes (2007)
Higginsia thielei	North Atlantic	HMA	NP	−0.042	−0.001			van Duyl *et al.* (2008)
Rossella nodastrella	North Atlantic	HMA	NP	−0.002	−2.5 10^{-5}			van Duyl *et al.* (2008)
μmol h^{-1} L^{-1}								
Agelas conifera	Florida keys	HMA	Low Chl a	**49**	−170	–	–	Southwell (2007)

(continued)

Table 3.4 *(continued)*

Sponge	Location	Host type	Phototrophic components	NH_4^+	NO_x^-	PO_4^{3-}	DON	Source
Aplysina archeri	Florida keys	HMA	High Chl a	−40	−100	–	–	Southwell (2007)
Aplysina lacunosa	Florida keys	HMA	High Chl a	−19	−120	–	–	Southwell (2007)
Ircinia strobilina	Florida keys	HMA	Low Chl a	−18	−370	–	–	Southwell (2007)
Niphates digitalis	Florida keys	LMA	Low Chl a	−260	NP	–	–	Southwell (2007)
Xestospongia muta	Florida keys	HMA	High Chl a	**2**	−170	–	–	Southwell (2007)
Aplysina cauliformis	Florida keys	HMA	High Chl a	−44	−160	–	–	Southwell (2007)
Aplysina fistularis	Florida keys	HMA	High Chl a	−68	−24	–	–	Southwell (2007)
Amphimedon compressa	Florida keys	LMA	Low Chl a	−150	0	–	–	Southwell (2007)
Callyspongia vaginalis	Florida keys	LMA	Low Chl a	−290	−2	–	–	Southwell (2007)
Ircinia campana	Florida keys	HMA	High Chl a	−220	−86	–	–	Southwell (2007)
Ircinia felix	Florida keys	HMA	High Chl a	−41	−270	–	–	Southwell (2007)
Niphates erecta	Florida keys	LMA	Low Chl a	−140	−5	–	–	Southwell (2007)
Smenospongia aurea	Florida keys	HMA	Low Chl a	−5	−230	–	–	Southwell (2007)

HMA, high-microbial abundance species; LMA, low-microbial abundance species; DON, dissolved organic nitrogen; Low and high Chl a, refers to Chl a measurements as indication of phototrophic components; DGGE, denaturing gradient gel electrophoresis used to detect phototrophic components; NP, not present; –, not measured; NS, not significant. Positive values (bolded) mean removal, negative values mean excretion; negative values are effluxes; *1, April measurements; *2, May-September measurements. Grey colour highlights LMA species.

of pumping arrest. Conversion to nitrate would be a solution for eliminating these compounds, which can be toxic to the sponge if accumulated above a certain concentration threshold (Steger *et al.*, 2008).

The most evident nitrogen flux in sponges mediating nitrification is a net production of NO_x^- (i.e. either nitrite and/or nitrate). Production of NO_x^- is a common feature in HMA species, irrespective of habitat and biogeographic area, which is in contrast with the low rates or even lack of nitrification in LMA species (Table 3.4, Fig. 3.3). In this sense, Southwell *et al.* (2008a) showed high rates of nitrification in HMA sponges of coral reefs systems, in agreement with previous work on a limited number of Caribbean species (Corredor *et al.*, 1988; Díaz and Ward, 1997). Beside NO_x^- production, nitrification can also result in NH_4^+ uptake (Fig. 3.4), being this combined nitrogen flux (sink of NH_4^+ and source of NO_x^-) specially evident in HMA species (Table 3.4). A seasonal variability can exist in those fluxes, as reported in the Mediterranean *A. aerophoba* (Bayer *et al.*, 2008). In this species, it was also demonstrated that ammonia addition stimulated nitrification, the sponge becoming an "ammonium source" during summer months. Nitrate release (excretion) was also significantly stimulated by ammonium in the spring months, while in August, addition of ammonium did not result in enhanced nitrate release (Bayer *et al.*, 2008). NH_4^+ uptake by sponges can also be related to photoautotrophy, as suggested by the presence of phototrophic microorganisms in the sponge exhibiting that ability (Table 3.4). This is mostly evident in HMA sponges.

3.1.3. Nitrogen fixation

This involves the conversion of ambient dinitrogen (N_2) to ammonia by sponge-associated cyanobacteria (Wilkinson and Fay, 1979) and/or heterotrophic microorganisms (Shieh and Lin, 1994). Sponge-mediated nitrogen fixation was first reported in several species in the Red Sea (Wilkinson and Fay, 1979) and years later in a tropical species of *Halichondria* sp. (Shieh and Lin, 1994). Detection of nitrogen fixing bacteria expressing *nifH* genes has been described in several sponges from Key Largo, Florida (Mohamed *et al.*, 2008). Nitrogen fixation by sponge-associated microorganisms may contribute to complete sponge nitrogen needs in environments with low nitrogen availability (Taylor *et al.*, 2007). Nitrogen fixation results in a net removal of nitrogen from the water column and it is mostly detected in HMA species (Fig. 3.3).

In situ studies showed that nitrogen fixation rates in several sponges of Florida Bay were very low relative to the ambient water, also compared to DIN (NH_4^+ and NO_x^-) production rates (Southwell, 2007). The studied species (*Aplysina cauliformis, A. conifera,* and *Ircinia felix*) are all HMA. The former one showed a maximum N_2 fixation about 2.5 μg g^{-1} day^{-1}. Yet this figure was much lower than DIN release rates (430 μg g^{-1} day^{-1}). Southwell (2007) concluded that sponge-mediated nitrogen fixation in Caribbean reefs probably contributes significantly neither to the sponge nutrition nor to nitrogen inputs to the reef. Nevertheless, the result may be different in other geographical regions with stronger nitrogen limitation.

3.1.4. Denitrification and anammox

Denitrification is the conversion of nitrate to N_2 by associated bacteria or archaea. Denitrification in sponges was recently discovered in *G. barretti* (Hoffmann *et al.*, 2009), then in the Mediterranean species *D. avara* and *Chondrosia reniformis* (Schläppy *et al.*, 2010). Recovery of two groups of *NirS* gene (encoding for nitrite reductase) suggest that denitrification activity is generally performed by more than a group of bacteria (Hoffmann *et al.*, 2009). Anammox (anaerobic denitrification) reaction is the conversion of ammonia and nitrite to N_2. Microorganisms that carry out this reaction are bacteria from the phylum *Planctomycetales*. First evidence of anammox activity in sponges was described in *G. barretti* and associated anammox bacteria in its tissue (Hoffmann *et al.*, 2009). Anammox bacteria have also been detected in *Ircinia strobilina* and *M. laxissima* (Mohamed *et al.*, 2010). Denitrification occurs under aerobic conditions but anammox requires hypoxic conditions. Combination of both aerobic and anaerobic mesohyl portions are common in HMA species and Hoffmann *et al.* (2009) showed evidence of anammox and denitrification in the same animal host (i.e. *G. barretti* from North Atlantic). The resulting nitrogen flux is then a net N_2 production. This production could be fuelled by providing the microorganisms with sponge metabolic waste products (a way to clean the sponge tissue from noxious compounds), and/or with inorganic nitrogen (NH_4^+) taken up from the environment, adding to the effect of nitrification and photoautotrophy in HMA species (Fig. 3.3). Reported nitrogen release due to denitrification and anammox in *G. barretti* was 1 or 2 orders of magnitude lower than that by nitrification: nitrification = 0.024, denitrification = 0.0038, and anammox = 0.00013 μmol cm^{-3} h^{-1} (Hoffmann *et al.*, 2009).

Few studies have included the measurement of dissolved organic nitrogen in marine sponge nitrogen fluxes. Data available for seven sponges including LMA and HMA species from the Mediterranean, North Pacific and Western Australia suggest that LMA species are characterized by either DON excretion or a neutral balance between inhaled and exhaled water, while a net DON uptake appears to characterize HMA species (Table 3.4). Interestingly, average DON incorporation by *H. caerulea* is 0.001 mmol cm^{-3} h^{-1}. This sponge harbours nitrifiers, but only shows average net nitrification after ammonia addition (Fleur C. van Duyl, in preparation).

3.2. Phosphorous fluxes in sponges

Phosphorous is an essential nutrient required by all organisms for biological synthesis and energy transfer processes (Tyrreell, 1999; Benitez-Nelson, 2000). In aquatic environment, budgets of total phosphorous results from the addition of three major phosphorous compounds: (1) dissolved inorganic phosphate, which is usually measured as phosphate (PO_4^{3-}); (2) dissolved organic phosphorous (DOP); and (3) particulate phosphorous. DOP is the dominant form (Dyhrman *et al.*, 2007). Since Taylor *et al.* (2007) devoted just a couple of lines to phosphorus cycling in marine sponges in their extensive review on sponge-associated microorganisms, not much advance has been attained in the understanding of sponge-mediated P fluxes. The few available measures come from some Mediterranean and North Pacific species, in which fluxes of PO_4^{3-} have been assessed (Table 3.4). Both LMA and HMA species acted as net PO_4^{3-} sources. This outcome is consistent with Taylor and co-workers' assumption that sufficient P is obtained by the sponges from the diet.

Recent work on *Dendrilla nigra* has shown that actinomycetes isolates were able to solubilize phosphate compounds, making this element available both for the host and favouring a faster P cycling (Sabarathnam *et al.*, 2010). However, "*in vivo*" studies are needed to evaluate the potential role of such microbe-mediated mechanisms in phosphate accumulation and solubilization by marine sponges.

3.3. Ecological significance of nitrogen and phosphorous use

Marine sponges are able to act either as sink or source of a variety of nitrogen compounds, which is more diverse in HMA species. Some general patterns can be summarized. Marine sponges act as a sink of POM. Their grazing impact on the small plankton fraction can be important to the point that picoplankton depletion layers have been registered above benthic communities where sponges were abundant (Savarese *et al.*, 1997). As a result of POM oxidation, release of inorganic dissolved compounds, such as ammonia and phosphate plus organic dissolved compounds, is expected. Nevertheless, the pathways followed after oxidation of organic matter is highly dependent on the abundance and type of the associated microbial communities. LMA sponges act as a net source of NH_4^+, PO_4^{3-}, and DOM. The scenario in HMA species is more complex due to the larger variety of the hosted microbial partners. Despite only few available data, it can be suggested that HMA sponges share the ability to act as a PO_4^{3-} source, irrespective of the amount and type of associated microbes. HMA sponges also represent a source of NO_x^-, although the magnitude of this flux appears to be substantially affected by seasonal variability. On the other side, HMA sponges are a sink of NH_4^+, as a result of either nitrification processes or photoautotrophy. The few data available on DON fluxes suggest that HMA species are a sink of these nitrogen compounds. To date, the available data indicate that sponges are neither a significant N_2 source nor a sink.

Being that sponges are one of the most abundant and widespread group of benthic organisms in a wide variety of marine systems, their activity can potentially increase local concentrations of NH_4^+, PO_4^{3-}, and DON, nutrients often limiting for primary production and bacterial growth. This effect would be more likely to occur in LMA-dominated sponge assemblages. In contrast, the metabolic activity of HMA sponges does not appear to represent a potential source of compounds stimulating primary production or bacterioplankton growth. Rather, these sponges will compete with planktonic organisms for limiting nutrients and DON. Most of studies relating bacterial productivity to inorganic and organic nutrient inputs derived from benthic activity have been performed in coral reef systems. Differences are found between studies, depending on the nutrient limiting bacterial growth at each site. DOM produced through mucus released by corals was responsible of enhancing activity in the bacterioplankton adjacent to corals (Ferrier-Pages *et al.*, 2000; Van Duyl and Gast, 2001; Wild *et al.*, 2008). Inorganic nutrients from the remineralization of ingested compounds by suspension feeders living in the reef also enhanced bacterial activity in reef crevices (Gast *et al.*, 1998; Torreton *et al.*, 2000; Richter *et al.*, 2001; Van Duyl and Gast, 2001; Scheffers *et al.*, 2005). A lack of studies from other geographical sites (including temperate and cold systems) prevents any generalization on the effects of sponge activity on plankton productivity.

4. Use of Silicon by Sponges

4.1. Biological role of silicon structures

About 75% of the extant sponges species (most Demospongiae and all known Hexactinellida) use Si to build a siliceous skeleton. It consists of discrete pieces (known as spicules; Fig. 3.4A and B), pieces fused into rigid networks or a combination of both elements. The siliceous skeleton may represent up to a 90–95% of the body dry weight, depending on the species (e.g. Barthel, 1995; Maldonado *et al.*, 2010a). Spicules range in size from 1 μm to about 2–3 m, depending on the species. They also show a large variety of shapes and ornamentation details (reviewed in Hartman, 1981; Uriz *et al.*, 2003; Uriz, 2006). Often small spicules ($\sim < 100$ μm) are termed microscleres, being larger spicules known as megascleres. Both shape and size of spicules are relatively constant within species, a condition also conserved through the geological time scale, as revealed by the fossil record (e.g. Hinde and Holmes, 1892; Mostler, 1976; Wiedenmayer, 1994; Vacelet *et al.*, 2009).

Figure 3.4 (A) Star-like spicules of the demosponge *Chondrilla nucula*. (B) Needle-like spicules of the demosponge *Petrosia ficiformis*. (C) Sclerocyte (sc) of *P. ficiformis* elaborating a siliceous spicule (sp) within an intracytoplasmic vesicle limited by the plasmalemma (pm) membrane. (D) Epithelial cell (ec) of the homosclerophorid *Corticium candelabrum* in the process of extruding to the mesohyl a sponge spicule (sp) that have been produced intracellularly. (E) Transversal section of a young spicule of *P. ficiformis*, showing a hexagonal axial filament (af) surrounded by a moderate amount of silica (si). (F) Transversal section of a young spicule of the demosponge *Crambe crambe*, showing a triangular axial filament (af) surrounded by an incipient silica layer (si).

Conserved spicule sizes and shapes could have resulted from particular adaptations to specific functions, which in most cases, have not been elucidated yet. Spicules are assumed to strengthen the mesohyl, helping to support the growth of soft parts and reinforcing the epithelia (pinacoderms) to avoid collapse of the aquiferous subectosomal spaces and canals. In many sponges, spicules form either a dense cortex at the surface or sharpened protruding palisades, thought to provide mechanical defences that deter potential predators. Spicules protruding from the body are also believed to protect the inhalant areas of the sponge surface from silt accumulation and clogging, particularly in deep-sea habitats. Some sponges also use giant spicules (centimetres to metres) to form rooting systems and stalks through which sponges attach to the seafloor, particularly in soft bottoms. In carnivorous sponges, siliceous micro-hooks that are finely and diversely micro-ornamented protrude from the body to hook the appendices of the microcrustaceans that will serve as food to these non-filtering sponges

(Vacelet and Duport, 2004). Additionally, siliceous spicules are excellent light transmitters that facilitate illumination of the subectosomal body regions, allowing phototrophic symbionts (i.e. cyanobacteria and zooxanthellae) to grow successfully in relatively deep regions of the sponge body for enhanced mutualistic or symbiotic relationship (Cattaneo-Vietti *et al.*, 1996; Brummer *et al.*, 2008). In sponge larval stages, spicules are thought to be involved in control of weight mass distribution and buoyancy (reviewed in Maldonado, 2006). Nevertheless, beyond the above generic functions postulated for sponge spicules, it remains largely unknown whether adaptation is the reason for the highly conserved nature of spicule size, shape, and ornamentation.

4.2. Cytology of silicon use: Intracellular versus intercellular modes

Siliceous skeletal pieces are elaborated by sponge cells (Fig. 3.4C and D), which are somehow able to take up the silicic acid (Si $(OH)_4$) dissolved in seawater and polymerize it into biogenic silica (SiO_2). Therefore, Si fluxes through sponges, unlike those of other dissolved nutrients, are thought to be exclusively mediated by sponge cells, being unrelated to the composition and density of sponge-associated microbial populations. Hereafter, we will refer to silicic acid as dissolved silicon (DSi) and to biogenic silica as BSi. In the past decades, the understanding of the mechanisms involved in spicule production has improved notably, but many unknowns still remain.

Plenty of observational evidence through light and electron microscopy during the twentieth century has corroborated that siliceous spicules start being elaborated intracellularly by amoeboid, free-moving cells that occur in the sponge mesohyl, known as sclerocytes (Simpson and Vaccaro, 1974; Simpson, 1978, 1984; Wilkinson and Garrone, 1980; Imsiecke *et al.*, 1995; Custodio *et al.*, 2002; Leys, 2003). Spiculogenesis begins within an intracellular vesicle, with the synthesis of a protein filament (termed axial filament; Fig. 3.4D and E) that is subsequently mineralized by condensation of silica around it. The protein making the axial filament is mostly silicatein, an enzyme that somehow directs polymerization of silicic acid into biogenic silica (Shimizu *et al.*, 1998; Cha *et al.*, 1999). The membrane of the vesicle, known as silicalemma (Simpson, 1984), has long been suspected to mediate in the condensation of silica around the axial filament. In most cases, each sclerocyte secretes only one spicule. Nevertheless, there are reports of sclerocytes secreting two or several spicules simultaneously (Imsiecke *et al.*, 1995; Custodio *et al.*, 2002), or even multiple spicules arranged in parallel in dense packs (Wilkinson and Garrone, 1980).

By investigating homosclerophorid sponges, it was recently found that spicules are produced intracellularly not only in the sclerocytes (Fig. 3.4C) but also within the epithelial cells (Fig. 3.4D), that is, exopinacocytes and, to a lesser degree, endopinacocytes (Maldonado and Riesgo, 2007). The unconventional epithelial origin and the particular internal structure of those homosclerophorid spicules—with two relatively thick, concentric organic layers intercalated among silica layers in addition to a central axial filament—strongly suggests that silicification in homosclerophorid sponges does not follow the standard described for other siliceous sponges. Therefore, since it cannot be discarded that the ability for silicification may have evolved independently more than once in the phylum, a careful re-examination of the hitherto monophyletic state attributed to biosilicification within the phylum Porifera is needed.

Most of studies reporting intracellular production of sponge spicules come from observations on microscleres or small megascleres (Simpson and Vaccaro, 1974; Simpson, 1978; Wilkinson and Garrone, 1980; Imsiecke *et al.*, 1995; Custodio *et al.*, 2002; Maldonado and Riesgo, 2007). Therefore, it remains unclear how the large spicules (several mm to cm) characterizing many demosponges and hexactinellids could be produced intracellularly and it has long been suspected that some sort of extracellular BSi condensation could occur during the building of the largest megascleres (Simpson, 1984). In at least some Hexactinellida, it has been corroborated that relatively large spicules are formed intracellularly, within multinucleate giant sclerocytes, named sclerosyncytia (Mackie and Singla, 1983). From TEM observations in the demosponge *Crambe crambe,* it has been proposed the alternative view that production of megascleres in most demosponges could occur intercellularly (Uriz *et al.*, 2000). This claim was based on the realization that axial filaments of *C. crambe* were initially produced intracellularly in a membrane bounded vesicle of the sclerocytes and subsequently exocytosed to the intercellular medium. There, DSi condensation would take place somehow within an external "pocket" formed around the silicatein filament by pseudopodia emitted by the sclerocyte membrane (i.e. the plasmalemma).

More recently, another mechanism has been proposed, mostly from accumulative research on silicification in primmorphs of the demosponge *Suberites domuncula.* It postulates that silicification starts intracellularly and is completed extracellularly through mechanisms different from the "sclerocyte pockets" suggested by Uriz and co-workers (reviews by Schröder *et al.*, 2008; Wang *et al.*, 2011b). This latter model vindicates that, at the cellular level, sponge silicification involves three major steps: (1) intracellular production of the axial filament and initiation of BSi condensation within the silicalemma-enveloped vesicle, (2) exocytosis of the unfinished spicule to the intercellular milieu (the mesohyl), and (3) completion of BSi deposition onto the external surface of the exocytosed spicule. This secondary BSi deposition is hypothesized to result from the release of concentrated DSi and silicatein enzymes previously stored in vesicles by one or several sclerocytes congregated in the vicinity of the exocytosed spicules. Such an intercellular silicification phase would occur directly onto the spicule surface without the need of intimate contact between the growing spicule and the membrane of nearby sclerocytes. A fourth extracellular silicification step has recently been proposed. It postulates that at least one of the sclerocytes in the vicinity of the newly exocytosed spicule emits a pseudopodium-like evagination of its membrane that enters the axial canal and deposits a silica layer, reducing considerably the lumen of the axial canal (Wang *et al.*, 2011b).

4.3. Molecular aspects of silicon use

As at the cellular level, many molecular aspects of DSi use by sponges still remain obscure. The available information has been somewhat repetitively reviewed in the past years (Schröder *et al.*, 2008; Müller *et al.*, 2009; Wang *et al.*, 2011a), because new findings that complicate the global picture are published continuously (Schloßmacher *et al.*, 2011; Wiens *et al.*, 2011). A major step in this field was the realization that the axial filament of spicules mostly consists of an enzymatic protein (silicatein) that directs polymerization of DSi into BSi (Shimizu *et al.*, 1998; Cha *et al.*, 1999). So far, two isoforms of silicatein (alpha and beta) have identified (Cha *et al.*, 1999). It has been shown *in vitro* that silicatein can use synthetic tetraethoxylene as substrate (Cha *et al.*, 1999), but, given that this type of compound is not naturally available to the sponges, the natural substrate for the enzyme remains enigmatic. It has preliminarily been suggested that silicic acid–sugar complexes might be possible natural candidate substrates for silicatein (Schröder *et al.*, 2008). The particulars of reaction mechanisms leading to the production of BSi from DSi in sponges are still under debate. A model postulating the basic reactions of enzymatically mediated BSi deposition starting from a synthetic alkoxide substrate was originally proposed by Cha *et al.* (1999), but no significant progress on the putative reactions regarding natural substrates has been achieved since.

In contrast, some progress has been attained relative to the mechanisms through which silicatein is produced. The silicatein gene renders an inactive 36-kDa translation product, which is cleaved twice before becoming a mature 23-kDa enzyme (Müller *et al.*, 2005). Pos-translational modifications also include multiple phosphorilations and a dehydroxilation step (Müller *et al.*, 2005; Schröder *et al.*, 2006). Silicatein monomers are able to self-assemble into oligomers, then into longer protein filaments by mechanisms still under debate (Murr and Morse, 2005; Schröder *et al.*, 2008).

Silicatein is not only the main component of axial filaments. It has also be demonstrated to occur on the spicule surface of intercelullar spicules (Müller *et al.*, 2006). Extracellular aggregation of silicatein onto the surface of "unfinished" spicules is postulated to control the extracellular silica condensation required to complete the spicule formation, according to the Müeller–Schröder's silicification model (Müller *et al.*, 2005; Schröder *et al.*, 2006). Although silicateins are the predominant proteins related with sponge silicification, a number of additional proteins, such as Galectin-2, Silintaphin-1, and Silintaphin-2, have been identified so far. Galectin is thought to serve as a structural matrix for the assembly of silicatein molecules (Schröder *et al.*, 2006, 2008). Silintaphin-1 appears to facilitate the assemble of silicatein-alpha multimers and somehow binds to those multimers to enhance BSi condensation around the silicatein scaffold (Wiens *et al.*, 2009; Schloßmacher *et al.*, 2011). The role of the recently discovered Silintaphin-2 (Wiens *et al.*, 2011) still remains unclear.

It also remains little understood the mechanism responsible for binding and transporting DSi from seawater to the cytoplasm of the silica-secreting cells. To date, the only enzymes properly identified in connection with DSi condensation belong to the silicatein family (Shimizu *et al.*, 1998; Cha *et al.*, 1999), but neither this enzyme nor other known components involved in the process of DSi polymerization inside or outside the sclerocytes cells are having a demonstrated role in internalization of DSi from ambient seawater. Suggestions have been made that a sodium-bicarbonate co-transporting system could somehow be involved in taking up silicic acid from seawater. Although active DSi uptake in diatoms has been shown to be supplemented by some passive diffusion across the cell membrane (Thamatrakoln and Hildebrand, 2008), a similar diffusion process is unlikely in sponges, because silica-secreting cells of sponges (sclerocytes) occur typically at the inner mesohyl regions of the sponge body. As a result, silicifying cells are typically isolated from ambient seawater by epithelial cell layers and dense intercellular deposits of collagen and other macromolecules.

4.4. Ecology of silicon use by sponges

Most DSi in the ocean is thought to be consumed by marine plankton (i.e. diatoms, radiolarians, choanoflagellates, silicoflagellates), particularly by diatoms (Nelson *et al.*, 1995; Tréguer *et al.*, 1995; Sarmiento and Gruber, 2006). Diatoms use DSi to build their extracellular siliceous skeleton (frustule) of BSi, which supports cell growth. High levels of DSi in seawater typically stimulate growth of diatom populations, which in turn decreases DSi, nitrate, and phosphate levels in surface waters and facilitate transfer of atmospheric carbon dioxide to the ocean, hence connecting silicon to carbon, phosphate, and nitrogen cycles (Siever, 1991; Harrison, 2000; Planavsky *et al.*, 2010). Because of these cycle interconnections, there is strong interest in predicting the interplay between DSi and BSi budgets and many efforts have been made during the last decades to unravel the "diatom" route of Si through the oceans.

There had always been suspicions that the contribution of sponges to the marine Si cycle, even if never quantified in global terms owing to its technical complexity, could be of some importance (e.g. Harriss, 1966; Siever, 1991). A small group of studies investigating the contribution by siliceous sponges (Rützler and Macintyre, 1978; Conley and Schelske, 1993; Reincke and Barthel, 1997; Maldonado *et al.*, 1999, 2005, 2010a) has already alerted the notion of a Si cycle exclusively revolving around diatoms may incur an unadvisable oversimplification. Here, we review the knowledge on Si fluxes through sponges, highlighting recent developments.

4.4.1. Experimental silicon uptake rates

The most serious limit to our understanding of the magnitude of Si turnover through sponges is the current lack of information regarding uptake kinetics. There are only two available studies to date. One study was conducted on the demosponge *H. panicea* (Reincke and Barthel, 1997), the other on several species in the genus *Axinella* spp. (Maldonado *et al.*, 2011). Both studies have agreed that DSi uptake rates increase with increasing ambient DSi availability according to a hyperbolic function (Fig. 3.5), which is the model characterizing ligand-binding kinetics with one site saturation (i.e. Michaelis–Menten kinetics). That kinetics is also in full agreement with the independent discovery of silicatein as the enzyme responsible for silica deposition in sponges (Cha *et al.*, 1999).

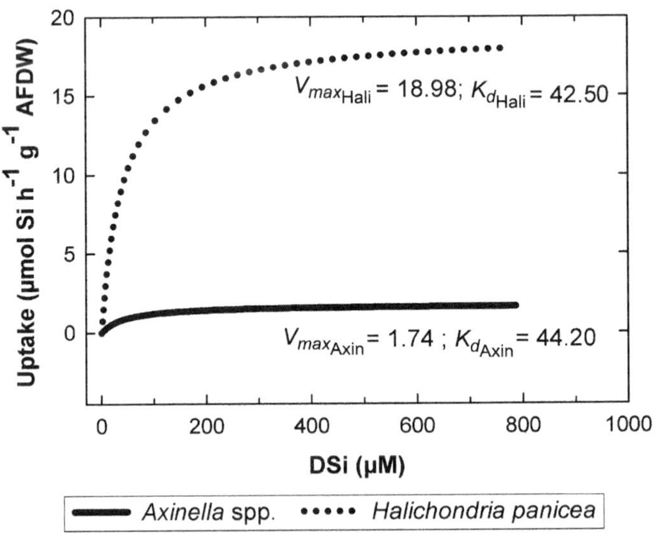

Figure 3.5 Comparative DSi uptake kinetics resulting from a reanalysis of raw experimental data for the demosponges *Halichondria panicea* (from Reincke and Barthel, 1997) and several *Axinella* species (from Maldonado *et al.*, 2011). For comparative purposes, uptake is expressed as μmol Si hour^{-1} and ash-free dry weight gram of sponge. Note that the half-saturation value (K_d) is relatively similar in both sponge genera, while maximum velocity of DSi transport (V_{max}) is much greater in *H. panicea*. This latter difference is assumed to result from uptake rates of *Axinella* spp. being measured on histologically undisturbed, complete individuals, while *H. panicea* uptake rates were measured on body fragments cut down for the experiment and inadvertently induced to regenerate much soft tissues and BSi skeleton at accelerate rates, probably biasing the uptake experiment.

According to Reincke and Barthel's (1997) experiments (Fig. 3.5), *H. panicea* has an uptake kinetics with a half-saturation constant (K_d) of 46.41 μM DSi and a saturated uptake rate (V_{max}) of 19.33 μmol DSi h^{-1} g^{-1} of sponge ash-free dry weight (AFDW). Experiments on *Axinella* spp. (Fig. 3.5) by Maldonado *et al.* (2011) indicated a K_d of 74.47 μM and a V_{max} of either 0.13 μmol h^{-1} sponge ml^{-1} or 1.74 μmol h^{-1} g^{-1}, if expressed as AFDW for comparative purposes. Surprisingly, saturated uptake rate for *H. panicea* was 19.33 μmol Si h^{-1} g^{-1} AFDW, an order of magnitude higher than that measured for *Axinella* spp. Such differences in V_{max} could be explained by both sponge genera having very different affinity by DSi. An alternative explanation is that *H. panicea* uptake measurements were not taken from complete sponge individuals but from explants, that is, regenerating pieces that had been obtained for the experiment by fragmenting larger sponges (Reincke and Barthel, 1997). It is well known that sponges regenerate rapidly large portions of their body (including the corresponding production of new silica skeleton), accomplishing regeneration at rates that are up to 2900 times the undisturbed growth rates (Ayling, 1983).

It is also worth noting that the *H. panicea's* uptake system saturates at ambient DSi concentrations around 100 μM DSi and that of *Axinella* spp. does it at about twice higher concentration (i.e. 200 μM), indicating that affinity for DSi may vary between sponges species. Interestingly, DSi uptake in both *H. panicea* and *Axinella* spp. becomes more efficient when DSi is provided as much higher concentrations (1 to 2 orders of magnitude higher) than the natural values (1–10 μM DSi) characterizing the sublittoral ecosystems where these sponges live. It has also been shown (Maldonado *et al.*, 2012) that DSi uptake rates by the intertidal demosponge *H. perlevis* increases when exposed to increasing DSi concentrations, with average DSi uptake rate (μmol Si mL^{-1} fresh sponge h^{-1}) at 70 μM being significantly higher (0.39 ± 0.07) than those at 40 and 25 μM (0.18 ± 0.04 and 0.15 ± 0.07, respectively), which in turn were not significantly different from each other, but were higher than the uptake rate at 10 μM (0.06 ± 0.02). In agreement with *H. panicea* and *Axinella* spp., these results support that the uptake system of *H. perlevis* performs with significantly higher efficiency when DSi concentrations are at least two- to fourfold higher than the DSi maximum expected in the natural habitat under the most conservative conditions (i.e. 10–15 μM). Consequently, all investigated sublittoral sponges are suffering a severe, chronic limitation by DSi (<10 μM) in their natural habitats. Another example of severe DSi limitation has been experimentally provided by long-term exposure of the shallow-water demosponge *C. crambe* to DSi concentrations higher (30 and 100 μM) than those in its natural habitats (about <2 μM). Those high concentrations induced secretion of not only longer and thicker spicules but also additional spicule types that are never produced in wild populations

(Maldonado *et al.*, 1999). Such a response suggests that the genetic systems controlling Si uptake and silicification are up-regulated by threshold DSi concentrations higher than those naturally available to the sponge.

Altogether, these findings on DSi uptake kinetics open the possibility that extant sponge species belonging to lineages that diversified before the decline in DSi availability caused by the ecological expansion of diatoms during the late Cretaceous and Early Tertiary (about 100 to 50 mya) may still silicify through uptake systems originally suited to deal with the high DSi concentrations that seemingly characterized pre-Tertiary oceans (Maldonado *et al.*, 2011). From an eco-physiological point of view, it means that the skeletal growth in the sublittoral populations of many Recent demosponges is chronically limited by DSi availability. The idea that low DSi concentrations in shallow waters of Recent oceans might limit sponges was originally disregarded (e.g. Hartman, 1981; Vacelet, 1988; Pisera, 1997), as it was assumed that their uptake systems had progressively adapted during the post-Cretaceous DSi crisis. In contrast, it seems that such an adaptation was never achieved by many sponges. Consequently, the reduced DSi availability characterizing modern oceans still operates as an important environmental pressure probably favouring evolutionary changes in these sponges towards either weakening their siliceous skeletons (Maldonado *et al.*, 1999; Maldonado, 2009) or tuning their silicification systems for higher efficiency at low DSi concentrations.

4.4.2. Silicon field demands

Information on DSi field demands by sponge populations is extremely scarce because of the technical complexity of the approaches. Any estimate of yearly DSi demands by sponge populations have to be taken as tentative, because calculations will be affected by several sources of variability for which corrections are difficult to implement. For instance, laboratory estimates of sponge DSi uptake on hourly or daily basis may not incorporate appropriately the biological variability inherent to the uptake process. It is likely that DSi uptake by sponges occurs neither as a continuous process nor at a constant rate, even if ambient DSi concentrations would be kept constant. DSi uptake is also linked to processes governing sponge growth, which in turn are strongly affected by seasonality (e.g. Stone, 1970a,b; Elvin, 1976; Turon *et al.*, 1998), individual age and size (e.g. Dayton *et al.*, 1974; Barthel, 1989), incidence of predation and grazing (e.g. Ayling, 1983), food availability (e.g. Frøhlich and Barthel, 1997), and species-specific "BSi/organic tissue" ratios. The effects of some of these parameters on DSi uptake rates may be dramatic. For instance, Ayling (1981) observed that some encrusting sponges, which had experimentally been injured to simulate predation, regenerated the lost tissues at rates between 22 and 2900 times the undisturbed growth rate, depending on species. Therefore, Si uptake and BSi production will be very different in two conspecific populations experiencing different levels of predation or grazing. Similarly, Frøhlich and Barthel (1997) demonstrated that sponges starved for 1 week showed Si uptake rates only 15% as large as those shown by non-starved sponges.

Despite problems, some tentative estimates have been produced. Field surveys along a relatively oligotrophic, sponge-poor Mediterranean rocky sublittoral revealed that siliceous sponges average 0.34 ± 0.52 L m^{-2} and ambient DSi 0.73 ± 0.44 μM (Maldonado *et al.*, 2011). At that ambient DSi concentration, sponge uptake rate is predicted to average $1.31 \pm 0.79 \times 10^{-3}$ μmol Si h^{-1} sponge ml^{-1} (according to equation for Mediterranean sponges in Fig. 3.5). It means that the sponge fauna per m^2 of habitat at that portion of Mediterranean rocky coast use yearly about 3.9 ± 5.9 mmol DSi, which represents yearly about 21.4 ± 32.7% of the average DSi available in a 30-m overlying water column and about 10.7 ± 16.3% in a 50-m deep-water column of that coastal ecosystem, if there is no replenishment. Similarly, it has been estimated that the abundance of siliceous sponges across a Mesoamerican continental shelf (Belize) averages 2.6 ± 14.3 L m^{-2} of bottom (Maldonado *et al.*, 2010a). Mean yearly DSi concentration in the 25-m deep-water column of such shelf ecosystem is about 3.6 ± 0.6 μM, and the sponge communities are predicted to consume DSi at an average rate of about $14.5 \pm 0.35 \times 10^{-3}$ μmol h^{-1} sponge ml^{-1} (according to the equation in Fig. 3.5). It means that yearly sponge uptake is about 332 ± 1826 mmol DSi m^{-2}, and that it would virtually deplete the shallow Belizean shelf of DSi once every 98.9 ± 83.7 days, if there is no DSi replenishment.

For a comparison (Table 3.5), the estimated Si consumption rate by the poor (in volume) sponge fauna of the Mediterranean sublittoral would be about 0.01 ± 0.01 mmol Si m^{-2} day^{-1}, while that of the richer Caribbean sponge assemblages would be 0.90 ± 5.00 mmol Si m^{-2} day^{-1}. By using available DSi uptake data for seasonal Baltic populations of the sponge *H. panicea* (Reincke and Barthel, 1997) and the mean biomass (20 ml m^{-2}) in the less favourable sponge habitat (Barthel, 1988), a conservative average consumption of 0.44 mmol Si m^{-2} day^{-1} during Baltic summer months may also be arrived at. From raw data in Dayton *et al.* (1974) and Dayton (1979), Maldonado *et al.* (2005) deduced that BSi production by the demosponge *Homaxinella balfouriensis* during an astonishing population bloom from 1968 to 1975 occurred at rates of about 22.94 mmol Si m^{-2} day^{-1} (Table 3.5). Surprisingly, this high value is not far behind the annual BSi production rate of the whole diatom assemblage in the Ross Sea, estimated between 37 and 93 mmol Si m^{-2} day^{-1}, according to mean and maximum production rates, respectively (Ragueneau *et al.*, 2000). Importantly, according to Dayton's abundance data (1979), the BSi production rate of the sponge *H. balfouriensis* would have experienced an additional twofold increase in the study area from 1975 to 1977, which, if considered in the calculations, would have lead to an even higher annual average rate of Si consumption.

Table 3.5 Rates of Si use by communities of planktonic diatoms and sublittoral sponge communities in various marine systems

Habitat/system	Mean Si demand ($mmol\ Si\ m^{-2}\ day^{-1}$)
Planktonic diatoms	
Coastal upwelling	90
Other coastal conditions	15
Southern ocean	15
Deep ocean	2.3
World ocean average	1.6–2.1
Sublittoral sponges	
Outer shelf of Belize	0.9
Baltic sublittoral bottoms	0.44
Mediterranean rocky bottoms	0.01
Antartic demosponge population bloom	22.9
Bathyal sponges	
Canadian hexactinellid reefs	3.52

Diatom Si demands were originally measured as BSi production rates (Nelson *et al.*, 1995; Tréguer *et al.*, 1995; Ragueneau *et al.*, 2000), demands by sublittoral non-Antarctic sponges have been derived from Si uptake rates (Reincke and Barthel, 1997; Maldonado *et al.*, 2011), demands by Antarctic sublittoral demosponges inferred by Maldonado *et al.* (2005) from Dayton (1979), and demands by bathyal hexactinellids inferred from estimated from growth rates (Chu *et al.*, 2011).

At present, it is impossible to estimate with any accuracy global Si consumption by sponges in the world oceans—as will be the case for decades to come owing to the extremely variable distribution of sponge biomass on the continental margins at depths that prohibit extensive measurements of individual Si contents per bottom area. With the exception of the transitory population bloom of the Antarctic demosponge *H. balfourensis*, the Si consumption rates inferred for sponges to date (Table 3.5) are somewhat smaller than the average BSi production by diatoms in the global ocean, estimated at 1.6–2.1 $mmol\ Si\ m^{-2}\ day^{-1}$ (Nelson *et al.*, 1995; Ragueneau *et al.*, 2000). They are also lower than the average diatom demand estimated in diverse marine systems. Nevertheless, sponge abundance is so important in some sublittoral areas that, despite their chronic DSi limitation, DSi incorporation at the community level is anticipated to have some relevance, particularly at local or regional level.

If the most conservative DSi consumption rates determined for sublittoral, non-Antarctic demosponges (Table 3.5) are standardized per unit area of bottom (according to the sponge abundances measured at the Mediterranean, the Baltic, and the Caribbean communities) and these figures extrapolated over the entire continental shelf ($22 \times 10^6\ Km^2$) of the earth, a first, very tentative estimate of the global Si consumption by sponges can be inferred, falling somewhere between 8.6×10^{10} and 7.3×10^{12} $mol\ Si\ year^{-1}$

(Maldonado *et al.*, 2011). This figure is still about 2– 4 orders of magnitude smaller than the 2.0 to 2.8×10^{14} mol Si year^{-1} estimated for diatoms (Nelson *et al.*, 1995; Tréguer *et al.*, 1995). Nevertheless, should DSi consumption by the dense sublittoral Antarctic populations and the many sponge communities that are being discovered at bathyal depths following the advent of ROVs and manned oceanographic submersibles (Richardson and Young, 1987; Rice *et al.*, 1990; Genin *et al.*, 1992; Maldonado and Young, 1996; Leys *et al.*, 2004; Maldonado *et al.*, 2005; Chu *et al.*, 2011) incorporated into the calculations, a generous—though unlikely to be ever accurately quantified—increase of the yearly global Si sponge consumption might be arrived at.

The information available to infer sponge DSi demands in bathyal and abyssal communities (where ambient DSi is not limiting) is even scarcer than that on sublittoral bottoms. To our knowledge, only the study by Chu *et al.* (2011) has addressed this issue, using a combination of field sampling and surveys with ROVs to investigate field populations of large and heavily skeletonized hexactinellid sponges (mostly *A. vastus*). These sponges are able to grow on top of non-dissolved skeletons of dead conspecifics, creating large siliceous mounds up to 21 m in height and about 9000 year old (Conway *et al.*, 1991, 2001; Chu and Leys, 2010). These sponge reefs are unique epibathyal habitats, so far known only from British Columbia (Canada), where they discontinuously extend over an area greater than 700 km^2 (Whitney *et al.*, 2005; Conway *et al.*, 2007). Because in this Pacific area annual DSi levels fluctuate in surface waters (<30 m) from 1 to 70 μM, but remain higher (about 50 μM) throughout depths (150–240) where sponge reefs occur, DSi is thought not be limiting to the sponge reefs. Under the assumption of a 1-cm year^{-1} growth rate in only the vertical dimension of reefs, the sponge populations of the 700-km^2 reef system were estimated to incorporate Si into their skeletons at an average rate of 1.3×10^6 mol Si km^{-2} year^{-1} (Table 3.5), leading to a current standing stock of 1.4×10^8 mol Si km^{-2} accumulated as BSi skeletons. These figures are thought to represent an important pool at the regional scale of the northeast Pacific continental shelf (Chu *et al.*, 2011).

4.4.3. Silicon in silica standings stocks

Only recently has the magnitude of sponge BSi standing stocks started being evaluated. A first assessment for a Caribbean sublittoral population of *Chondrilla nucula* in reef habitats and for a sublittoral Mediterranean population of *C. crambe* revealed that either population accumulated amounts of Si in the form of skeletal BSi that were about 200–300 times those available as DSi in the water column of their respective habitats (Maldonado *et al.*, 2005). The local population of *C. nucula* contained about 67.4 tons BSi km^2 of habitat and *C. crambe* about 81 kg BSi per linear km of rocky sublittoral.

To date, only a study has compared the relative contribution of BSi standing stocks by sponges and diatoms in a continental shelf ecosystem and it concluded, contrary to expectations, that sponges (and not diatoms) comprise the largest standing stock of BSi in the regional pool (Maldonado *et al.*, 2010a). Collectively, BSi in the various sponge communities was about 88.6% of the total Si pool in the studied Mesoamerican shelf ecosystem, while diatoms represented only 4.2%, being the remaining Si (7.2%) as ambient DSi. Such sponge dominance was not attributable to impoverished environments in terms of either nutrients or phytoplankton, but to high sponge abundance. On average, siliceous sponges in that shelf showed a mean BSi content of $0.3\,6 \pm 2.7$ kg m^{-2}, but large individuals of some species were estimate to contain up to 28 kg BSi each.

The few available, deep-sea studies strongly suggest that below the mixed layer of the ocean, where diatoms do not proliferate, sponge standing BSi stocks may also be significant. It has been shown that Si standing stocks by dense populations of bathyal hexactinellid *Sericolophus hawaiicus* average 12.6 g Si m^{-2}, which means that the sponges retain an amount of Si per m^2 of bottom that is equivalent to that contained in 2.6×10^6 l of ambient water from their habitat, an amount equivalent several times to that contained in the entire 400-m high water column above the sponge population (Maldonado *et al.*, 2005). These data are a very conservative estimate of total BSi content trapped in that bathyal sponge bed, because the abundant masses of large spicules (30–40 cm) released to the seafloor after sponge death were not considered. More importantly, the population of *S. hawaiicus* is not an isolated case of sponge abundance at bathyal depths. Recent surveys of 9-km abyssal transects at a depth of 4100 m off the California coast have shown occurrence of 2418 heavily silicified hexactinellid stalks at densities of 0.18–0.33 m^{-2} (Beaulieu, 2001). Dense populations of heavily silicified demosponges and hexactinellids are also known from continental slopes of several North Atlantic locations (Rice *et al.*, 1990; e.g. Barthel *et al.*, 1996; Maldonado and Young, 1996), and many of these populations are reported to accumulate on the bottom large amounts of BSi skeletons released after sponge death (Bett and Rice, 1992; Barthel and Tendal, 1993). Siliceous spicules are also extremely abundant in both Arctic and Antarctic bottoms at diverse depths, where they form mats up to 2 m thick (Koltun, 1968; Dayton *et al.*, 1974; Van Wagoner *et al.*, 1989). Of particular interest to illustrate the issue of BSi retention by sponge skeletons below the photic layer are the epibathyal reefs of hexactinellid sponges in the Hecate Strait and the Strait of Georgia in British Columbia (Conway *et al.*, 2001; Krautter *et al.*, 2001). In these habitats, the abundance of living hexactinellids can reach 240 individuals in 10 m^2 (Leys *et al.*, 2004). Because discrete reefs can be up to 19 m high and 2–10 km in extension and they are discontinuously scattered over an area greater than 700 km^2 (Conway *et al.*, 2001; Krautter *et al.*, 2001; Leys *et al.*, 2004), it can easily be deduced that these sponge systems function as huge Si traps. Estimates of BSi content in only three discrete reefs reported values ranging from 7 to 11 kg m^{-2}, amounting to a total of 915 tons of BSi locked in only the exposed portion of reefs, which extended for about 0.047 km^2 (Chu *et al.*, 2011). Because discrete reefs are discontinuously scattered over an area greater than 700 km^2, it was deduced that glass sponge reefs equate to 65% of the DSi reservoir (3.6×10^9 mol Si) in that area and represent a substantial silicon sink in the continental shelf waters of the northeastern Pacific Ocean.

It is also worth stressing that the skeletal Si incorporated by most kinds of sponges will be locked in their bodies for their entire lifetime, a period which may extend from decades to millennia, in striking contrast with the retention period by the short-lived diatoms, which is only days.

4.4.4. Dissolution and burial of silicon through sponges

Because seawater is highly undersaturated with Si, it is predicted that any siliceous skeleton exposed to seawater will readily dissolve into silicic acid, following a thermodynamically favoured reaction. The rapid postmortem dissolution observed for diatom frustules (e.g. Bidle and Azam, 1999) clearly fits such a theoretical prediction. At the global scale, diatoms are estimated to produce yearly about 240×10^{12} mol of BSi in the world oceans (Tréguer *et al.*, 1995). Nevertheless, after a few days of planktonic life, diatoms die and their BSi skeletons dissolve, releasing DSi. Dissolution is so rapid that about 50% of the skeletal BSi produced yearly by diatoms dissolves back into DSi before sinking below the photic layer (~200 m). Dissolution continues once frustule fragments have reached the ocean bottom, so that it is estimated that only about 3% of the skeletal BSi produced yearly by diatoms is finally buried in the marine sediment and exported to the geological cycle. Because the features of diatom BSi favours a rapid turnover between DSi and BSi states, the idea of a rapid BSi–DSi re-cycling has traditionally been extrapolated to the functioning of the global Si cycle (Nelson *et al.*, 1995; Tréguer *et al.*, 1995; Sarmiento and Gruber, 2006). Nevertheless, several experimental approaches have revealed that dissolution dynamics of sponge BSi are far slower than that of diatom BSi. For instance, by exposing acid-cleaned frustules and a variety of sponge spicule to abiotic seawater for 8 months, Maldonado *et al.* (2005) demonstrated that frustules dissolve about 75%. In contrast, three types of sponge spicules resisted dissolution similarly (0–5%), despite having marked differences in surface area. Therefore, surface area by itself cannot account for the drastic differences in dissolution dynamics between frustules and spicules. Studies by Katamani (1971) and Chu *et al.* (2011) have consistently agreed that major differences in dissolution rates between diatom frustules and sponge spicules also occur when the solvent is either alkaline solutions or non-aseptic seawater.

The reasons for these large difference in the dissolution dynamics remain unclear. Katamani (1971) reported infrared absorption near 940 cm^{-1} by diatom frustules, an absorption peak that is missing in sponge spicules. This peak indicates either water molecules trapped in the interstitial voids of the BSi polymeric network or hydroxyl groups in various states of association. Nevertheless, such minor differences, even if favouring dissolution of frustules relative to spicules, would be insufficient to explain the large differences detected in dissolution kinetics (Katamani, 1971). Substantial incorporation of fourfold-coordinated Al into the sponge BSi during silicification could be responsible for the very different dissolution kinetics. This element is known to drastically reduce dissolution rate of BSi skeletal remnants when it is incorporated in trace amounts from the sediment (Lewin, 1961; Dixit *et al.*, 2001). More recently, structural polysaccharides, such as chitin, have been shown to be incorporated into the concentric BSi layers of the sponge spicules (Ehrlich *et al.*, 2007), and it could be at least partially responsible for stabilizing the BSi matrix against dissolution. Chitin itself is naturally insoluble in water (Hock, 1940; Austin *et al.*, 1981) and enhances exoskeleton insolubility when imbedded into protein complexes of other marine invertebrates (Hunt, 1970; Weiner *et al.*, 1983).

Whatever the mechanism that sponges have developed to prevent passive dissolution of their spicules, such ability has resulted in multiple adaptive advantages. Spicule resistance to dissolution allow sponges to project large spicules out of the body forming "roots" for attachment, long stalks that elevate the sponge body for enhanced filter feeding, velvety surfaces that smooth water flow around the body, lacerating walls that deter predators, long palisades that prevent filtering surfaces from clogging, and minute hooks that capture microinvertebrates.

The resistance of sponge spicules to dissolution may also have major implications for Si cycling dynamics. According to the available data on spicule dissolution, it appears that BSi cycles into DSi through sponges over the timescale of decades to centuries, while that cycling takes only days to years when going through diatom frustules. The diatom-based model of the Si cycle postulates that 32–39% of the total burial of BSi is concentrated along continental margins owing to coastal diatom populations (Ragueneau *et al.*, 2000; DeMaster, 2002). Nevertheless, the persisting neglect of the impressive sponge populations characterizing continental shelves and upper slopes strongly suggests that global Si burial values may be underestimated. For instance, Rützler and MacIntyre (1978) reported that sponge spicules (and not diatom frustules) are the main component of particulate silica in Caribbean reef sediments. Likewise, sponge spicules have been shown to dominate the sediments of some Mediterranean coastal areas (Bavestrello *et al.*, 1996; Bertolino *et al.*, 2012). In Arctic and Antarctic bottoms at various depths, sponge spicules form mats up to 2 m thick (Koltun, 1968; Dayton *et al.*, 1974; Van Wagoner *et al.*, 1989). Dense populations of heavily silicified demosponges and hexactinellids are known from continental slopes of several North Atlantic locations (Rice *et al.*, 1990; Barthel *et al.*, 1996; Maldonado and Young, 1996), and many of these populations are reported to accumulate on the bottom large amounts of BSi skeletons released after death of the sponges (Bett and Rice, 1992; Barthel and Tendal, 1993).

The available evidence suggests that the magnitude of Si burial through sponges at the global scale may be more important than traditionally thought, even if no global quantification can be provided currently because of insufficient knowledge of global patterns of sponge BSi content in sediments. Unfortunately, most on-going research efforts towards BSi quantification in sediments and burial rates are guided by the extended notion that DSi removal on continental margins derives from production and burial of BSi primarily by diatoms and radiolarians. In contrast, the recent findings relative to sponges support the idea that these organisms may account for puzzling gaps in the current Si cycle model, particularly missing Si sinks on continental margins. For instance, after the recent realization that early estimates of BSi accumulation in the Southern Ocean and Antarctic deep sea were about 35% overestimated (DeMaster, 2002), the widely accepted model of steady-state balance for the marine Si cycle became unbalanced, because a Si sink equivalent to approximately one quarter of the global BSi burial is now missing. Because in order to bring back the cycle into its assumed balance an additional BSi sink should be identified, it has been proposed that BSi accumulation by diatoms on continental margins should account for most of the "missing" BSi burial (DeMaster, 2002; Ragueneau *et al.*, 2010). Admittedly, reliable direct estimates of diatom Si retention on continental margins are still lacking (Ragueneau *et al.*, 2010). Current data on sponge DSi demands, along with those already available on BSi standing stocks in sponge populations, suggest that much of the "missing" BSi could correspond to BSi stocks in extant populations of long-live sponges and in skeletons being buried in continental margins.

4.5. The role of sponges in the ocean silicon cycle

The current model of the marine Si cycle is summarized in Fig. 3.6. In brief, it describes that about 2.4×10^{14} Si mol of DSi are removed yearly from the photic ocean layer. This consumption is exclusively attributed to Si uptake of diatoms, which use it to build their BSi frustules. The model also predicts that about 50% of the BSi produced in the photic layer rapidly dissolves, being returned to the photic ocean as DSi for subsequent diatom new BSi production. The remaining 50% of BSi is exported to the non-photic ocean. Therefore, in a steady-state Si cycle, it can be assumed that the diatom BSi standing stock is about 1.2×10^{14} mol Si. Approximately 75% (9×10^{13}) of the sinking BSi that escapes from the photic layer upon diatom death, dissolves as DSi during its slow gravity transit through the aphotic water column, with only about 25% of the sinking BSi reaching the bottom as particulate material (BSi rain). From the BSi arriving at the bottom (i.e.

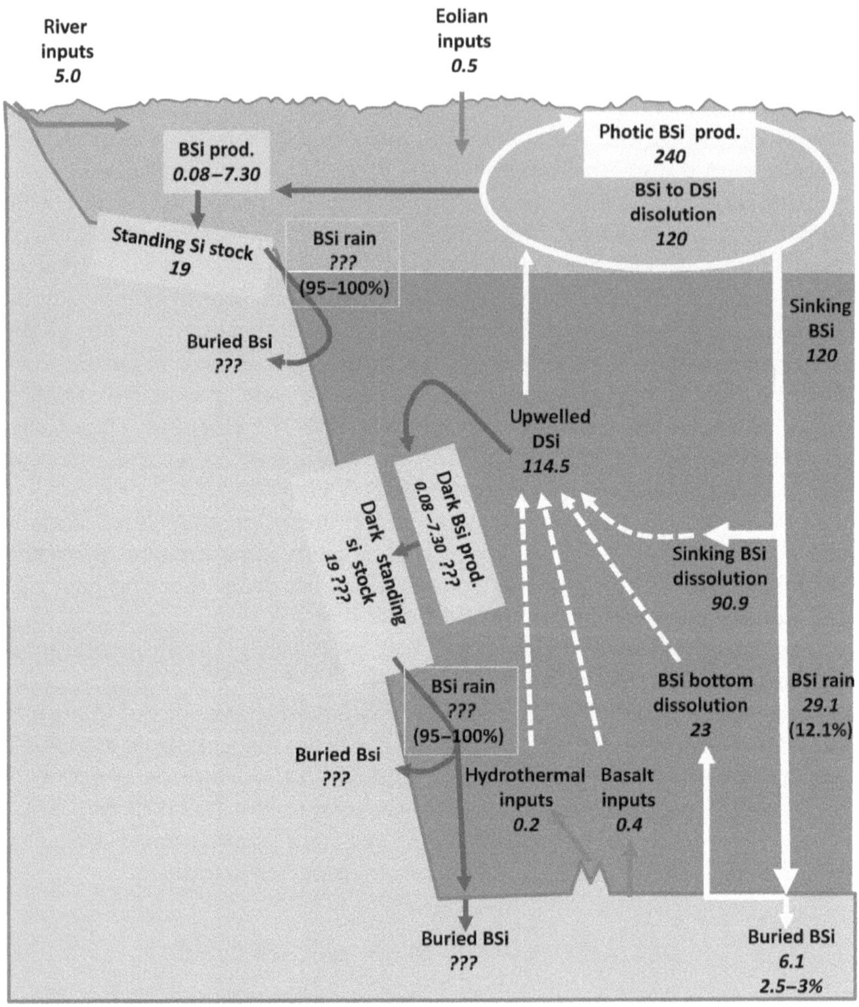

Figure 3.6 Global biogeochemical cycle of Si in the world ocean (as redrawn from Tréguer *et al.*, 1995), in which putative sponge contributions have been incorporated. Purple arrows refer to input Si fluxes to the ocean. Yellow arrows and boxes refer to diatom-mediated budgets and fluxes. Sponge budgets are given in orange boxes and putative fluxes referred by red arrows. Numbers in italics refer to budgets and fluxes calculated on a global yearly basis and expressed in mol Si $\times 10^{12}$. "Percentage values" for BSi rain and burial refer to total yearly BSi production estimated for either diatoms or sponges. Unknown or doubtful budgets and fluxes are indicated by a question mark (?). Note that most sponge-mediated processes concentrate around continental margins. (For interpretation of the references to colour in this figure legend, the reader is referred to the online version of this chapter.)

25% of sinking BSi or about 12% of the annual BSi production in the photic layers, i.e. 29.1×10^{12} mol Si), about four-fifths (i.e. about 10% of the BSi production in photic layers) dissolves and is returned as DSi to the deep-ocean water. Therefore, only about 3% (6.1×10^{12} mol Si) of the yearly BSi production in the photic ocean is finally buried in the sediments and exported to the geological cycle (Nelson *et al.*, 1995; Tréguer *et al.*, 1995; Sarmiento and Gruber, 2006). Because this model is largely based on diatom contribution, it assumes that: (1) DSi removal takes place exclusively in the photic ocean and deep-BSi production does not occur; (2) BSi rapidly cycles into DSi, because frustules dissolved readily when exposed to seawater after diatom death; and (3) only about 12% of the yearly BSi production arrives in particulate form to the marine sediments and only a small fraction (3%) is finally buried.

However, these cycle assumptions are not met when DSi is processed through sponges, for two major reasons: (1) Important populations of heavily silicified sponges occur at bathyal depths on continental margins (and to a lesser extent on abyssal plains), accounting for some DSi removal and BSi production in aphotic environments; (2) rates of sponge BSi return into DSi is predicted to be dramatically slower than that estimated for diatoms, because of greater sponge longevity and greater resistance of spicules to dissolution. It is likely that a large (> 3%) percentage of the yearly sponge BSi production becomes finally buried in the sediments. It could also be that a significant fraction of the Si burial attributed to diatom frustules indeed corresponds to sponge spicules in sediments, particularly on continental margins.

Figure 3.6 provides a summary of sponge-mediated Si fluxes that can tentatively be estimated under the current state of knowledge. The Si standing stock in siliceous sponges of the relatively sponge-poor rocky Mediterranean sublittoral averages 0.27 ± 0.94 mol Si m^{-2}, while it is 5.00 ± 45.07 mol Si m^{-2} in the sponge-reef Caribbean sublittoral (Maldonado *et al.*, 2010a, 2011). A preliminary tentative Si standing stock of about 19×10^{12} mol Si in sponges can be arrived at for earth continental shelves. This figure would be based on the tentative, conservative assumptions that (1) only about a third of the earth continental shelf area (22×10^6 km^2) harbours sponge populations and (2) abundances in half of the sponge-populated area are equivalent to those in the poor Mediterranean and, in the other half, equivalent to those in the Caribbean. Likewise, DSi demand by sponges on shelves has tentatively been estimated to fall somewhere from 0.08 to 7.30 Tmol Si (Maldonado *et al.*, 2011) and similar figures could be expected globally for sponges at bathyal and abyssal communities. Given that sponge BSi is highly reluctant to dissolution in seawater, nearly 95–100% of the BSi yearly released upon sponge death (whatever its magnitude), is assumed to reach the sediments. The rates of sponge BSi burial remain unknown, but a large fraction of the spicules reaching the bottom is likely to be buried before experiencing significant dissolution (e.g. Bertolino *et al.*, 2012), except in particular, highly carbonated environments, such as coral reefs (e.g. Rützler and Macintyre, 1978).

Therefore, the role of sponges is likely to affect current estimates of several Si partial budgets in the global Si cycle model, namely:

(1) DSi demands on continental shelves are slightly larger than currently predicted.
(2) There is "dark DSi consumption" due to sponges growing on the aphotic regions of the continental margins, where no diatom BSi production occurs.
(3) Large amounts (nearly 100%) of produced BSi is delivered directly to the sediments (BSi rain) upon sponge death.
(4) Rates of Si burial on continental margins are likely greater than predicted. The magnitude of this burial could even challenge the idea that the marine Si cycle is at a perfect steady balance.

More importantly, the ecological importance of sponge DSi demands, BSi production, and Si burial does not derive strictly from the magnitude of these figures in the global cycle, but rather from the fact that these processes largely concentrate on continental margins. Therefore, if we are to understand adequately Si fluxes in some coastal regions, we may need a new conceptual Si model, involving three major Si pools: (1) Si in rocks and deep sediment layers that cycles on geological time scales; (2) Si in diatoms that cycles comparatively rapidly and is connected to primary production processes; and (3) Si in sponges that cycles apart from primary production and does so at a much slower (but still undetermined) rate as compared with that of diatoms.

It also deserves attention that the fact that all DSi used by sponges on continental shelves is at the expense of the stock available for diatoms. The DSi progressively accumulated into sponge BSi on continental shelves (and slopes) is taken away from the primary production circuit for a long period, given the longevity of most sponges and the low levels of dissolution characterizing sponge skeletons. By disregarding this sponge DSi-sequestering process while considering exclusively diatom-related DSi and BSi stocks, the current biogeochemical models may be overestimating the real levels of connection between Si and C cycles on continental margins.

5. Concluding Remarks

- Sponges are opportunistic suspension feeders that process rapidly large volumes of water and incorporate particulate and dissolved nutrients.

They are able to harvest carbon as POC, DOC, and inorganic carbon. They are also able to incorporate organic nitrogen and organic phosphorous through feeding on planktonic cells and detritus. Siliceous sponges (i.e. most Demospongiae and all Hexactinellida) are also able to incorporate inorganic silicon from its dissolve silicic acid form. While Si fluxes are exclusively related to the elaboration of their skeleton of biogenic silica, C, N, and P fluxes derive from a complex combination of metabolic processes that include feeding, respiration, egestion, excretion, as well as hosting of large microbial populations within the sponge body. Sponge-associated microorganisms mediate in most C and N fluxes. Such a metabolic integration complicates the assessment of C and N balances through the sponges, particularly in sponges hosting high microbial density. In the case of N fluxes, the situation may even be more complex, for aerobic and anaerobic microbial processes co-occurring in the same sponge. Because of these major differences between the metabolic pathways, through which each of the concerned elements (i.e. C, N, P, Si) is processed, the available information on the sponge-meditated fluxes is very different in each case. For C and Si, there have been attempts to establish individual balances to infer subsequently fluxes at the population or community level and further assess the role of sponges in element cycling at local, regional, or global scales. For N and P, the situation is more complicated, as most studies are still attempting to unravel the basic metabolic pathways, as well as understanding the basis of environmental and between-species variability in fluxes. Therefore, the available information does not allow scaling up to large ecological sceneries.

- The study of C flux clearly shows that sponges have a C metabolism that differs in many aspects from the typical known in other invertebrates. This probably results from sponges being simple and ancient multicellular organisms that are evolutionarily close to the colonial organization stage. Due to the action of associated microbes, it is not surprising that C ingestion rates are often exceeded by C respiration rates. The most likely explanation is that total carbon ingestion rates may have been underestimated somehow. DOM is now an acknowledged food source for sponges that have to be incorporated into C balances and diet studies. Sponge cells along with sponge-associated microorganisms have been reported to assimilate DOC. HMA sponges, however, appear to rely more on DOM as a carbon source than LMA sponges. Whether sponges consume DOM probably depends on the naturally available amounts of DOM versus picoplankton cells. Most of the carbon assimilated by sponges appears to be used for maintenance and is respired. Costs for maintenance are relatively high in sponges compared to the ingestion, leaving not much energy for net growth in size. Moreover, the costs of growth also appear to be higher than in other invertebrates. These are probably constrains maintaining sponge growth usually at slow rates, despite the potential for growth being surprisingly high in these organisms. The impact that sponge populations have on their environment in terms of C flux remains to be accurately quantified. The qualitative effects of their activity are a net withdrawal of organic carbon from the passing water, accumulation of biomass, and excretion and biodeposition of detrital organic matter. Additionally, DOM consuming sponges may convert DOM into POM conserving energy within the system.

- Regarding N, the role of sponges is much related to the activity of their microbial associates. Microbes with the potential to mediate in each of the steps known in N cycling (i.e. nitrogen fixation, nitrification, denitrification, and anammox) have been molecularly identified in marine sponges. Nevertheless, there is a serious lack of studies quantifying the resulting N fluxes. Sponges, irrespective of being HMA or LMA, act as a source of NO_x^- (due to nitrification). HMA species appear to act consistently as a DON and NH_4^+ sink (potentially related to phototrophic activity and nitrification). Nevertheless, the role of HMA species in N_2 cycling is not that consistent, as they can be a N_2 sink (through nitrogen fixation) or a N_2 source (through denitrification or anammox). HMA sponges may theoretically compete with the phytoplankton and the bacterioplankton for limiting nutrients, such as DON and NH_4^+, while LMA species are more likely to release nutrients that are limiting to primary production. Nevertheless, preliminary assessments of these processes suggest that they contribute significantly neither to sponge nutrition nor to induce relevant changes in concentration of N compounds in the ambient water of the sponge habitats.
- The few available data on PO_4^{3-} fluxes indicate that marine sponges generate a net efflux of this nutrient, irrespective of their HMA or LMA condition. This suggests that regular heterotrophic feeding is providing the sponges with more than enough P to cover their metabolic needs. Despite phosphate being a limiting nutrient for phytoplankton in many local environments, the ecological relevance of the net P efflux by sponges remains uninvestigated.
- Regarding Si, the available information for shallow-water demosponges reveals that they incorporate Si following a Michaelis–Menten kinetic, a process that operates with maximum efficiency when ambient Si concentrations are about 1–2 orders of magnitude higher than natural availability. Consequently, all sponges investigated to date are strongly limited by Si ambient availability. This condition reveals a marked lack of adaptation of the sponge Si uptake system to the low ambient Si concentration characterizing modern oceans, a situation compensated for by other better-adapted Si consuming organisms, such as diatoms. The exact molecular and cellular pathway through which Silicon is transported to the inner of the silica-secreting cells for polycondensation remain poorly known, but it appears to use mechanisms different from those in diatoms

and other Si using organisms. The yearly Si consumption by sponges on continental shelves has tentatively been estimated somewhere between 8.6×10^{10} and 7.3×10^{12} mol Si year^{-1} and the accumulation of Si in their skeletons at 19×10^{12} mol Si. These figures may double when the contributions by sponge populations on continental slopes and deeper bottoms are incorporated into calculations. Therefore, sponges represent a modest, though non-negligible, contribution (probably around 15%) to the global marine Si budget. Because sponges may live for decades, centuries, and even millennia, and because the dissolution of their siliceous skeletons is extremely slowly, sponge populations on continental shelves and slopes function as transitory traps that slow down Si re-cycling around continental margins, favouring also a significant net export to the sediments that awaits quantification. The ecological importance of sponge Si consumption and net export does not derive strictly from the magnitude of their rates in global terms, but rather from the fact that the effects are accumulative and largely concentrated on continental margins. All Si incorporated by sponges on continental shelves and slopes is at the expense of the stock available for diatoms, being taken away from the primary production circuit for a long period. The lack of quantitative data on this sponge Si-sequestering process favours a conceptual biogeochemical model overestimating the real levels of connection between Si and C cycles on continental margins.

ACKNOWLEDGEMENTS

This research was supported by MINYC grants from the Spanish Government to M. M. (BFU2008-00227/BMC and BFU2011-24936) and to M. R. (CGL2010-18466), also by a grant to from "Generalitat de Catalunya" (Catalan Government) to M. R. (2009SGR142).

REFERENCES

Alber, M., and Valiela, I. (1995). Organic aggregates in detrital food webs: Incorporation by bay scallops Argopecten irradians. *Marine Ecology Progress Series* **121,** 117–124.

Austin, P. R., Brine, C. J., Castle, J. E., and Zikakis, J. P. (1981). Chitin: New facets of research. *Science* **212,** 749–753.

Ayling, A. M. (1981). The role of biological disturbance in temperate subtidal encrusting communities. *Ecology* **62,** 830–847, Brooklyn.

Ayling, A. L. (1983). Growth and regeneration rates in thinly encrusting demospongiae from temperate waters. *The Biological Bulletin* **165,** 343–352.

Ayukai, T. (1995). Retention of phytoplankton and planktonic microbes on coral reefs within the Great Barrier Reef, Australia. *Coral Reefs* **14,** 141–147.

Barthel, D. (1988). On the ecophysiology of the sponge *Halichondria panicea* in Kiel Bight. II. Biomass, production, energy budget and integration in environmental processes. *Marine Ecology Progress Series* **43,** 87–93.

Barthel, D. (1989). Growth of the sponge Halichondria panicea in the North Sea habitat. In "Proceedings of the 21st European Marine Biology Symposium." pp. 23–30.

Barthel, D. (1995). Tissue composition of sponges from the Weddell Sea, Antarctica: Not much meat on the bones. *Marine Ecology Progress Series* **123,** 149–153.

Barthel, D., and Tendal, O. S. (1993). Sponge spicules in abyssal and bathyal sediments of the NE Atlantic. *Deep-Sea Newsletter* **20,** 15–18.

Barthel, D., and Theede, H. (1986). A new method for the culture of marine sponges and its application for experimental studies. *Ophelia* **25,** 75–82.

Barthel, D., Tendal, O. S., and Thiel, H. (1996). A wandering population of the Hexactinellid sponge *Pheronema carpenteri* on the continental slope off Morocco, Northwest Africa. *Marine Ecology* **17,** 603–616.

Bavestrello, G., Cattaneo-Vietti, R., Cerrano, C., Cerutti, S., and Sarà, M. (1996). Contribution of sponge spicules to the composition of biogenic silica in the Ligurian Sea. *Marine Ecology* **17,** 41–50.

Bayer, K., Schmitt, S., and Hentschel, U. (2007). Microbial nitrification in Mediterranean sponges: Possible involvement of ammonia-oxidizing Betaproteobacteria. In "Porifera Research: Biodiversity, Innovation and Sustainability" (M. C. Custódio, G. Lôbo-Hajdu, E. Hajdu and G. Muricy, eds), pp. 165–171. Museu Nacional do Rio de Janeiro, Rio de Janeiro.

Bayer, K., Schmitt, S., and Hentschel, U. (2008). Physiology, phylogeny and in situ evidence for bacterial and archaeal nitrifiers in the marine sponge *Aplysina aerophoba*. *Environmental Microbiology* **10,** 2942–2955.

Beaulieu, S. E. (2001). Life on glass houses: Sponge stalk communities in the deep sea. *Marine Biology* **138,** 803–817.

Behringer, D. C., and Butler, M. J. IV (2006). Stable isotope analysis of production and trophic relationships in a tropical marine hard-bottom community. *Oecologia* **148,** 334–341.

Bell, J. J. (2008). Sponges as agents of biological disturbance. *Marine Ecology Progess Series* **368,** 127–135.

Benitez-Nelson, C. R. (2000). The biogeochemical cycling of phosphorus in marine systems. *Earth-Science Reviews* **51,** 109–135.

Bertilsson, S., Berglund, O., Karl, D., and Chisholm, S. (2003). Elemental composition of marine Prochlorococcus and Synechococcus: Implications for the ecological stoichiometry of the sea. *Limnology and Oceanography* **48,** 1721–1731.

Bertolino, M., Calcinai, B., Capellacci, S., Cerrano, C., Lafratta, A., Pansini, M., Penna, A., and Bavestrello, G. (2012). *Posidonia oceanica* meadows as sponge spicule traps. *Italian Journal of Zoology* (in perss), 1–8.

Bett, B. J., and Rice, A. L. (1992). The influence of hexactinellid sponge spicules on the patchy distribution of macrobenthos in the Porcupine Seabight (bathyal NE Atlantic). *Ophelia* **36,** 217–226.

Bidle, K. D., and Azam, F. (1999). Accelerated dissolution of diatom silica by marine bacterial assemblages. *Nature* **397,** 508–512.

Boury-Esnault, N., De Vos, L., Donadey, C., Vacelet, J., and Rützler, K. (1990). Ultrastructure of choanosome and sponge classification. In "New Perspectives in Sponge Biology," pp. 237–244. Smithsonian Institution Press, Washington, D.C./London.

Brummer, F., Pfannkuchen, M., Baltz, A., Hauser, T., and Thiel, V. (2008). Light inside sponges. *Journal of Experimental Marine Biology and Ecology* **367,** 61–64.

Cattaneo-Vietti, R., Bavestrello, G., Cerrano, C., Sara, M., Benatti, U., Giovine, M., and Gaino, E. (1996). Optical fibres in an Antarctic sponge. *Nature* **383,** 397–398.

Cha, J. N., Shimizu, K., Zhou, Y., Christiansen, S. C., Chmelka, B. F., Stucky, G. D., and Morse, D. E. (1999). Silicatein filaments and subunits from a marine sponge direct the polymerization of silica and silicones in vitro. *Proceedings of the National Academy of Sciences of the United States of America* **96,** 361–365.

Chu, J. W. F., and Leys, S. P. (2010). High resolution mapping of community structure in three glass sponge reefs (Porifera, Hexactinellida). *Marine Ecology Progress Series* **417,** 97–113.

Chu, J. W. F., Maldonado, M., Yahel, G., and Leys, S. P. (2011). Glass sponge reefs as a silicon sink. *Marine Ecology Progress Series* **441,** 1–14.
Clausen, I., and Riisgård, H. U. (1996). Growth, filtration and respiration in the mussel *Mytilus edulis*: No evidence for physiological regulation of the filter-pump to nutritional needs. *Marine Ecology Progress Series* **141,** 37–45.
Coma, R., Zabala, M., and Gili, J. M. (1995). Reproduction and cycle of gonadal development in the Mediterranean gorgonian *Paramuricea clavata. Marine Ecology Progress Series* **117,** 173–183.
Coma, R., Ribes, M., Gili, J.-M., and Hughes, R. N. (2001). The ultimate opportunists: Consumers of seston. *Marine Ecology Progress Series* **219,** 305–308.
Coma, R., Ribes, M., Gili, J.-M., and Zabala, M. (2002). Seasonality of in situ respiration rate in three temperate benthic suspension feeders. *Limnology and Oceanography* **47,** 324–331.
Conley, D. J., and Schelske, C. L. (1993). Potencial role of sponge spicules in influencing the silicon biogeochemistry of Florida lakes. *Canadian Journal of Fisheries and Aquatic Sciences* **50,** 296–302.
Conway, K. W., Barrie, J. V., Austin, W. C., and Luternauer, J. L. (1991). Holocene sponge bioherms on the western Canadian continental shelf. *Continental Shelf Research* **11,** 771–790.
Conway, K. W., Krautter, M., Barrie, J. V., and Neuweiler, M. (2001). Hexactinellid sponge reefs on the Canadian continental shelf: A unique "living fossil". *Geoscience Canada* **28,** 71–78.
Conway, K. W., Barrie, J. V., Hill, P. R., Austin, W. C., and Pickard, K. (2007). Mapping sensitive benthic habitats in the Strait of Georgia, coastal British Columbia: Deep-water sponge and coral reefs. *Geological Survey of Canada* **A2,** 1–6.
Corredor, J. E., Wilkinson, C. R., Vicente, V. P., Morell, J. M., and Otero, E. (1988). Nitrate release by Caribbeaan reef sponges. *Limnology and Oceanography* **33,** 114–120.
Custodio, M. R., Hajdu, E., and Muricy, G. (2002). In vivo study of microsclere formation in sponges of the genus *Mycale* (Demospongiae, Poecilosclerida). *Zoomorphology* **121,** 203–211.
Davy, S. K., Trautman, D. A., Borowitzka, M. A., and Hinde, R. (2002). Ammonium excretion by a symbiotic sponge supplies the nitrogen requirements of its rhodophyte partner. *The Journal of Experimental Biology* **205,** 3505–3511.
Dayton, P. K. (1979). Observations of growth, dispersal and population dynamics of some sponges in McMurdo Sound, Antarctica. *Colloques Internationaux du Centre National de la Recherche Scientifique* **291,** 271–282.
Dayton, P. K., Robilliard, G. A., Paine, R. T., and Dayton, L. B. (1974). Biological accommodation in the benthic community at the McMurdo Sound, Antarctica. *Ecological Monographs* **44,** 105–128.
de Caralt, S., Uriz, M. J., and Wijffels, R. H. (2008). Grazing, differential size-class dynamics and survival of the Mediterranean sponge *Corticium candelabrum. Marine Ecology Progress Series* **360,** 97–106.
De Goeij, J. M., and Van Duyl, F. C. (2007). Coral cavities are sinks of dissolved organic carbon (DOC). *Limnology and Oceanography* **52,** 2608–2617.
De Goeij, J. M., Moodley, L., Houtekamer, M., Carballeira, N. M., and van Duyl, F. C. (2008a). Tracing 13C-enriched dissolved and particulate organic carbon in the bacteria containing coral reef sponge *Halisarca caerulea*: Evidence for DOM-feeding. *Limnology and Oceanography* **53,** 1376–1386.
De Goeij, J. M., Van de Berg, H., Van Oostveen, M., Epping, E. H. G., and Van Duyl, F. C. (2008b). Major bulk dissolved organic carbon (DOC) removal by encrusting coral reef cavity sponges. *Marine Ecology Progress Series* **357,** 139–151.
De Goeij, J. M., De Kluijver, A., Van Duyl, F. C., Vacelet, J., Wijffels, R. H., De Goeij, A. F. P. M., Cleutjens, J. P. M., and Schutte, B. (2009). Cell kinetics of the

marine sponge *Halisarca caerulea* reveal rapid cell turnover and shedding. *Journal of Experimental Biology* **212,** 3892–3900.

Del Giorgio, P. A., and Cole, J. J. (2000). Bacterial energetics and growth efficiency. In "Microbial Ecology of the Oceans" (D. L. Kirchman, ed.), pp. 289–325. Wiley-Liss, New York.

DeMaster, D. J. (2002). The accumulation and cycling of biogenic silica in the Southern Ocean: Revisiting the marine silica budget. *Deep Sea Research Part II: Topical Studies in Oceanography* **49,** 3155–3167.

Diaz, M. C., and Rützler, K. (2001). Sponges: An essential component of Caribbean coral reefs. *Bulletin of Marine Science* **69,** 535–546.

Díaz, M. C., and Ward, B. B. (1997). Sponge-mediated nitrification in tropical benthic communities. *Marine Ecology Progess Series* **156,** 97–107.

Dixit, S., Van Capellen, P., and van Bennekom, A. J. (2001). Processes controlling solubility of biogenic silica and pore water build-up of silicic acid in marine sediments. *Marine Chemistry* **73,** 333–352.

Duckworth, A. R. (2003). Effect of wound size on the growth and regeneration of two temperate subtidal sponges. *Journal of Experimental Marine Biology and Ecology* **287,** 139–153.

Duckworth, A., and Battershill, C. N. (2003). Sponge aquaculture for the production of biologically active metabolites: The influence of farming protocols and environment. *Aquaculture* **221,** 311–329.

Dyhrman, S. T., Ammerman, J. W., and van Mooy, B. A. S. (2007). Microbes and the marine phosphorus cycle. *Oceanography* **20,** 110–116.

Ehrlich, H., Krautter, M., Hanke, T., Simon, P., Knieb, C., Heinemann, S., and Worch, H. (2007). First evidence of the presence of chitin in skeletons of marine sponges. Part II. Glass sponges (Hexactinellida: Porifera). *Journal of Experimental Zoology. Part B, Molecular and Developmental Evolution* **308,** 473–483.

Elvin, D. W. (1976). Seasonal growth and reproduction of an intertidal sponge, *Haliclona permollis* (Bowerbank). *The Biological Bulletin* **151,** 108.

Erwin, P. M., and Thacker, R. W. (2007). Incidence and identity of photosynthetic symbionts in Caribbean coral reef sponge assemblages. *Journal of the Marine Biological Association of the United Kingdom* **87,** 1683–1692.

Ferrier-Pages, C., Leclercq, N., Jaubert, J., and Pelegri, S. (2000). Enhancement of pico- and nanoplankton growth by coral exudates. *Aquatic Microbial Ecology* **21,** 203–209.

FrØhlich, H., and Barthel, D. (1997). Silica uptake on the marine sponge *Halichondria panicea* in Kiel Bight. *Marine Biology* **128,** 115–125.

Frost, T. M. (1976). Sponge feeding: A review with a discussion of some continuing research. In "Aspects of Sponge Biology" (F. W. Harrison and R. R. Cowden, eds), pp. 283–298. Academic Press, New York.

Frost, T. M. (1978). In situ measurement of clearance rates for the freshwater sponge *Spongilla lacustris*. *Limnology and Oceanography* **23,** 1034–1039.

Frost, T. M. (1980). Selection in sponge feeding processes. In "Nutrition in the Lower Metazoa" (D. C. Smith and Y. Tiffon, eds), pp. 33–44. Pergamon Press, Oxford.

Garrabou, J., and Zabala, M. (2001). Growth dynamics in four Meditarranean demosponges. *Estuarine, Coastal and Shelf Science* **52,** 293–303.

Gast, G. J., Wiegman, S., Wieringa, E., van Duyl, F. C., and Bak, R. P. M. (1998). Bacteria in coral reef water types: Removal of cells, stimulation of growth and mineralization. *Marine Ecology Progess Series* **167,** 37–45.

Gatti, S., Brey, T., Mueller, W. E. G., Heilmayer, O., and Host, G. (2002). Oxygen microoptodes: A new tool for oxygen measurements in animal ecology. *Marine Biology* **140,** 1075–1085.

Geider, R. J., and La Roche, J. (2002). Redfield revisited: Variability of C: N: P in marine microalgae and its biochemical basis. *European Journal of Phycology* **37,** 1–17.

Genin, A., Paull, C. A., and Dillon, W. P. (1992). Anomalous abundances of deep-sea fauna on a rocky bottom exposed to strong currents. *Deep-Sea Research I* **39**, (2), 293–302.

Genin, A., Monismith, S. G., Reidenbach, M. A., Yahel, G., and Koseff, J. R. (2009). Intense benthic grazing of phytoplankton in a coral reef. *Limnology and Oceanography* **54**, 938–951.

Gerrodette, T., and Flechsig, A. O. (1979). Sediment-induced reduction in the pumping rate of the tropical sponge *Verongia lacunosa*. *Marine Biology* **55**, 103–110.

Gili, J.-M., and Coma, R. (1998). Benthic suspension feeders: Their paramount role in littoral marine food webs. *Trends in Ecology & Evolution* **13**, 316–321.

Granek, E. F., Compton, J. E., and Philips, D. L. (2009). Mangrove-exported nutrient incorporation by sessile coral reef invertebrates. *Ecosystems* **12**, 462–472.

Haas, A. F., Jantzen, C., Naumann, M. S., Iglesias-Prieto, R., and Wild, C. (2010). Organic matter release by the dominant primary producers in a Caribbean reef lagoon: Implication for in situ O2 availability. *Marine Ecology Progess Series* **409**, 27–39.

Hadas, E., Marie, D., Shpigel, M., and Ilan, M. (2006). Virus predation by sponges is a new nutrient-flow pathway in coral reef food webs. *Limnology and Oceanography* **51**, 1548–1550.

Hadas, E., Ilan, M., and Shpigel, M. (2008). Oxygen consumption by a coral reef sponge. *Journal of Experimental Biology* **211**, 2185–2190.

Hadas, E., Shpigel, M., and Ilan, M. (2009). Particulate organic matter as food source for a coral reef sponge. *Journal of Experimental Biology* **212**, 3643–3650.

Hanson, C. E., McLaughlin, M. J., Hyndes, G. A., and Strzelecki, J. (2009). Selective uptake of prokaryotic picoplankton by a marine sponge (*Callyspongia* sp.) within an oligotrophic coastal system. *Estuarine, Coastal and Shelf Science* **84**, 289–297.

Harrison, K. G. (2000). The role of increased marine silica input on paleo-pCO_2 levels. *Paleoceanography* **15**, 292–298.

Harriss, R. C. (1966). Biological buffering of oceanic silica. *Nature* **212**, 275–276.

Hartman, W. D. (1981). Form and distribution of silica in sponges. In "Silicon and Siliceous Structures in Biological Systems" (T. L. Simpson and B. E. Volcani, eds), pp. 453–493. Springer Verlag, New York.

Hatcher, A. I. (1994). Nitrogen and phosphorous turnover in some benthic marine invertebrates: implications for the use of C:N ratios to assess food quality. *Marine Biology* **121**, 161–166.

Hentschel, U., Fieseler, L., Wehrl, M., Gernert, C., Steinert, M., Hacker, J., and Horn, M. (2003). Microbial diversity of marine sponges. In "Molecular Marine Biology of Sponges" (W. E. G. Mueller, ed.), pp. 60–88. Springer Verlag, Heidelberg.

Hentschel, U., Usher, K. M., and Taylor, M. W. (2006). Marine sponges as microbial fermenters. *FEMS Microbial Ecology* **55**, 167–177.

Hinde, G. J., and Holmes, W. M. (1892). On the sponge remains in the Lower Tertiary strata near Oamaru, New Zealand. *Journal of the Linnean Society of London, Zoology* **24**, 177–262.

Hochmuth, T., Niederkruger, H., Gernert, C., Siegl, A., Taudien, S., Platzer, M., Crews, P., Hentschel, U., and Piel, J. (2010). Linking chemical and microbial diversity in marine sponges: Possible role for Poribacteria as producers of methyl-branched fatty acids. *ChemBioChem* **11**, 2572–2578.

Hock, C. W. (1940). Decomposition of chitin by marine bacteria. *Biological Bulletin, Marine Biological Laboratory, Woods Hole* **79**, 199–206.

Hoffmann, F., Larsen, O., Thiel, V., Rapp, H. T., Pape, T., Michaelis, W., and Reitner, J. (2005). An anaerobic world in sponges. *Geomicrobiology Journal* **22**, 1–10.

Hoffmann, F., Radax, R., Woebken, D., Holtappels, M., Lavik, G., Rapp, H. T., Schlappy, M.-L., Schleper, C., and Kuypers, M. M. (2009). Complex nitrogen cycling in the sponge *Geodia barretti*. *Environmental Microbiology* **11**, 2228–2243.

Hoppe, W. F. (1988). Growth, regeneration and predation in three species of large coral reef sponges. *Marine Ecology Progress Series* **50,** 117–125.

Hunt, S. (1970). Polysaccharide-Protein Complexes in Invertebrates. Academic Press, New York.

Imsiecke, G., Steffen, R., Custodio, M., Borojevic, R., and Müller, W. E. G. (1995). Formation of spicules by sclerocytes from the freshwater sponge *Ephydatia muelleri* in short-term cultures in vitro. *In Vitro Cellular & Developmental Biology. Animal* **31,** 528–535.

Jaeckle, W. B. (1995). Transport and metabolism of alanine and palmitic acid by field-collected larvae of *Tedania ignis* (Porifera, Demospongiae): Estimated consequences of limited label translocation. *Biological Bulletin, Marine Biological Laboratory, Woods Hole* **189,** 159–167.

Jiménez, E. (2011). Nutrient fluxes in marine sponges: Methodology, geographical variabilit and the role of associated microorganisms. PhD thesis. Universitat Politècnica de Catalunya, p 140.

Jiménez, E., and Ribes, M. (2007). Sponges as a source of dissolved organic nitrogen: Nitrification mediated by temperate sponges. *Limnology and Oceanography* **52,** 948–958.

Jørgensen, C. B., Larsen, P. S., and Riisgård, H. U. (1990). Effects of temperature on the mussel pump. *Marine Ecology Progress Series* **64,** 89–97.

Karchenko, T. A., and Lyashenko, A. V. (1986). Oxygen consumption of fresh-water sponges. *Hydrobiological Journal* **22,** 99–102.

Katamani, A. (1971). Physical and chemical characteristics of biogenous silica. *Marine Biology* **8,** 89–95.

Koltun, V. M. (1968). Spicules of sponges as an element of the bottoms sediments of the Antarctic. In "Symposium on Antarctic Oceanography," pp. 121–123. Scott Polar Research Institute, Cambridge.

Koopmans, M., and Wijffels, R. H. (2008). Seasonal growth rate of the sponge *Haliclona oculata* (Demospongiae: Haplosclerida). *Marine Biotechnology* **10,** 502–510.

Koopmans, M., Martens, D., and Wijffels, R. H. (2010). Growth efficiency and carbon balance for the sponge *Haliclona oculata. Marine Biotechnology* **12,** 340–349.

Kowalke, J. (2000). Ecology and energetics of two Antarctic sponges. *Journal of Experimental Marine Biology and Ecology* **247,** 85–97.

Krautter, M., Conway, K. W., Barrie, J. V., and Neuweiler, M. (2001). Discovery of a "living Dinosaur": Globally unique modern hexactinellid sponge reefs off British Columbia, Canada. *Facies* **44,** 265–282.

Lesser, M. P. (2006). Benthic-pelagic coupling on coral reefs: Feeding and growth of Caribbean sponges. *Journal of Experimental Marine Biology and Ecology* **328,** 277–288.

Lewin, J. C. (1961). The dissolution of silica from diatom walls. *Geochimica et Cosmochimica Acta* **21,** 182–198.

Leys, S. P. (2003). Comparative study of spiculogenesis in demosponge and hexactinellid larvae. *Microscopy Research and Technique* **62,** 300–311.

Leys, S. P., and Eerkes-Medrano, D. I. (2006). Feeding in a calcareous sponge: Particle uptake by pseudopodia. *Biological Bulletin, Marine Biological Laboratory, Woods Hole* **211,** 157–171.

Leys, S. P., and Lauzon, R. J. (1998). Hexactinellid sponge ecology: Growth rates and seasonality in deep water sponges. *Journal of Experimental Marine Biology and Ecology* **230,** 111–129.

Leys, S. P., Wilson, K., Holeton, C., Reiswig, H. M., Austin, W. C., and Tunnicliffe, V. (2004). Patterns of glass sponge (Porifera, Hexactinellida) distribution in coastal waters of British Columbia, Canada. *Marine Ecology Progress Series* **283,** 133–149.

Lopez-Legentil, S., Song, B., McMurray, S. E., and Pawlik, J. R. (2008). Bleaching and stress in coral reef ecosystems: *hsp70* expression by the giant barrel sponge *Xestospongia muta. Molecular Ecology* **17,** 1840–1849.

Mackie, G. O., and Singla, C. L. (1983). Studies on hexactinellid sponges. I. Histology of *Rhabdocalyptus dawsoni* (Lambe, 1873). *Philosophical Transactions of the Royal Society of London* **301,** 365–400.

Magnino, G., Sara, A., Lancioni, T., and Gaino, E. (1999). Endobionts of the coral reef sponge *Theonella swinhoei* (Porifera, Demospongiae). *Invertebrate Biology* **118,** 213–220.

Maldonado, M. (2006). The ecology of the sponge larva. *Canadian Journal of Zoology (Journal Canadien de Zoologie)* **84,** 175–194.

Maldonado, M. (2007). Intergenerational transmission of symbiotic bacteria in oviparous and viviparous demosponges, with emphasis on intracytoplasmically-compartmented bacterial types. *Journal of the Marine Biological Association of the United Kingdom* **87,** 1701–1713.

Maldonado, M. (2009). Embryonic development of verongid demosponges supports independent acquisition of spongin skeletons as alternative to the siliceous skeleton of sponges. *Biological Journal of the Linnean Society* **97,** 427–447.

Maldonado, M., and Riesgo, A. (2007). Intra-epithelial spicules in a homosclerophorid sponge. *Cell and Tissue Research* **328,** 639–650.

Maldonado, M., and Young, C. M. (1996). Bathymetric patterns of sponge distribution on the Bahamian slope. *Deep-Sea Research I* **43,** 897–915.

Maldonado, M., and Young, C. M. (1998). Limits on the bathymetric distribution of keratose sponges: A field test in deep water. *Marine Ecology Progress Series* **174,** 123–139.

Maldonado, M., and Young, C. M. (1999). Effects of the duration of the larval life on post-larval stages of the demosponge *Sigmadocia caerulea*. *Journal of Experimental Marine Biology and Ecology* **232,** 9–21.

Maldonado, M., Carmona, M. C., Uriz, M. J., and Cruzado, A. (1999). Decline in Mesozoic reef-building sponges explained by silicon limitation. *Nature* **401,** 785–788.

Maldonado, M., Carmona, M. C., Velásquez, Z., Puig, A., Cruzado, A., López, A., and Young, C. M. (2005). Siliceous sponges as a silicon sink: An overlooked aspect of the benthopelagic coupling in the marine silicon cycle. *Limnology and Oceanography* **50,** 799–809.

Maldonado, M., Riesgo, A., Bucci, A., and Rützler, K. (2010a). Revisiting silicon budgets at a tropical continental shelf: Silica standing stocks in sponges surpass those in diatoms. *Limnology and Oceanography* **55,** 2001–2010.

Maldonado, M., Sánchez-Tocino, L., and Navarro, C. (2010b). Recurrent disease outbreaks in corneous demosponges of the genus *Ircinia*: Epidemic incidence and defense mechanisms. *Marine Biology* **157,** 1577–1590.

Maldonado, M., Zhang, X., Cao, X., Xue, L., Cao, H., and Zhang, W. (2010c). Selective feeding by sponges on pathogenic microbes: A reassessment of potential for abatement of microbial pollution. *Marine Ecology Progress Series* **403,** 75–89.

Maldonado, M., Navarro, L., Grasa, A., González, A., and Vaquerizo, I. (2011). Silicon uptake by sponges: A twist to understanding nutrient cycling on continental margins. *Scientific Reports* **1,** 1–8.

Maldonado, M., Heng, C., Xupeng, C., Yuefan, S., Yi, Q., and Zhang, W. (2012). Experimental silicon demand by the sponge *Hymeniacidon perlevis* reveals chronic limitation in field populations. *Hydrobiologia* **687,** 251–257.

McClintock, J. B., Amsler, C. D., Baker, B. J., and van Soest, R. M. W. (2005). Ecology of Antarctic marine sponges: An overview. *Integrative and Comparative Biology* **45,** 359–368.

McMurray, S. E., Blum, J. E., and Pawlik, J. R. (2008). Redwood of the reef: Growth and age of the giant barrel sponge *Xestospongia muta* in the Florida Keys. *Marine Biology* **155,** 159–171.

Mohamed, N. M., Colman, A. S., Tal, Y., and Hill, R. T. (2008). Diversity and expression of nitrogen fixation genes in bacterial symbionts of marine sponges. *Environmental Microbiology* **10,** 2910–2921.

Mohamed, N. M., Saito, K., Tal, Y., and Hill, R. T. (2010). Diversity of aerobic and anaerobic ammonia-oxidizing bacteria in marine sponges. *ISME Journal* **4,** 38–48.

Monismith, S. G., Davi, K. A., Shellenbarger, G. G., Hench, J. L., Nidzieko, N. J., Santoro, A. E., Reidenbach, M. A., Rosman, J. H., Holtzman, R., Martens, C. S., Lindquist, N. L., Southwell, M. W., and Genin, A. (2010). Flow effects on benthic grazing on phytoplankton by a Caribbean reef. *Limnology and Oceanography* **55,** 1881–1892.

Mostler, H. (1976). Poriferenspiculae der alpinen Trias. *Geologische-Paläontologische Mitteilungen Innsbruck* **6,** 1–42.

Moya, A., Pereto, J., Gil, R., and Latorre, A. (2008). Learning how to live together: Genomic insights into prokaryote-animal symbioses. *Nature Reviews. Genetics* **9,** 218–229.

Müller, W. E. G., Rothenberger, M., Boreiko, A., Tremel, W., Reiber, A., and Schröder, H. C. (2005). Formation of siliceous spicules in the marine demosponge *Suberites domuncula. Cell and Tissue Research* **321,** 285–297.

Müller, W. E. G., Belikov, S. I., Tremel, W., Perry, C. C., Gieskes, W. W. C., Boreiko, A., and Schröder, H. C. (2006). Siliceous spicules in marine demosponges (example *Suberites domuncula*). *Micron* **37,** 107–120.

Müller, W., Wang, X., Cui, F.-Z., Jochum, K., Tremel, W., Bill, J., Schröder, H., Natalio, F., Schloßmacher, U., and Wiens, M. (2009). Sponge spicules as blueprints for the biofabrication of inorganic–organic composites and biomaterials. *Applied Microbiology and Biotechnology* **83,** 397–413.

Murr, M. M., and Morse, D. E. (2005). Fractal intermediates in the self-assembly of silicatein filaments. *Proceedings of the National Academy of Sciences of the United States of America* **102,** 11657–11662.

Naumann, M., Haas, A. F., Struck, U., Mayr, C., el-Zibdah, M., and Wild, C. (2010). Organic matter release by dominant hermatypic corals of the Northern Red Sea. *Coral Reefs* **29,** 649–659.

Nelson, D. M., Tréguer, P., Brzezinski, M. A., Leynaert, A., and Quéguiner, B. (1995). Production and dissolution of biogenic silica in the ocean: Revised global estimates, comparison with regional data and relationship to biogenic sedimentation. *Global Biogeochemical Cycles* **9,** 359–372.

Nelson, C. E., Alldredge, A. L., McCliment, E. A., Amaral-Zettler, L. A., and Carlson, C. A. (2011). Depleted dissolved organic carbon and distinct bacterial communities in the water column of a rapid-flushing coral reef ecosystem. *ISME Journal* **5,** 1374–1387.

Newell, R. I. E., Hilbish, T. J., Koehn, R. K., and Newell, C. J. (1982). Temporal variation in the reproductive cycle of *Mytilus edulis* (Bivalvia, Mytildae) from localities on the East Coast of North America. *The Biological Bulletin* **162,** 299–310.

Nielsen, A. M., Eriksen, N. T., Iversen, J. J. L., and Riisgård, H. U. (1995). Feeding, growth and respiration in the polychaetes Nereis diversicolor (facultative filter-feeder) and N. virens (omnivorous)-a comparative study. *Marine Ecology Progress Series* **125,** 149–158.

Nishijima, M., Lindsay, D. J., Hata, J., Nakamura, A., Kasai, H., Ise, Y., Fisher, C. R., Fujiwara, Y., Kawato, M., and Maruyama, T. (2010). Association of thioautotrophic bacteria with deep-sea sponges. *Marine Biotechnology* **12,** 253–260.

Osinga, R., Tramper, J., and Wijfels, R. H. (1999). Cultivation of marine sponges. *Marine Biotechnology* **1,** 509–532.

Pansini, M., and Pronzato, R. (1990). Observations on the dynamics of a Mediterranean sponge community. In "New Perspectives in Sponge Biology" (K. Rützler, ed.), pp. 404–415. Smithsonian Institution Press, Washington, DC/London.

Petersen, J. K., and Riisgård, H. U. (1992). Filtration capacity of the ascidian *Ciona intestinalis* and its grazing impact in a shallow fjord. *Marine Ecology Progess Series* **88,** 9–17.

Petersen, J. K., Schou, O., and Thor, P. (1995). Growth and energetics in the ascidian Ciona intestinalis. *Marine Ecology Progress Series* **120,** 175–184.

Petersen, J. K., Bougrier, S., Smaal, A. C., Garen, P., Robert, S., Larsen, J. E. N., and Brummelhuis, E. (2004). Intercalibration of mussel Mytilus edulis clearance rate measurements. *Marine Ecology Progess Series* **267,** 187–194.

Pile, A. J. (1997). Finding Reiswig's missing carbon: quantification of sponge feeding using dual-beam flow cytometry. Proceedings of the 8th International Coral Reef Symposium, vol. 2, pp. 1403–1410.

Pile, A. J. (1999). Resource partitioning by caribbean coral reef sponges: Is there enough food for everyone? *Memoirs of the Queensland Museum* **44,** 457–461.

Pile, A. J., and Young, C. M. (2006). The natural diet of a hexactinellid sponge: Benthic-pelagic coupling in a deep-sea microbial food web. *Deep Sea Research Part I: Oceanographic Research Papers* **53,** 1148–1156.

Pile, A. J., Patterson, M. R., and Witman, J. D. (1996). In situ grazing on plankton < 10 μm by the boreal sponge *Mycale lingua. Marine Ecology Progess Series* **141,** 95–102.

Pile, A. J., Pattersen, M. R., Savarese, M., Chernykh, V. I., and Fialkov, V. A. (1997). Trophic effects of sponge feeding within Lake Baikal's littoral zone. II.Sponge abundance, diet, feeding efficiency, and carbon flux rates. *Limnology and Oceanography* **42,** 178–184.

Pile, A. J., Grant, A., Hinde, R., and Borowitzka, M. A. (2003). Heterotrophy on ultraplankton communities is an important source of nitrogen for a sponge-rhodophyte symbiosis. *Journal of Experimental Biology* **206,** 4533–4538.

Pisera, A. (1997). Upper Jurasic siliceous sponges from Swabian Alb: Taxonomy and paleoecology. *Palaeontologica Polonica* **57,** 1–216.

Planavsky, N. J., Rouxel, O. J., Bekker, A., Lalonde, S. V., Konhauser, K. O., Reinhard, C. T., and Lyons, T. W. (2010). The evolution of the marine phosphate reservoir. *Nature* **467,** 1088–1090.

Queric, N. V., Arrietta, J. M., Soltwedel, T., and Arntz, W. E. (2008). Prokaryotic community dynamics in the sedimentary microenvironment of the demosponge *Tentorium semisuberites* from deep Arctic waters. *Marine Ecology Progess Series* **370,** 87–95.

Ragueneau, O., Tréguer, P., Leynaert, A., Anderson, R. F., Brzezinski, M. A., DeMaster, D. J., Dugdale, R. C., Dymond, J., Fischer, G., François, R., Heinze, C., Maier-Reimer, E., Martin-Jézéquel, V., Nelson, D. M., and Quéguiner, B. (2000). A review of the Si cycle in the modern ocean: Recent progress and missing gaps in the application of biogenic opal as paleoproductivity proxy. *Global and Planetary Change* **26,** 317–365.

Ragueneau, O., Conley, D. J., DeMaster, D. J., Dürr, H. H., and Dittert, N. (2010). Biogeochemical transformations of silicon along the land–ocean continuum and implications for the global carbon Cycle. In "Carbon and Nutrient Fluxes in Continental Margins" (K.-K. Liu, L. Atkinson, R. Quiñones and L. Talaue-McManus, eds), pp. 515–527. Springer, Berlin, Heidelberg.

Reincke, T., and Barthel, D. (1997). Silica uptake kinetics of *Halichondria panicea* in Kiel Bight. *Marine Biology* **129,** 591–593.

Reiswig, H. M. (1971a). *In situ* pumping activities of tropical Demospongiae. *Marine Biology* **9,** 38–50.

Reiswig, H. M. (1971b). Particle feeding in natural populations of three marine demosponges. *Biological Bulletin, Marine Biological Laboratory, Woods Hole* **141,** 568–591.

Reiswig, H. M. (1973). Population dynamics of three Jamaican Demospongiae. *Bulletin of Marine Science* **23,** 191–226.

Reiswig, H. M. (1974). Water transport, respiration and energetics of three tropical marine sponges. *Journal of Experimental Marine Biology and Ecology* **14,** 231–249.

Reiswig, H. M. (1981). Partial carbon and energy budgets of the bacteriosponge *Verongia fistularis* (Porifera: Demopongiae) in Barbados. *Marine Ecology* **2,** 273–293.

Reiswig, H. M. (1990). In situ feeding in two shallow-water hexactinellid sponges. In "New Perspectives in Sponge Biology" (K. Rützler, ed.), pp. 504–510. Smithsonian Institution Press, Washington, D.C./London.

Ribes, M., Coma, R., and Gili, J. M. (1999a). Seasonal variation of particulate organic carbon, dissolved organic carbon and the contribution of microbial communities to the live particulate organic carbon in a shallow near-bottom ecosystem at the Northwestern Mediterranean Sea. *Journal of Plankton Research* **21,** 1077–1100.
Ribes, M., Coma, R., and Gili, J. M. (1999b). Natural diet and grazing rate of the temperate sponge *Dysidea avara* (Demospongiae, Dendroceratida) throughout an annual cycle. *Marine Ecology Progress Series* **176,** 179–190.
Ribes, M., Coma, R., Gili, J. M., Svoboda, A., Julia, A., and Parera, J. (2000). A 'semi-closed' recirculating system for the in situ study of feeding and respiration of benthic suspension feeders. *Scientia Marina* **64,** 265–275.
Ribes, M., Coma, R., Atkinson, M. J., and Kinzie, R. A. III (2003). Particle removal by coral reef communities: Picoplancton is a major source of nitrogen. *Marine Ecology Progess Series* **257,** 12–23.
Ribes, M., Coma, R., Atkinson, M. J., and Kinzie, R. A. III (2005). Sponges and ascidians control removal of particulate organic nitrogen from coral reef waters. *Limnology and Oceanography* **50,** 1480–1489.
Ribes, M., Jimenez, E., Yahel, G., Lopez-Sendino, P., Diez, B., Massana, R., Sharp, J. H., and Coma, R. (2012). Functional convergence of microbes associated with temperate marine sponges. *Environmental Microbiology* (in press). DOI: 10.1111/j.1462-2920.2012.02701.x.
Rice, A. L., Thurston, M. H., and New, A. L. (1990). Dense aggregations of a hexactinellid sponge, *Pheronema carpenteri*, in the Porcupine Seabight (northeast Atlantic Ocean), and possible causes. *Progress in Oceanography* **24,** 179–196.
Richardson, M. D., and Young, D. K. (1987). Abyssal benthos of the Venezuela Basin, Caribbean Sea: Standing stock considerations. *Deep Sea Research Part I: Oceanographic Research Papers* **34,** 145–164.
Richter, C., and Wunsch, M. (1999). Cavity-dwelling suspension feeders in coral reefs—A new link in reef trophodynamics. *Marine Ecology Progess Series* **188,** 105–116.
Richter, C., Wunsch, M., Rasheed, M., Kötter, I., and Badran, M. I. (2001). Endoscopic exploration of Red Sea coral reefs reveals dense populations of cavity-dwelling sponges. *Nature* **413,** 726–730.
Riisgård, H. U. (2001). On measurement of filtration rates in bivalves—The stony road to reliable data: Review and interpretation. *Marine Ecology Progress Series* **211,** 275–291.
Riisgård, H. U. (2004). Intercalibration of methods for measurement of bivalve filtration rates—A turning point. *Marine Ecology Progress Series* **276,** 307–308.
Riisgård, H. U., Thomassen, S., Jacobsen, H., Weeks, J. M., and Larsen, P. S. (1993). Suspension feeding in marine sponges *Halichondria panicea* and *Haliclona urceolus*: Effects of temperature on filtration rate and energy cost of pumping. *Marine Ecology Progress Series* **96,** 177–188.
Roditi, H. A., and Fisher, N. S. (1999). Rates and routes of trace element uptake in zebra mussels. *Limnology and Oceanography* **44,** 1730–1749.
Roy, S. O., Packard, T. T., Berdalet, E., and St-Amand, L. (1999). Impact of acetate, pyruvate, and physiological state on respiration and respiratory quotients in *Pseudomonas nautica*. *Aquatic Microbial Ecology* **17,** 105–110.
Rützler, K. (1970). Spatial competition among Porifera: Solution by epizoism. *Oecologia* **5,** 85–95.
Rützler, K. (1990). Associations between Caribbean sponges and photosynthetic organisms. In "New Perspectives in Sponge Biology," pp. 455–466. Smithsonian Institution Press, Washington, D.C..
Rützler, K., and Macintyre, I. G. (1978). Siliceous sponge spicules in coral reefs sediments. *Marine Biology* **49,** 147–159.

Sabarathnam, B., Manilal, A., Sujith, S., Kiran, G. S., Selvin, J., Thomas, A., and Ravji, R. (2010). Role of sponge associated actinomycetes in the marine phosphorous biogeochemical cycle. *American-Eurasian Journal of Agriculture and Environmental Sciences* **8,** 253–256.

Sarà, M. (1970). Competition and cooperation in sponge populations. In "The Biology of Porifera" (W. G. Fry, ed.), pp. 273–284. Academic Press, London.

Sarmiento, J., and Gruber, N. (2006). Ocean Biogeochemical Dynamics. Princeton University Press, Princeton.

Savarese, M., Patterson, M. R., Chernykh, V. I., and Fialkov, V. A. (1997). Trophic effects of sponge feeding within Lake Baikal's littoral zone. I. In situ pumping rates. *Limnology and Oceanography* **42,** 171–178.

Scheffers, S. R., Nieuwland, G., Bak, R. P. M., and van Duyl, F. C. (2004). Removal of bacteria and nutrient dynamics within the coral reef framework of Curaçao (Netherlands Antilles). *Coral Reefs* **23,** 413–422.

Scheffers, S. R., Bak, R. P. M., and van Duyl, F. C. (2005). Why is bacterioplankton growth in coral reef framework cavities enhanced? *Marine Ecology Progress Series* **299,** 89–99.

Schlappy, M. L., Hoffmann, F., Roy, H., Wijffels, R. H., Mendola, D., Sidri, M., and de Beer, D. (2007). Oxygen dynamics and flow patterns of *Dysidea avara* (Porifera: Demospongiae). *Journal of the Marine Biological Association of the United Kingdom* **87,** 1677–1682.

Schlappy, M. L., Schottner, S. I., Lavik, G., Kuypers, M. M. M., de Beer, D., and Hoffmann, F. (2010). Evidence of nitrification and denitrification in high and low microbial abundance sponges. *Marine Biology* **157,** 593–602.

Schloßmacher, U., Wiens, M., Schröder, H. C., Wang, X., Jochum, K. P., and Müller, W. E. G. (2011). Silintaphin-1—Interaction with silicatein during structure-guiding bio-silica formation. *FEBS Journal* **278,** 1145–1155.

Schmidt, I. (1970). Phagocytose et pincytose chez les *Spongillidae. Etude in vivo de l'ingestion de bactéries et de protéines marquées à l'aide d'un colorant fluorescent en lumière ultra-violette. Zeitschrift für Vergleichende Physiologie* **66,** 398–420.

Schmitt, S., Wehrl, M., Bayer, K., Siegl, A., and Hentschel, U. (2007). Marine sponges as models for commensal microbe-host interactions. *Symbiosis* **44,** 43–50.

Schönberg, C. H. L., De Beer, D., and Lawton, A. (2005). Oxygen microsensor studies on zooxanthellate clionaid sponges from the Costa Brava, Mediterranean Sea. *Journal of Phycology* **41,** 774–779.

Schröder, H. C., Boreiko, A., Korzhev, M., Tahir, M. N., Tremel, W., Eckert, C., Ushijima, H., Müller, I. M., and Müller, W. E. G. (2006). Co-expression and functional interaction of silicatein with galectin: Matrix-guided formation of siliceous spicules in the marine demosponge *Suberites domuncula. Journal of Biological Chemistry* **281,** 12001–12009.

Schröder, H. C., Wang, X., Tremel, W., Ushijima, H., and Muller, W. E. G. (2008). Biofabrication of biosilica-glass by living organisms. *Natural Product Reports* **25,** 455–474.

Shieh, W. Y., and Lin, Y. M. (1994). Association of heterotrophic nitrogen-fixing bacteria with a marine sponge of *Halichondria* sp.. *Bulletin of Marine Science* **54,** 557–564.

Shimizu, K., Cha, J. N., Stucky, G. D., and Morse, D. E. (1998). Silicatein alpha: Cathepsin L-like protein in sponge biosilica. *Proceedings of the National Academy of Sciences of the United States of America* **95,** 6234–6238.

Siegl, A., Bayer, K., Kozytska, S., Hentschel, U., and Schmitt, S. (2008). Sponges and microbes-new frontiers in ancient symbiosis. *Vie et Milieu* **58,** 165–174.

Siegl, A., Kamke, J., Hochmuth, T., Piel, J., Richter, M., Liang, C., Dandekar, T., and Hentschel, U. (2011). Single-cell genomics reveals the lifestyle of Poribacteria, a candidate phylum symbiotically associated with marine sponges. *ISME Journal* **5,** 61–70.

Siever, R. (1991). Silica in the oceans: Biological-geochemical interplay. In "Scientists on Gaia" (S. H. Scheneider and P. J. Boston, eds), pp. 287–295. MIT Press, Cambridge, MA.

Simpson, T. L. (1978). The biology of the marine sponge Microciona prolifera (Ellis and Solander). III. Spicule secretion and the effect of temperature on spicule size. *Journal of Experimental Marine Biology and Ecology* **35,** 31–42.
Simpson, T. L. (1984). The Cell Biology of Sponges. Springer Verlag, New York.
Simpson, T. L., and Vaccaro, C. A. (1974). An ultrastructural study of silica deposition in the freshwater sponge Spongilla lacustris. *Journal of Ultrastructure Research* **47,** 296–309.
Southwell, M. W. (2007). Sponge impacts on coral reef nitrogen cycling, Key Largo, Florida. (Ph.D. thesis, Univ. of North Carolina at Chapel Hill).
Southwell, M. W., Popp, B. N., and Martens, C. S. (2008a). Nitrification controls on fluxes and isotopic composition of nitrate from Florida Keys sponges. *Marine Chemistry* **108,** 96–108.
Southwell, M. W., Weisz, J. B., Martens, C. S., and Lindquist, N. (2008b). In situ fluxes of dissolved inorganic nitrogen from the sponge community on Conch Reef, Key Largo, Florida. *Limnology and Oceanography* **53,** 986–996.
Steger, D., Ettinger-Epstein, P., Whalan, S., Hentschel, U., de Nys, R., Wagner, M., and Taylor, M. W. (2008). Diversity and mode of transmission of ammonia-oxidizing archaea in marine sponges. *Environmental Microbiology* **10,** 1087–1094.
Stephens, G. C., and Schinske, R. A. (1961). Uptake of amino acids by marine invertebrates. *Limnology and Oceanography* **6,** 175–181.
Stone, A. R. (1970a). Seasonal variation of spicule size in *Hymeniacidon perleve*. *Journal of the Marine Biological Association of the United Kingdom* **50,** 343–348.
Stone, A. R. (1970b). Growth and reproduction of *Hymeniacidon perleve* (Montagu) in Langstone Harbour, Hampshire. *Journal of Zoology* **161,** 443–459.
Stuart, V., and Klumpp, D. W. (1984). Evidence for food-resource partitioning by kelp-bed filter feeders. *Marine Ecology Progress Series* **16,** 27–37.
Tanaka, K. (2002). Growth dynamics and mortality of the intertidal encrusting sponge *Halichondria okadai* (Demospongiae, Halichondrida). *Marine Biology* **140,** 383–389.
Taylor, M. W., Schupp, P. J., Dahllof, I., Kjelleberg, S., and Steinberg, P. D. (2004). Host specificity in marine sponge-associated bacteria, and potential implications for marine microbial diversity. *Environmental Microbiology* **6,** 121–130.
Taylor, M. W., Radax, R., Steger, D., and Wagner, M. (2007). Sponge-associated microorganisms: Evolution, ecology, and biotechnological potential. *Microbiology and Molecular Biology Reviews* **71,** 295–347.
Thacker, R. W. (2005). Impacts of shading on sponge-cyanobacteria symbioses: A comparison between host-specific and generalist associations. *Integrative and Comparative Biology* **45,** 369–376.
Thacker, R. W., Becerro, M. A., Lumbang, W. A., and Paul, V. J. (1998). Allelopathic interactions between sponges on a tropical reef. *Ecology* **79,** 1740–1750.
Thamatrakoln, K., and Hildebrand, M. (2008). Silicon uptake in diatoms revisited: A model for saturable and nonsaturable uptake kinetics and the role of silicon transporters. *Plant Physiology* **146,** 1397–1407.
Thomassen, S., and Riisgård, H. U. (1995). Growth and energetics of the sponge *Halichondria panicea*. *Marine Ecology Progress Series* **128,** 239–246.
Topçu, N., Pérez, T., Grégori, G., and Harmelin-Vivien, M. (2010). In situ investigation of *Spongia officinalis* (Demospongiae) particle feeding: Coupling flow cytometry and stable isotope analysis. *Journal of Experimental Marine Biology and Ecology* **389,** 61–69.
Torreton, J. P., Talbot, V., and Garcia, N. (2000). Nutrient stimulation of bacterioplankton growth in Tuamotu atoll lagoons. *Aquatic Microbial Ecology* **21,** 125–137.
Tréguer, P., Nelson, D. M., Bennekom, A. J. V., DeMaster, D. J., Leynaert, A., and Quéguiner, B. (1995). The silica balance in the world ocean: A reestimate. *Science* **268,** 375–379.

Trussell, G. C., Lesser, M. P., Patterson, M. R., and Genovese, S. J. (2006). Depth-specific differences in growth of thr reef sponge *Callyspongia vaginalis*: Role of bottom-up effects. *Marine Ecology Progress Series* **323,** 149–158.
Turon, X., Galera, J., and Uriz, M. J. (1997). Clearance rates and aquiferous systems in two sponges with contrasting life-history strategies. *The Journal of Experimental Zoology* **278,** 22–36.
Turon, X., Tarjuelo, I., and Uriz, M. J. (1998). Growth dynamics and mortality of the encrusting sponge *Crambe crambe* (Poecilosclerida) in two contrasting habitats: Correlation with population structure and investment in defence. *Functional Ecology* **12,** 631–639.
Tyrreell, T. (1999). The relative influences of nitrogen and phosphorus on oceanic primary production. *Nature* **400,** 525–531.
Uriz, M. J. (2006). Mineral skeletogenesis in sponges. *Canadian Journal of Zoology-Revue Canadienne De Zoologie* **84,** 322–356.
Uriz, M. J., Turon, X., and Becerro, M. A. (2000). Silica deposition in demosponges: Spiculogenesis in *Crambe crambe. Cell and Tissue Research* **301,** 299–309.
Uriz, M.-J., Turon, X., Becerro, M. A., and Agell, G. (2003). Siliceous spicules and skeleton frameworks in sponges: Origin, diversity, ultrastructural patterns, and biological functions. *Microscopy Research and Technique* **62,** 279–299.
Usher, K. M. (2008). The ecology and phylogeny of cyanobacterial symbionts in sponges. *Marine Ecology Progress Series* **29,** 178–192.
Vacelet, J. (1967). Les cellules a inclusions de l'éponge cornée *Verongia cavernicola* Vacelet. *Journal de Microscopie* **6,** 237–240.
Vacelet, J. (1988). Indications de profondeur donnés par les Spongiaires dans les milieux benthiques actuels. *Géologie Méditerranéenne* **15,** 13–26.
Vacelet, J., and Donadey, C. (1977). Electron Microscope study of the association between some sponges and bacteria. *Journal of Experimental Marine Biology and Ecology* **30,** 301–314.
Vacelet, E., and Donadey, C. (1987). A new species of Halisarca (Porifera, Demospongiae) from the Caribbean, with remarks on the cytology and affinities of the genus. (W. C. Jones, ed.). European Contributions to the Taxonomy of Sponges, Vol. 1. Sherkin Island Marine Station, Midleton, Co. Cork, Ireland.
Vacelet, J., and Duport, É. (2004). Prey capture and digestion in the carnivorous sponge *Asbestopluma hypogea* (Porifera: Demospongiae). *Zoomorphology* **123,** 179–190.
Vacelet, J., Boury-Esnault, N., Fiala-Medioni, A., and Fisher, C. R. (1995). A methanotrophic carnivorous sponge. *Nature* **377,** 296.
Vacelet, J., Fiala-Medioni, A., Fisher, C. R., and Boury-Esnault, N. (1996). Symbiosis between a methane-oxidizing bacteria and a deep-sea carnivoroous cladorhizid sponge. *Marine Ecology Progress Series* **145,** 77–85.
Vacelet, J., Kelly, M., and Schlacherhoenlinger, M. (2009). Two new species of *Chondrocladia* (Demospongiae: Cladorhizidae) with a new spicule type from the deep south Pacific, and a discussion of the genus *Meliiderma. Zootaxa* **2073,** 57–68.
Vahl, O. (1984). The relationship between specific dynamic action (SDA) and growth in the common starfish, Asterias rubens L. *Oecologia* **61,** 122–125.
Van de Vyver, G., Vray, B., Belaouane, S., and Toussaint, D. (1990). Efficiency and selectivity of microorganism retention by *Ephydatia fluviatilis*. In "New Perspectives in Sponge Biology" (K. Rützler, ed.), pp. 511–515. Smithsonian Institution Press, Washington D.C./London.
Van Duyl, F. C., and Gast, G. J. (2001). Linkage of small-scale spatial variations in DOC, inorganic nutrients and bacterioplankton growth with different coral reef water types. *Aquatic Microbial Ecology* **24,** 17–26.
Van Duyl, F. C., Gast, G. J., Steinhoff, W., Kloff, S., Veldhuis, M. J. W., and Bak, R. P. M. (2002). Factors influencing the short-term variation in phytoplankton composition and biomass in coral reef waters. *Coral Reefs* **21,** 293–306.

Van Duyl, F. C., Hegeman, J., Hoogstraten, A., and Maier, C. (2008). Dissolved carbon fixation by sponge–microbe consortia of deep water coral mounds in the northeastern Atlantic Ocean. *Marine Ecology Progress Series* **358,** 137–150.

Van Duyl, F. C., Moodley, L., Nieuwland, G., Van Ijzerloo, L., Van Soest, R. W. M., Houtekamer, M., Meesters, E. H., and Middelburg, J. J. (2011). Coral cavity sponges depend on reef-derived food resources: Stable isotope and fatty acid constraints. *Marine Biology* **158,** 1653–1666.

Van Wagoner, N. A., Mudie, P. J., Cole, F. E., and Daborn, G. (1989). Siliceous sponge communities, biological zonation and Recent sea-level change on the Arctic margin: Ice Islands results. *Canadian Journal of Earth Sciences/Revue Canadienne des Sciences de la Terre* **26,** 2341–2355.

Vogel, S. (1977). Current-induced flow through living sponges in nature. *Proceedings of the National Academy of Sciences of the United States of America* **74,** 2069–2071.

Vogel, S. (1978). Organisms that capture currents. *Scientific American* **239,** 128–135.

Wafar, M., Wafar, S., and David, J. J. (1990). Nitrification in reef corals. *Linmology & Oceanography* **35,** 725–730.

Walters, K. D., and Pawlik, J. R. (2005). Is there a trade-off between wound-healing and chemical defenses among Caribbean reef sponges? *Integrative and Comparative Biology* **45,** 352–358.

Wang, X.-H., Wang, S.-F., Gan, L., Schloßmacher, U., Zhou, F., Jochum, K. P., Wiens, M., Schröder, H. C., and Müller, W. E. G. (2011a). Advances in research on siliceous sponge spicules: Novel insight into the understanding of biomineralization mechanisms and bionic applications. (in Chinese). *Acta Geoscientica Sinica* **32,** 129–141.

Wang, X., Wiens, M., Schröder, H. C., Schloßmacher, U., Pisignano, D., Jochum, K. P., and Müller, W. E. G. (2011b). Evagination of cells controls bio-silica formation and maturation during spicule formation in sponges. *PLoS One* **6,** e20523.

Ward, B. B., Capone, D. G., and Zehr, J. P. (2007). What's new in the nitrogen cycle? *Oceanography* **20,** 101–109.

Wehrl, M., Steinert, M., and Hentschel, U. (2007). Bacterial Uptake by the Marine Sponge *Aplysina aerophoba. Microbial Ecology* **53,** 355–365.

Weiner, S., Traub, W., and Lowenstam, H. A. (1983). Organic matrix in calcified exoskeletons. In "Biomineralization and Biological Metal Accumulation" (P. Westbroek and E. W. de Jong, eds), pp. 205–224. Reidel, Amsterdam.

Weissenfels, N. (1976). Bau und Funktion des süßwasserschwamms *Ephydatia fluviatilis* L. (Porifera). III. Nahrungsaufnahme, verdauung und defäkation. *Zoomorphologie* **85,** 73–88.

Weisz, J.B. (2006). Measuring impacts of associated microbial communities on Caribbean reef sponges: searching for symbiosis. PhD thesis. University of North Carolina at Chapell Hill. pp 165.

Weisz, J. B., Lindquist, N., and Martens, C. S. (2008). Do associated microbial abundances impact marine demosponge pumping rates and tissue densities? *Oecologia* **155,** 367–376.

Weisz, J. B., Massaro, A. J., Ramsby, B. D., and Hill, M. S. (2010). Zooxanthellar symbionts shape host sponge trophic status through translocation of carbon. *The Biological Bulletin* **219,** 189–197.

Wheeler, P. A., and Kirchman, D. L. (1986). Utilization of inorganic and organic nitrogen by bacteria in marine systems. *Limnology and Oceanography* **31,** 998–1009.

Whitney, F., Conway, K., Thomson, R., Barrie, V., Krautter, M., and Mungov, G. (2005). Oceanographic habitat of sponge reefs on the Western Canadian Continental Shelf. *Continental Shelf Research* **25,** 211–226.

Widdows, J., and Johnson, D. (1988). Physiological energetics of Mytilus edulis: Scope for Growth. *Marine Ecology Progress Series* **46,** 113–121.

Wiedenmayer, F. (1994). Contributions to the knowledge of post-Paleozoic neritic and archibental sponges (Porifera). *Schweizerische Paläontologidche Abhandlungen* **116,** 1–147.

Wiens, M., Bausen, M., Natalio, F., Link, T., Schlossmacher, U., and Müller, W. E. G. (2009). The role of the silicatein-[alpha] interactor silintaphin-1 in biomimetic biomineralization. *Biomaterials* **30,** 1648–1656.

Wiens, M., Schröder, H.-C., Wang, X., Link, T., Steindorf, D., and Müller, W. E. G. (2011). Isolation of the Silicatein-α Interactor Silintaphin-2 by a Novel Solid-Phase Pull-Down Assay. *Biochemistry* **50,** 1981–1990.

Wild, C., Hüttel, M., Klüter, A., Kremb, S. G., Rasheed, M. Y. M., and Jørgensen, B. B. (2004). Coral mucus functions as an energy carrier and particle trap in the reef ecosystem. *Nature* **428,** 66–70.

Wild, C., Mayr, C., Wehrmann, L., Schottner, S. I., Naumann, M., Hoffmann, F., and Rapp, H. T. (2008). Organic matter release by cold water corals and its implication for fauna-microbe interaction. *Marine Ecology Progress Series* **372,** 67–75.

Wilkinson, C. R. (1978). Microbial Associations in Sponges. 2. Numerical-analysis of sponge and water bacterial populations. *Marine Biology* **49,** 169–176.

Wilkinson, C. R. (1979). Nutrient translocation from symbiotic cyanobacteria to coral reef sponges. (C. Lèvi and N. Boury-Esnault, eds), Biologie des spongiaires. Colloques Internationaux du CNRS, Vol. 291, pp. 373–380. Editions du CNRS, Paris.

Wilkinson, C. R. (1983). Net primary productivty in coral reef sponges. *Science* **219,** 410–412.

Wilkinson, C. R., and Cheshire, A. C. (1990). Comparison of Sponge populations across the Barrier Reefs of Australia and Belize: Evidence for higher productivity in the Caribbean. *Marine Ecology Progress Series* **67,** 285–294.

Wilkinson, C. R., and Fay, P. (1979). Nitrogen fixation in coral reef sponges with symbiotic cyanobacteria. *Nature* **279,** 527–529.

Wilkinson, C. R., and Garrone, R. (1979). In "Nutrition in the Lower Metazoa" (D. C. Smith and Y. Tiffon, eds), pp. 157–161. Pergamon Press, Oxford, Caen, France.

Wilkinson, C. R., and Garrone, R. (1980). Ultrastructure of siliceous spicules and microsclerocytes in the marine sponge *Neofibularia irata* s. sp.. *Journal of Morphology* **166,** 51–63.

Wilkinson, C. R., and Vacelet, J. (1979). Transplantation of marine sponges to different conditions of light and current. *Journal of Experimental Marine Biology and Ecology* **37,** 91–104.

Wilson, S. K., Bellwood, D. R., Choat, J. H., and Furnas, M. J. (2003). Detritus in the epilithic algal matrix and its use by coral reef fishes. *Oceanography and Marine Biology* **41,** 279–309.

Witte, U., and Graf, G. (1996). Metabolism of deep-sea sponges in the Greenland-Norwegian Sea. *Journal of Experimental Marine Biology and Ecology* **198,** 223–235.

Witte, U., Brattegard, T., Graf, G., and Springer, B. (1997). Particle capture and deposition by deep-sea sponges from the Norwegian-Greenland Sea. *Marine Ecology Progress Series* **154,** 241–252.

Wolfrath, B., and Barthel, D. (1989). Production of faecal pellets by the marine sponge *Halichondria panicea* Pallas (1766). *Journal of Experimental Marine Biology and Ecology* **129,** 81–94.

Wulff, J. L. (1984). Sponge-mediated coral reef growth and rejuvenation. *Coral Reefs* **3,** 157–163.

Wulff, J. L. (2001). Assessing and monitoring coral reef sponges: Why and how? *Bulletin of Marine Science* **69,** 831–846.

Wulff, J. L. (2008). Collaboration among sponge species increases sponge diversity and abundance in a seagrass meadow. *Marine Ecology* **29,** 193–204.

Yahel, G., Post, A. F., Fabricius, K., Vaulot, D. M., Vaulot, D., and Genin, A. (1998). Phytoplankton distribution and grazing near coral reefs. *Limnology and Oceanography* **43,** 551–563.

Yahel, G., Sharp, J. H., Marie, D., Häse, C., and Genin, A. (2003). In situ feeding and element removal in the symbiont-bearing sponge *Theonella swinhoei*: Bulk DOC is the major source for carbon. *Limnology and Oceanography* **48,** 141–149.

Yahel, G., Marie, D., and Genin, A. (2005). InEx—a direct in situ method to measure filtration rates, nutrition and metabolism of active suspension feeders. *Limnology and Oceanography Methods* **3,** 46–58.

Yahel, G., Eerkes-Medrano, D. I., and Leys, S. P. (2006). Size independent selective filtration of ultraplankton by hexactinellid glass sponges. *Aquatic Microbial Ecology* **45,** 181–194.

Yahel, G., Whitney, F., Reiswig, H. M., Eerkes-Medrano, D. I., and Leys Sally, P. (2007). In situ feeding and metabolism of glass sponges (Hexactinellida, Porifera) studied in a deep temperate fjord with a remotely operated submersible. *Limnology and Oceanography* **52,** 428–440.

CHAPTER FOUR

Sponge Chemical Diversity: From Biosynthetic Pathways to Ecological Roles

Grégory Genta-Jouve and Olivier P. Thomas[1]

Contents

Abstract

Since more than 50 years, sponges have raised the interest of natural product chemists due to the presence of structurally original secondary metabolites. While the main objective were first to discover new drugs from the Sea, a large number of interrogations arose along with the isolation and structure elucidations of a wide array of original architectures and new families of natural products not found in the terrestrial environment. In this chapter, we focus

University of Nice-Sophia Antipolis, Institut de Chimie de Nice UMR 7272 CNRS, PCRE, Parc Valrose 06108 Nice, France
[1]Corresponding author: Email: oliver.thomas@unice.fr

Advances in Marine Biology, Volume 62
ISSN 0065-2881, DOI: 10.1016/B978-0-12-394283-8.00004-7

on the results obtained during this period on the following questions. A preliminary but still unresolved issue to be addressed will be linked to the role of the microbiota into the biosynthesis of these low-weight compounds. Our knowledge on the biosynthetic pathways leading to plant secondary metabolites is now well established, and this background will influence our comprehension of the biosynthetic events occurring in a sponge. But is the level of similarity between both metabolisms so important? We clearly need more experimental data to better assess this issue. This question is of fundamental interest because sponges have a long evolutionary history, and this will allow a better understanding on the transfer of the genetic information corresponding to the biosynthesis of secondary metabolites. After the how, the why! The question of the ecological role of these metabolites is also of high importance first not only because they can serve as synapomorphic characters but also because they may represent chemical cues in the water environment. Even if most of these compounds are considered as defensive weapons for these sessile invertebrates, they may also be linked to physiological characters as the reproduction. Finally, a metabolomic approach can appear as a complementary tool to give additional information on the sponge fitness. All the new developments in molecular biology and bioanalytical tools will open the way for a better comprehension on the complex field of sponge secondary metabolites.

Key Words: Metabolic pathway; chemical ecology; sponge; biosynthesis

1. Introduction

Even if sponges are considered as primitive organisms, located on the basis of the metazoan tree, our knowledge of these "simple" organisms is still surprisingly very limited. The first sponge genome has been reported in 2010 and will certainly open the way to a better understanding of these intriguing animals at the molecular level (Srivastava *et al.*, 2010). Among the reasons that justify scientific interests on this group of marine invertebrates culminates the fact that sponges are now recognized as producers of extremely diverse secondary metabolites with applications in the therapeutic field (Huyck *et al.*, 2011). Since the 1970s, several groups of natural product chemists have turned their attention to this group of animals and today more than 8000 sponge natural products have been referenced, mainly in the search for new bioactive compounds (Blunt *et al.*, 2011). Apart from the high economical value of these original compounds, more fundamental issues have been rarely addressed as, for example, how and why these compounds are produced? These questions serve as a basis not only for our understanding of the biochemical pathways leading to these metabolites but also for a better assessment of their ecological role in the marine environment. Two main reasons may explain our poor knowledge in these fields. Firstly, the taxonomic identification of marine organisms and especially sponges is a highly difficult science and, consequently, the characterization of a "specific" metabolome is often a subject of controversy (Erpenbeck and van Soest, 2007). Furthermore, the role of the associated or symbiotic bacteria in the production of these compounds is still a matter of debate and may be confusing for the interpretation of ecologically relevant data (Lee *et al.*, 2001; Piel, 2009).

Consequently, we will only focus this report on some families of natural products where at least part of the biochemical pathways have been postulated to originate from sponge cells. In a similar manner, the chemical ecological studies will be only detailed for these families of sponge secondary metabolites. Several reviews have described the ecological studies performed on marine invertebrates including sponges, and we will detail herein the main results on the cell localization, fluctuation in the production and the ecological activity of sponge natural products (Pawlik, 1993; Paul and Puglisi, 2004; Paul *et al.*, 2006, 2011; Paul and Ritson-Williams, 2008). Results obtained on the extracts will not be reported here because synergistic effects are often involved in the bioactivity and it seems of higher importance to identify an isolated compound associated with an ecological role and a biosynthetic pathway. The construction of these complex metabolites is the result of a long evolutionary history, where the selection of the biosynthetic genes has been realized under a complex set of interactions between these benthic invertebrates and their environment. The impact of the environment and the biotope on the modulation of these biosynthetic pathways is still largely unknown. Marine biosynthetic studies have also been the subject of several important reviews (Garson, 1989, 1993, 2001; Moore, 2006). We will detail herein the recent advances in the biosynthesis and the chemical ecology of the most important groups of sponge secondary metabolites organized by usual classes of natural compounds. This question is of high interest in order to assess the level of similarity between sponge biosynthetic pathways and their terrestrial counterparts in a context of a large microbial diversity.

2. Acetate Pathway

Polyketides constitute a very large class of natural products including fatty acids and their derivatives, polyacetylenes, anthraquinones, etc. Despite a large structural diversity, this important class of secondary metabolites originates from very simple building blocks and biochemical transformations. Indeed, successive Claisen-type condensations between acetyl, propionyl or butyryl precursors are responsible for the elongation of the polyketide carbon chains leading to poly-β-ketones through a non-processive pathway. In some cases, additional modules are present in the biosynthetic multienzymatic complex. The polyketide chemical diversity then emerges from the ability of these ketones to be reduced into alcohols, alkenes or even saturated alkanes which can undergo further cyclizations into complex macrolides for examples.

The processes leading to the simplest saturated fatty acids have now been well documented and follow a universal mechanism catalyzed by the enzyme fatty acid synthase. While in animals, this enzyme is a multifunctional protein involving all the iterative domains, in plants and bacteria, the catalytic domains are usually separated.

2.1. Fatty acid derivatives

As for other animals, the cell membranes of marine sponges are mostly constituted of a lipid bilayer where additional sterols or proteins play the role of stabilizers or receptors. The fatty acid composition of the phospholipids present in the lipid bilayer of the sponges has been studied earlier and revealed some interesting features. Original and long-chain (C_{16}–C_{34}) fatty acids have been identified during the 1970s in marine sponges of the class Demospongiae (Litchfield and Morales, 1976). Some of these long-chain fatty acids have been identified to possess two 5*Z* and 9*Z* unsaturations which is quite uncommon in other living organisms and especially in the terrestrial environment. Other original features of sponge phospholipids include the presence of cyclopropyl, α-acetoxy or -methoxy chemical entities but also branched methyls and the abundance of odd-chain fatty acids (Djerassi and Lam, 1991). Due to their presence in other marine organisms and even in terrestrial plants, the demospongic acids have lost their specificity but it remains clear that sponge cell membranes have striking features that render them unique in the living world (Kornprobst and Barnathan, 2010).

The originalities in the structure of sponge fatty acids may originate from a *de novo* biosynthesis, a modification of dietary intake or even from associated bacteria. The first biosynthetic studies on the Caribbean sponge *Clathria* (*Clathria*) *prolifera* using radiolebelled precursors showed that the demospongic acid present in this sponge was biosynthesized from a short-chain unsaturated fatty acid 14:0 and not from a *de novo* biosynthesis (Fig. 4.1) (Morales and Litchfield, 1977; Hahn *et al.*, 1988). The authors concluded that, on the basis of their fatty acid desaturation pathways, sponges possess both animal and plant characteristics.

In the same manner, the methyl-branched $\Delta^{5,9}$-anteiso-27:2 demospongic acid produced by the Australian sponge *Jaspis stellifera* was proved to be biosynthesized from a short-chain 15:0 fatty acid supposedly produced by an associated bacteria (Fig. 4.2) (Carballeira *et al.*, 1986).

An additional biosynthetic study on the Caribbean sponge *Aplysina fistularis* revealed that the methyl chains of the demospongic acids might not be produced by the sponge itself but more probably by an associated

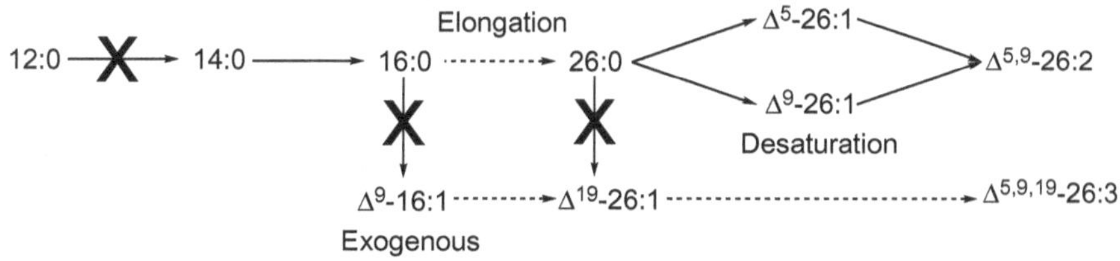

Figure 4.1 Biosynthesis of straight-chain demospongic acids in the marine sponge *Clathria (Clathria) prolifera* (Morales and Litchfield, 1977; Hahn *et al.*, 1988).

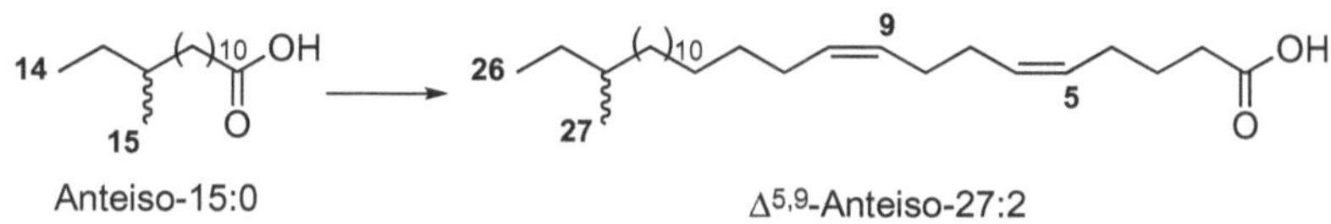

Figure 4.2 Incorporation of 1-^{14}C-labelled branched-chain fatty acids into the marine sponge *Jaspis stellifera* (Carballeira *et al.*, 1986).

bacteria providing the corresponding methyl-branched short-chain fatty acid (Raederstorff *et al.*, 1987). Finally, the group of Djerassi demonstrated that the Caribbean sponge *Xestospongia* sp. was not able to biosynthesize *de novo* odd-chain demospongic acids, but this sponge rather metabolizes odd-chain short fatty acids produced by associated bacteria. Unlike higher animals, they seem unable to incorporate propionate for the biosynthesis of odd-chain fatty acids (Djerassi and Lam, 1991). Functionalization of the alkyl chain, as for the bromination, was also proposed to occur after elongation and desaturation. Recently, Piel and co-workers confirmed that the "poribacteria" were likely to be the real producers of the methyl-branched mid-chain fatty acids, further metabolized into demospongic acids by cell sponges (Hochmuth *et al.*, 2010). They were able to identify a methyl transferase domain in these specific associated bacteria. Using a metagenomic approach, Schirmer and co-workers were able to isolate and characterize a polyketide synthase (PKS) gene cluster responsible for the biosynthesis of multimethyl branched fatty acids reminiscent of mycobacteria origin (Schirmer *et al.*, 2005). Furthermore, this gene cluster was also identified in a bacterial symbiont of the Caribbean sponge *Discodermia dissoluta*.

Phospholipids bearing two acyl chains are mainly constituents of cell membranes, but the less common lysophospholipids, lacking a chain at *sn*-1 or *sn*-2 of the glycerol, have been found to act as secondary messenger molecules in biological processes (Parrill, 2008). Two abundant ether and ester lysophospholipids were found in the Mediterranean sponge *Oscarella tuberculata* (Homoscleromorpha) (Fig. 4.3) (Ivanisevic *et al.*, 2011).

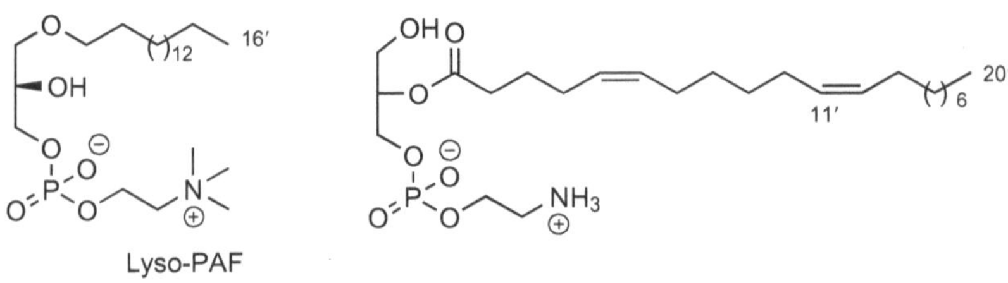

Figure 4.3 Lysophospholipids isolated from *Oscarella tuberculata* (Ivanisevic *et al.*, 2011).

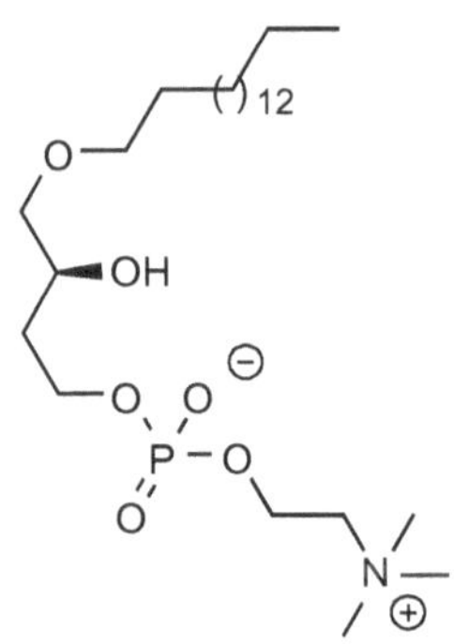

Figure 4.4 Lyso-PAF from *Suberites domuncula* (Müller *et al.*, 2004).

The highest expression level of both lysophospholipids was found to occur in summer during the period of embryogenesis and larval development. This observation suggests an important role of these compounds as secondary messengers during the reproduction of this sponge species. Fluctuation could then be related to physiological factors, a role that has been very little studied so far but which could be of high importance in the context of several theories on resource allocations.

Other important ecological studies have been performed by the group of Müller on some lyso-PAF (Platelet Activating Factor) type phosphoglycerolipids produced by the sponge *Suberites domuncula* (Fig. 4.4) (Müller *et al.*, 2004).

In this study, the authors were able to demonstrate that the presence of bacterial endotoxins lipopolysaccharides induced the biosynthesis of these antibacterial compounds. The biosynthetic processes of bioactive phosphoglycerolipids could then be regulated by toxic environmental factors. Key biosynthetic enzymes have further been identified in the host and not in the associated bacterial community which proved an adaptive response of the sponge to environmental stress.

Even if these compounds act as secondary metabolites, they are very closely related to the primary metabolism and the boundary between these two metabolisms is not well defined in these cases. Biosynthetically related derivatives named polyacetylenes is another family of fatty acid derivatives that have been largely found in marine sponges and they are more likely part of the secondary metabolites.

2.2. Polyacetylenes

Sponges have yielded a wide array of mid- and long-chain oxygenated and halogenated acetylenic fatty acid derivatives (Jung *et al.*, 2006). These compounds are often considered as chemotaxonomic markers of several genera of the order Haplosclerida like *Xestospongia*, *Pellina*, *Petrosia* and some *Haliclona*, even if they were also found recently in other groups of sponges (Erpenbeck and van Soest, 2007). In the order Hadromerida, the place of the genus *Diplastrella* was discussed in view of the presence of similar compounds in several species of this group which was also the case for the genus *Phakellia* from the order Halicondrida. Among the most important representatives of long oxygenated polyacetylenes, the petroformynes and the petrocortynes were isolated from *Petrosia* species (Fig. 4.5) (Cimino *et al.*, 1985a; Seo *et al.*, 1998; Curran and Sui, 2009). The osirisynes were later isolated from a *Haliclona* species (Shin *et al.*, 1998).

The biosynthesis of these very long-chain unsaturated polyols is very intriguing and only one hypothesis has been postulated so far (Minto and

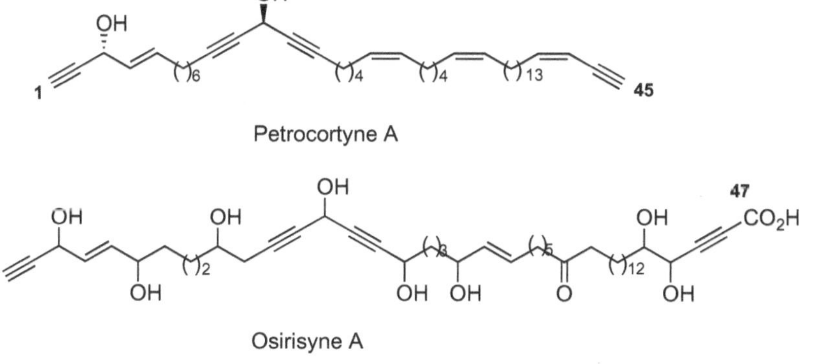

Figure 4.5 Structures of some long-chain acetylenic fatty acid derivatives (Cimino *et al.*, 1985a; Shin *et al.*, 1998; Curran and Sui, 2009).

Blacklock, 2008). It is based on the biosynthetic studies performed on mycolic acids, long-chain polyacetylenic compounds isolated from mycobacteria. A carboxylative process would allow the elongation of a second long alkyl chain. These compounds would therefore be formed by condensation of two long fatty acid chains. Formation of a central triple bond would occur via an elimination process (Fig. 4.6). The first acetylenase was characterized in a fungus and was proven to be distinct from usual desaturases found in plants or fungi (Blacklock *et al.*, 2010).

A wide array of shorter and sometimes halogenated analogues has been uncovered in several sponge species and could serve as precursors of longer derivatives (Dai *et al.*, 1996; Watanabe *et al.*, 2000; Hitora *et al.*, 2011a,b). Unfortunately, no ecological study has been conducted on these compounds so far and their role in the ecosystem is still largely unknown even if strong cytotoxicities have been revealed.

The biosyntheses of fatty acids and polyketides are closely related and both share an iterative process based on acetyl, propionyl or butyryl precursors, but the chemical diversity of polyketides was found to be much higher.

2.3. Polyketides

A large array of bioactive polyketides has been reported from marine sponges worldwide. As a highly important example, eribulin mesylate (Halaven; Eisai), a compound closely related to the sponge polyketide

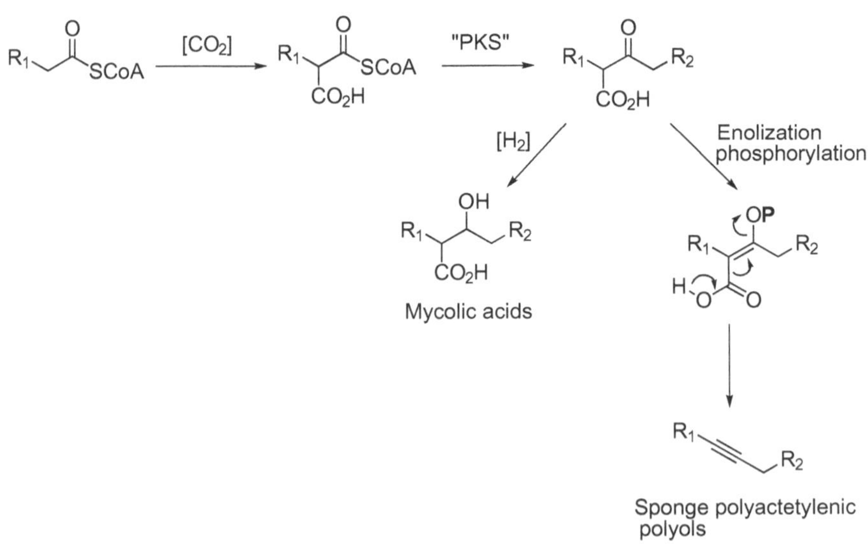

Figure 4.6 Biosynthetic hypothesis for the formation of sponge very long chain polyacetylenic polyols (Minto and Blacklock, 2008).

halichondrin B, was launched in 2010 for the treatment of metastatic breast cancer (Huyck *et al.*, 2011). The main features of this class of natural product are based on an iterative process using acetate (C2), propionate or butyrate building blocks, and a key Claisen-type condensation for the chain elongations. Important progress has been made in the comprehension of the metabolic pathways involved in this important class of natural products. The main reason is that the real producer of these compounds has usually been identified as an associated microorganism which in some cases has been amenable to *in vitro* culture (Nguyen *et al.*, 2008; Fisch *et al.*, 2009; Piel, 2009). The PKSs enzymes and the iterative process leading to the structural complexity of polyketides are among the most well-known metabolic pathways to date and they mostly originate from biosynthetic gene clusters present in microbes. The highly bioactive macrolides are common sponge polyketides usually produced by associated microorganisms, and we will give a small overview on this topic in the next section.

2.3.1. Macrolides

In a few cases, the biosynthetic genes have been clearly identified in the sponge-associated bacteria like for the polyketides theopederin and onnamide A in the sponge *Theonella swinhoei*, even if the symbiont is not cultivable (Piel *et al.*, 2004). This observation could explain the difficulty to find correlations of ecological relevance for the fluctuation of the macrolides concentration found in sponges. The variation in the concentration of the polyketide salicylihalamide A found in an Australian sponge *Haliclona* sp. could not be totally rationalized (Abdo *et al.*, 2007). Nevertheless, an increase of temperature has been correlated with a decrease in the production of the compound which could also be explained by a change in the sponge microfauna, an effect known as bleaching for the corals and its photosymbionts. Temporal and geographical variations have been described for mycalamide A, pateamine and peloruside A from the New Zealand *Mycale hentschelli* (Page *et al.*, 2005). The concentration of these compounds changed dramatically with the environment of the sponge, and this effect can also be attributed to the presence of associated bacteria and, in particular, the surrounding biotopes.

A specific class of endoperoxide polyketides has attracted much attention due its presence in several species of *Plakortis* distributed worldwide. Furthermore, the production by associated bacteria has not been proven to date, and on the contrary, the high concentration of these compounds in the sponge extracts would suggest a sponge origin or at least participation.

2.3.2. Endoperoxides

Endoperoxide polyketides of the plakortin, plakortolide, plakortide and plakortone families were described from several species of marine sponges and most specifically from species of the genus *Plakortis* (Fig. 4.7) (Fattorusso

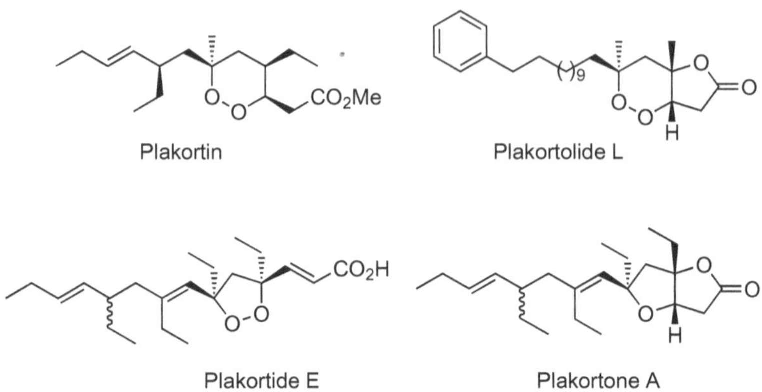

Figure 4.7 Structures of some endoperoxide derivatives isolated from *Plakortis* sponges (Fattorusso and Taglialatela-Scafati, 2010).

and Taglialatela-Scafati, 2010). Moreover, a high therapeutic potential as antimalarials has been detected for this family of compounds.

The unique study on the cell localization of these compounds was performed by the group of Fattorusso (Laroche *et al.*, 2007). After cell separation, they were able to identify a high concentration of plakortin in the bacterial cell fraction of the Caribbean sponge *Plakortis simplex* but this result cannot prove unambiguously a microbial origin for the compound. Bacterial cells could also serve as storage for these compounds.

Their biosynthesis has been the subject of several hypotheses and this issue is still a matter of debate. Even if an elegant Diels–Alder reaction between a diene and molecular oxygen could lead to the endoperoxide present in plakortin and plakortolide, this hypothesis is not in agreement with the stereochemical outcomes found for some of these compounds (Yong *et al.*, 2011). Garson and co-workers rather proposed the preliminary formation of a hydroperoxide similar to those observed in the prostaglandin biosynthesis starting from arachidonic acid. This hydroperoxide would then undergo an intramolecular Michael addition leading to the endoperoxide core (Fig. 4.8). A similar addition of molecular dioxygen and cyclization on the other double bond could give the plakortide family of compounds.

A reduction process would afford the plakortin family, while an electrophilic addition of the carboxylic acid on the cyclic double bond would lead to the plakortolides. An additional reduction of the peroxide could afford the seco derivatives of plakortides that were already described, and a final dehydration/Michael addition would yield the plakortones.

The ecological roles of these molecules were only little investigated but their pronounced antimalarial bioactivities suggest antimicrobial activities in the medium. Furthermore, the Homoscleromorpha sponge *Plakortis*

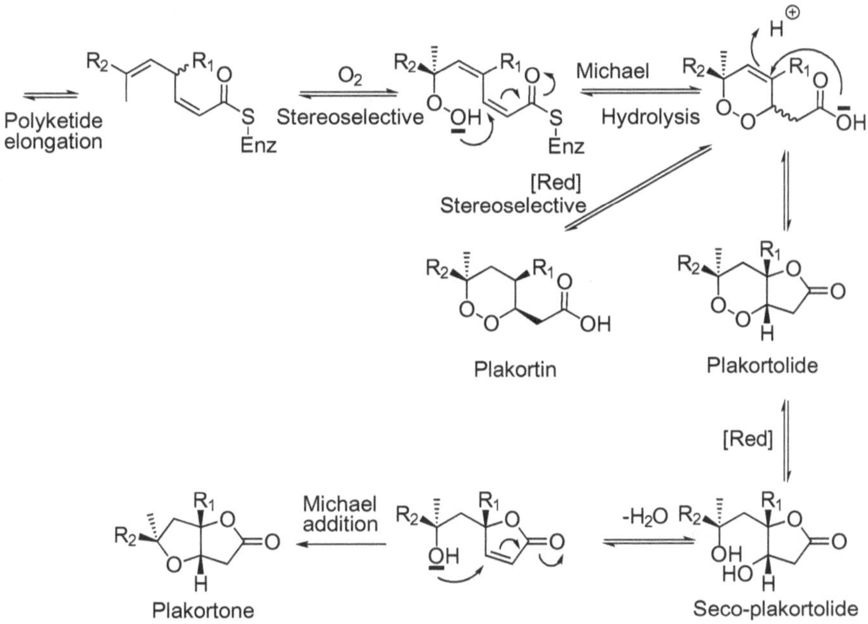

Figure 4.8 Putative biosynthesis of plakortin and plakortolide derivatives. Adapted from Yong *et al.* (2011).

halichondroides was observed to bleach corals in the Caribbean Sea (Porter and Targett, 1988). Because the concentrations of endoperoxides in this sponge were proved to be high, it is believed that the coral bleaching is due to the allelochemical properties of these compounds. The sponge could exude these compounds in the water environment which could have a detrimental effect on coral respiration and photosynthesis. A fouling activity of decan-2-one identified by cryo-trapping in a *Plakortis* sponge was also reported, suggesting that diverse classes of secondary metabolites may contribute to the interactions with the environment (Bowling *et al.*, 2010). This observation underlines the importance of the full chemical study of an extract in order to give reliable information at ecological levels.

3. Mevalonate Pathway

The terpenoids represent a large class of natural products produced by either a mevalonate or a non-mevalonate pathway. The C5 building blocks are the isopentenyl pyrophosphate or dimethylallyl pyrophosphate which can lead to terpenes in C10, sesquiterpenes in C15, diterpenes in C20, sesterterpenes in C25 or triterpenes in C30. The presence of carbocationic intermediates and the high number of asymmetric centres afford a large chemical diversity, and even if terpenoids are largely produced by plants, they are also found in marine sponges. Sponge diterpenoids are considered as chemotaxonomic markers of some genera of the orders Dictyoceratida and Dendrophorida (Keyzers *et al.*, 2006). Whereas the biosynthetic pathways are now largely described for plants, the few experiments performed to prove the biosynthetic pathways in sponges have not been conclusive and the construction of terpenoids in sponges could imply the non-mevalonate pathway discovered by the group of Rohmer (Garson, 1986; Garson *et al.*, 1988; Rohmer, 1999).

We will detail herein the main results obtained for four groups of terpenoids produced in large quantity by marine sponges: the isocyanide terpenes, sesquiterpene quinones, furanoterpenes and finally, steroids.

3.1. Isocyanide and related terpenes

The presence of the isocyanide group is quite rare in the terrestrial natural products, but a large quantity of sponge secondary metabolites was found to contain this functional group (Scheuer, 1992). Several biosynthetic and ecological studies have been undertaken early in this family of compounds, and most of the results have been described in two comprehensive reviews (Edenborough and Herbert, 1988; Garson and Simpson, 2004). This particular group of sesquiterpenes and diterpenes is found in very distinct groups of sponges thus precluding any chemical taxonomic marker.

The sponge *Amphimedon terpenensis* has been shown to produce a large variety of isocyanide terpenes and closely related derivatives. Because of the presence of a cyanobacterial symbiont in this sponge, some experiments have been undertaken to discriminate between a symbiotic or a host production for these compounds (Garson *et al.*, 1992). The group of Garson was able to demonstrate that a high content of isocyanide terpenes was associated with the sponge cell fraction and importantly, that these terpenes may be structural components of sponge cell membranes.

Whereas similar sesquiterpenes derivatives in C15 are much less frequently described, a large chemical diversity of sponge diterpenes derivatives in C20 was found in diverse sponge species, mainly as members of the kalihinane, amphilectane or even cycloamphilectane families which are characterized by two, three or four fused cyclohexanes, respectively. The putative biosynthetic pathway presented in Fig. 4.9 is mostly inspired from Simpson and Garson (2004).

Even if cyclization of terpene structural units is common among the terrestrial and marine environments, the presence of the isocyanide moieties or related functions as thiocyanate, isothiocyanate or formamide seems to be inherent to the marine environment. The issue of the origin of these

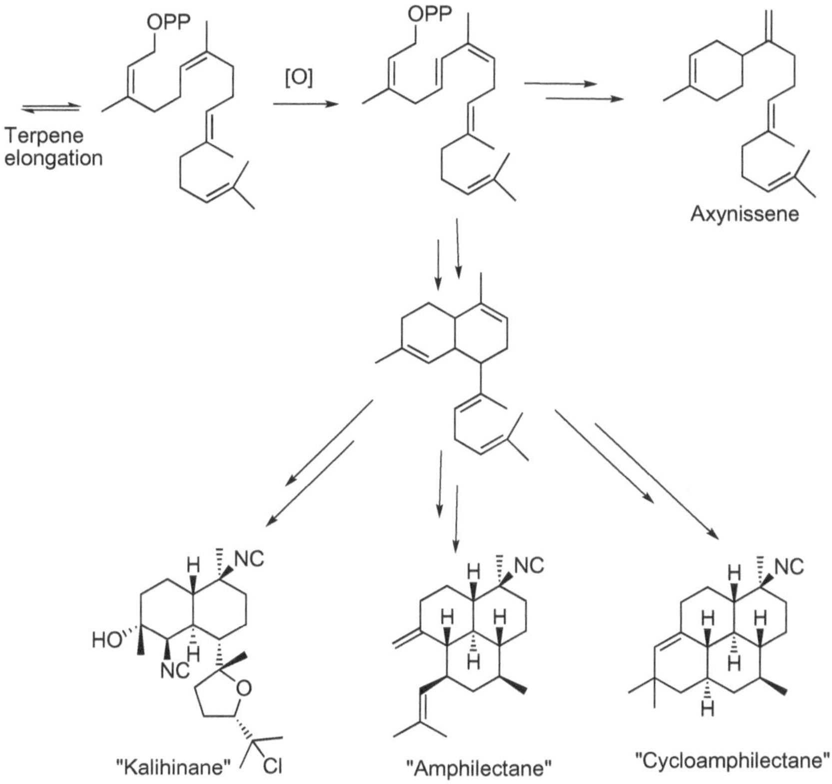

Figure 4.9 Putative biosynthetic pathways leading to sponge isocyanide diterpenes (Simpson and Garson, 2004).

functional groups has been addressed by several biosynthetic experiments. The first hypotheses on the origin of the isocyanide groups were based on a formamide intermediate, possibly formed by condensation between a primary amine and the usual C1 building block leading to the isocyanide function after a last dehydration step (Cable *et al.*, 1991). Nevertheless, several feeding experiments with labelled and radiolabelled cyanide proved unambiguously that both nitrogen and carbon atoms of the cyanides were incorporated intact into the isocyanides of sponge terpenes (Garson, 1986, 1993; Karuso and Scheuer, 1989). The subsequent interconversion of isocyanide into isothiocyanate was also demonstrated in both ways, thus unravelling the origin of these unusual sponge functional groups (Dumdei *et al.*, 1997; Simpson *et al.*, 1997; Simpson and Garson, 1998, 2004). Some enzymes have been already reported from other organisms to promote these interconversions. Additional biosynthetic experiments also proved the conversion of cyanide and thiocyanate into the dichloroimines function also characteristic of some sponge terpenes like the stylotellanes (Simpson *et al.*, 1997). Sponge-associated bacteria have been proposed to play a role in the production of the cyanide ions used for the biosynthesis of these secondary metabolites (Dekas *et al.*, 2009). A very interesting metabolite transfer from the marine sponge *Acanthella cavernosa* to the nudibranch *Phyllidiella pustulosa* was evidenced, thus suggesting that similar compounds also present in nudibranchs may originate from their diet (Dumdei *et al.*, 1997).

The ecological relevance of sponge isocyanide terpenes has been underlined in several reports. First, some sesquiterpenes isocyanides isolated from a nudibranch diet were found to deter goldfish, thus suggesting an important ecological role *in situ* for this class of compounds (Fig. 4.10) (Thompson *et al.*, 1982). Axisonitrile-1, the major component of the Mediterranean sponge *Axinella cannabina* (renamed *Acanthella cannabina*), was not only toxic but also exhibited deterrence activity on fishes (Fig. 4.10) (Cimino *et al.*, 1982). The fish toxicity detected for the extract of the Mediterranean sponge *Acanthella acuta* was attributed to 1-isocyanoaromadendrane, axisonitrile-3 being also present in the extract (Braekman *et al.*, 1987) (Fig. 4.10). Axisonitrile-3, also present in the Caribbean sponge *A. cavernosa*, was not associated with a deterrence activity, and a synergistic effect could explain the activity detected for the extract of this sponge (Dumdei *et al.*, 1997). Moreover, even if an unidentified sponge of the genus *Ircinia* was described to exude large quantities of methyl isocyanide and isothiocyanate in the water environment, these compounds were devoid of any toxicity or deterrence activity (Pawlik *et al.*, 2002).

Sponge cyanide terpenes have also proved to be associated with very interesting antifouling activities, especially against larvae settlements

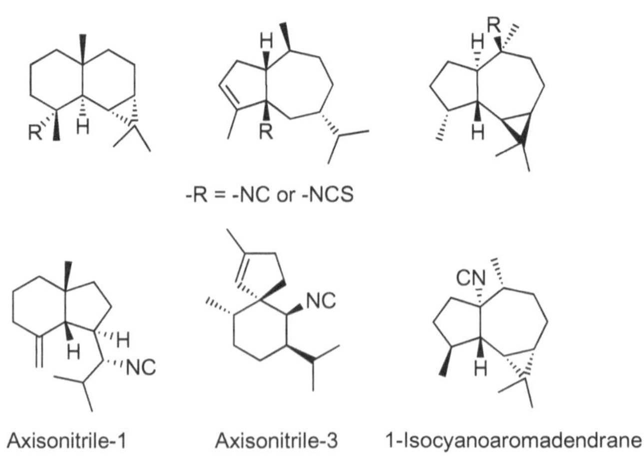

Figure 4.10 Ecologically relevant sponge sesquiterpenes.

(Fusetani, 1997). Additional important terpenes with an oxidized side chain have been largely encountered in several marine sponge species.

3.2. Oxygenated terpenes

3.2.1. Sesquiterpene hydroquinones

Sesquiterpene hydroquinones are largely found in sponges of the order Dictyoceratida (Sladic and Gasic, 2006). Avarol, first isolated from the Mediterranean sponge *Dysidea avara*, is an important representative of the sesquiterpene hydroquinones (Fig. 4.11) (Minale *et al.*, 1974). This compound features a sesquiterpene rearranged drimane skeleton substituted with a hydroquinone. Other closely related sponge sesquiterpene quinones include ilimaquinones (Luibrand *et al.*, 1979) and nakijiquinones (Fig. 4.11) (Shigemori *et al.*, 1994). This family of compounds is characterized by a high therapeutic potential that induced various biological and synthetic studies.

Despite the high potential of this class of compounds, only one cellular localization study has been performed by the group of Müller (Müller *et al.*, 1986). Avarol was localized in the spherulous cells of the sponge *D. avara*, and no production by its cultivable bacterial symbionts was observed. Furthermore, a primmorph culture of sponge cells led to the production of this compound (Müller *et al.*, 2000).

The biosynthetic pathways leading to these oxygenated sesquiterpenes would involve a cyclization of farnesyl diphosphate and a key carbocationic rearrangement leading to the bicyclic skeleton. A further substitution of the diphosphate group by a shikimate-derived quinone or hydroquinone would lead to the secondary metabolites (Fig. 4.12). Additional information could be given by the *de novo* biosynthesis of drimane sesquiterpenes from mevalonate in nudibranchs (Fontana *et al.*, 1999). Indeed, biosynthetic experiments are much easier to perform on nudibranchs, and the results are highly valuable to assess sponge biosynthetic pathways even if they

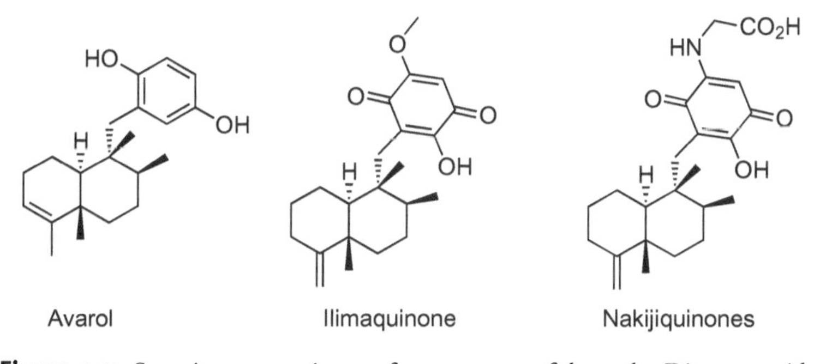

Figure 4.11 Sesquiterpene quinones from sponges of the order Dictyoceratida.

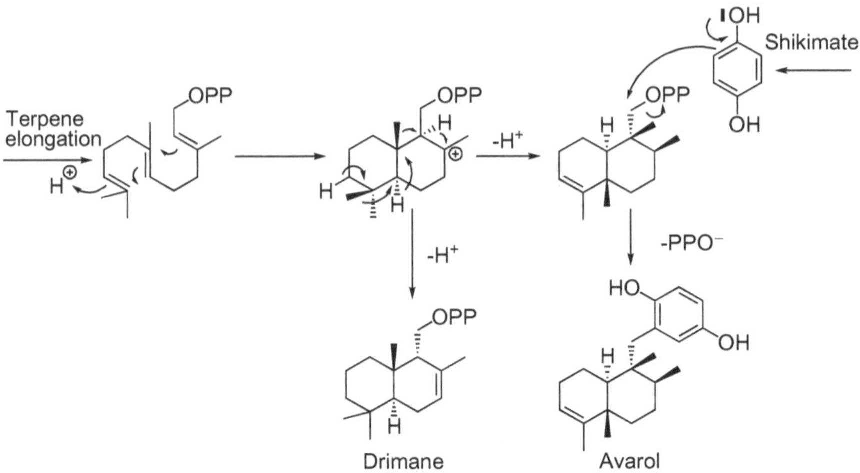

Figure 4.12 Putative biosynthetic pathway of sponge sesquiterpene quinones.

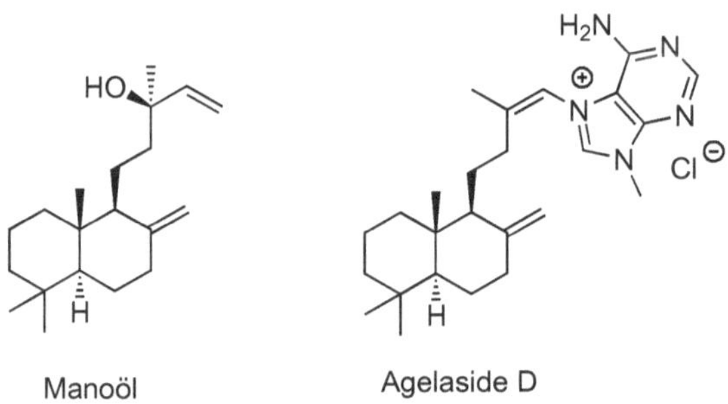

Figure 4.13 Oxygenated sesquiterpenes studied for chemical ecology purposes.

should be taken with caution. Unfortunately, no biosynthetic study has been reported for these compounds in sponges and no definitive conclusion can then be made.

From an ecological perspective, avarol and its derivatives exhibit not only interesting antifouling activity on the settlement of *Balanus amphitrite* but also antimicrobial activities on a large array of marine bacteria and fungi (Tsoukatou *et al.*, 2007). The groups of Uriz and Cimino were able to demonstrate that avarol was indeed toxic against marine bacteria using a standardized Microtox® assay (Marti *et al.*, 2003). This bioassay was also proved to be efficient for the assessment of the intraspecific variability of the sponge toxicity. A close drimane derivative manoöl present in the Bahamian sponge *Aplysilla glacialis* was deterrent against some fishes (Fig. 4.13)

(Bobzin and Faulkner, 1992). The concentration of the metabolites was monitored and found to be relatively stable in space and time. A nitrogenated drimane analogue, agelaside D, isolated from a sponge *Agelas* sp., inhibited the settlement but did not kill larvae of *Balanus improvisus* (Fig. 4.13) (Sjögren *et al.*, 2008).

In some cases, the terpene chain is not cyclized and linear hydroquinones have been identified as chemotaxonomic markers of some genera of the order Dictyoceratida, like the genus *Sarcotragus* (Erpenbeck and van Soest, 2007). Some linear prenylated hydroquinones isolated from *Ircinia spinosula* showed antimicrobial activity as well as ichthyotoxicity (De Rosa *et al.*, 1994). The corresponding quinones were less active. Prenylated hydroquinones found in the Mediterranean sponge *Ircinia spinosulus* were found to be active against marine bacteria and fungi, and they also inhibited the attachment of microalgae (Tsoukatou *et al.*, 2002).

3.2.2. Oxidized di- and sester-terpenes

Sponge diterpenoids have been largely studied and reviewed extensively as characteristic chemomarkers of the orders Dictyoceratida and Dendrophorida (Keyzers *et al.*, 2006). The Australian sponge *Luffariella variabilis* was monitored for temporal and spatial variability in the production of monocyclized diterpenes of the maoalide family (Fig. 4.14) (Ettinger-Epstein *et al.*, 2008). Both sponge species *Hyrtios erecta* and *Hyrtios altum* from Guam were

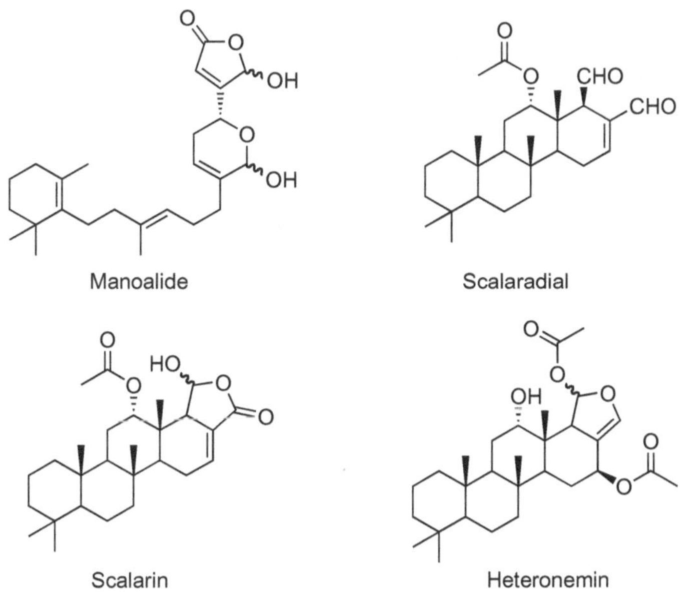

Figure 4.14 Examples of sponge di- and sester-terpenes.

investigated for the defensive properties of their secondary metabolites (Rogers and Paul, 1991). Scalaradial, scalarin and heteronemin were identified as major components of these sponges, but the results about their antifeedant activity were confusing and no clear conclusion could be given.

Intraspecimen variation in the concentration of sponge diterpenes was evaluated for the Pacific sponge *Cacospongia* sp. (Becerro *et al.*, 1998). A higher concentration of scalaradial has been identified in the tips of this sponge while structural components were mostly found in the base of the sponge but variations among specimens were relatively high. The extracts of this sponge also deterred feeding by reef fishes.

For a large number of sponge linear terpenes, the oxidized side chain is cyclized into one or several furans moieties. Furanosesterterpenes (C25) are identified as chemotaxonomic markers of some genera of the order Dictyoceratida but mostly for the genus *Ircinia* (Erpenbeck and van Soest, 2007). No biosynthetic study has been reported on this family of compounds in sponges even if some interesting results have been obtained with analogues produced by nudibranchs (Cimino *et al.*, 1985b).

Intraspecific variations have been very little studied in general for sponge secondary metabolites, and a preliminary report has been published on the Australian marine sponge *Rhopaloeides odorabile* and its main furanoterpenes metabolites (Thompson *et al.*, 1987). The tetracyclic furanoditerpenes were mostly found on the surface of the sponge and, because of a low content of sterols, they were supposed to play a role in the cell membrane stability (Fig. 4.15). The authors were also able to evidence a light-induced production of these compounds. A recent geographical study was performed on the Mediterranean sponge *Spongia lamella* which produced nitenin as well as scalarin analogues (Noyer *et al.*, 2011) (Fig. 4.15). The main results, based on a metabolomic approach, indicated that variations in the concentrations can be related to the geographical collection site of the individuals. A good correlation was further noticed between the geographical distance and the dissimilarity in the metabolomic profiles.

Screening for antiderrent activity of Californian sponges, a series of furanoterpenes have been identified as responsible for the activity against the feeding of some predatory fishes (Thompson *et al.*, 1985) (Fig. 4.16).

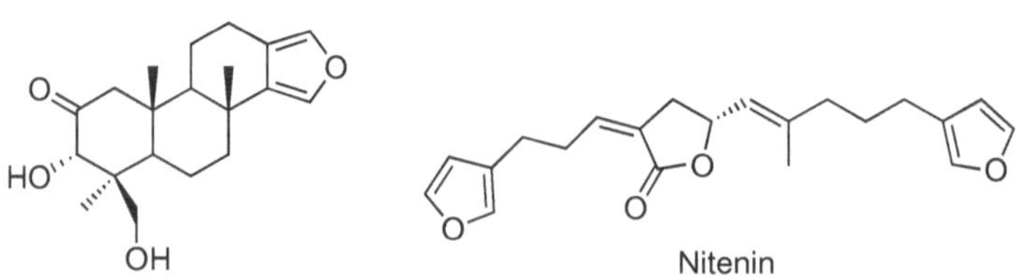

Figure 4.15 Sponge furanoditerpenes studied for their intraspecific variability.

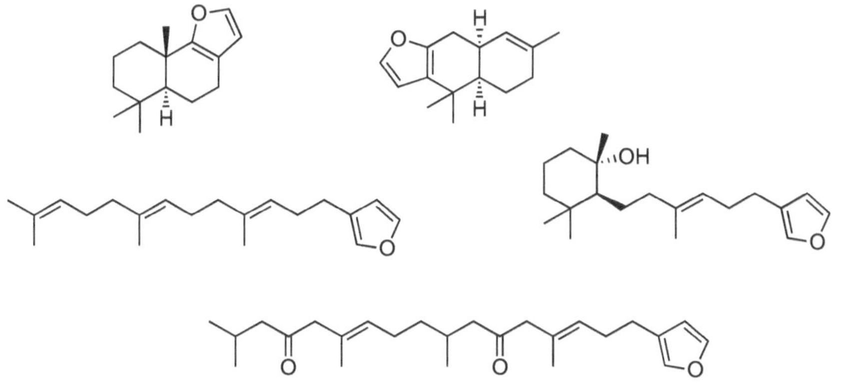

Figure 4.16 Antiderrent furanosesqui-, di- and sester-terpenes from Californian sponges (Thompson *et al.*, 1985).

In the nudibranch *Dendrodoris grandiflora*, it was proven that furanoterpenes had a dietary origin and more precisely after feeding on a sponge (Cimino *et al.*, 1985b). They were found to exhibit antifeedant activity except for fishes. Variabilin and ircinin-1 and -3, isolated from several species of the genus *Ircinia*, exhibited antifeedant activity against the wrasse *Thalassoma bifasciatum* (Fig. 4.17) (Pawlik *et al.*, 2002; Tsoukatou *et al.*, 2002). Finally, palinurin was identified as the toxic metabolite of the Mediterranean sponge *Ircinia variabilis* in the Microtox® assay (Fig. 4.17) (Marti *et al.*, 2003).

After an overview on true secondary metabolites of terpene origin, we will end up this chapter with the more complex steroids (C30-x) usually considered at the interface between the primary and the secondary metabolisms.

3.3. Triterpenes and steroids

3.3.1. Steroids

The first chemical studies on sponge sterols date back to the late 1940s with the work of Bergman, (Bergmann, 1949; Bergmann and Feeney, 1949), and these metabolites have been the subject of intense ecological and biosynthetic studies since then. Original sponge sterols have early been recognized as chemotaxonomic markers of some taxons. The issue of their biosynthetic pathways has also triggered a high number of studies focusing on the modification of either the rings or the side chain (Bergquist *et al.*, 1991; Djerassi and Silva, 1991; Giner, 1993; Sarma *et al.*, 2005).

Cellular localization of the natural products present in the Caribbean sponge *P. simplex* was performed using cell separation techniques (Laroche *et al.*, 2007). Not surprisingly, the large bacteriohopanoids isolated from this sponge were only present in the bacterial fraction, thus suggesting a

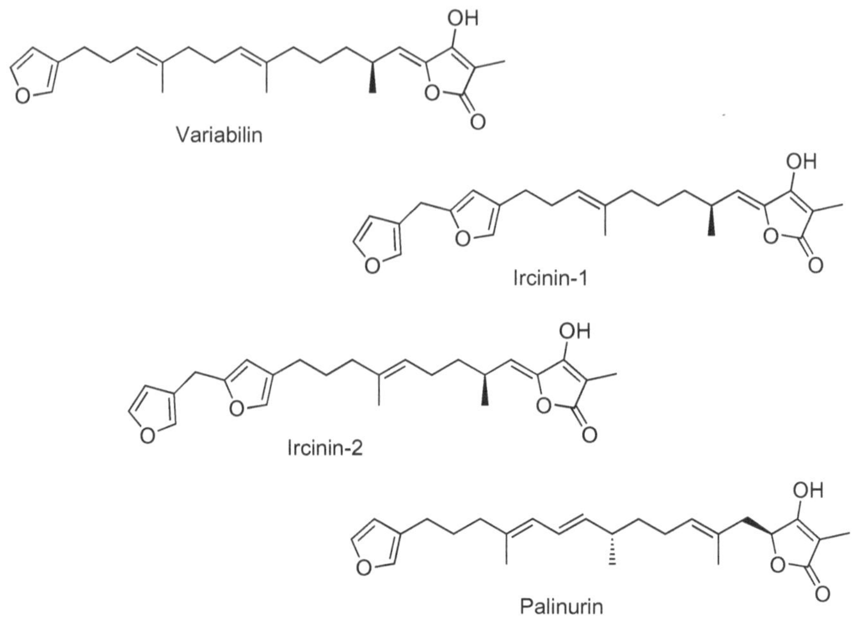

Figure 4.17 Furanosesterterpenes from sponges of the genus *Ircinia*.

microbial origin for these compounds. Nevertheless, no other study has been reported on the role of the associated microorganisms in the production of steroid derivatives.

The question of the biosynthetic origin of the sponge sterols was addressed in the late 1980s by the group of Djerassi through feeding experiments using labelled precursors. The very low incorporations of early precursors like acetate, methionine and mevalonate prevented any definitive conclusion on the *de novo* biosynthesis of sterols by marine sponges. Nevertheless, with lipid precursors like squalene, they were able to demonstrate that sponges are capable of the *de novo* biosynthesis of sterols (Silva *et al.*, 1991). It soon appeared that sponges produce an extremely high diversity of original sterols, much more than any other taxonomic group. The sterols differ largely in their structure from the sterols found in plants due to the presence of four characteristic features: multiply alkylated side chain, cyclopropane side chains, unsaturated ring B and the presence of a nor-ring A and 19-nor derivatives (Fig. 4.18). The loss of a methylene unit at C-19 or in ring A was explained by a specific oxidation at this position. The incorporation of multiple alkyl groups and a cyclopropane on the side chain has been demonstrated to occur from *S*-adenosyl methionine (SAM) (Catalan *et al.*, 1985).

Original sterols have been determined as chemotaxonomic markers of sponges of the order Haplosclerida (Erpenbeck and van Soest, 2007).

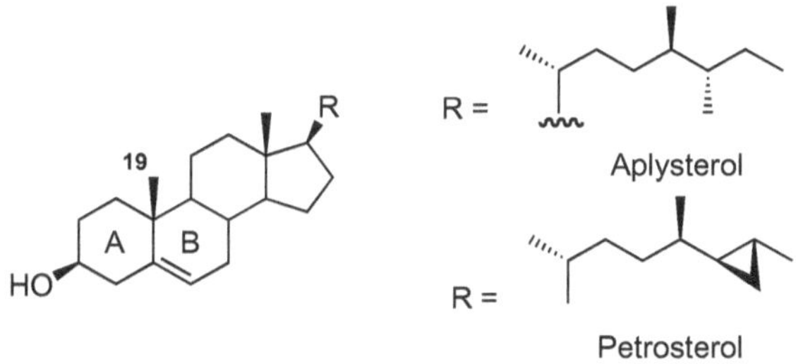

Figure 4.18 Examples of usual sponge sterol side-chains.

A unified biosynthetic scheme linking these metabolites was proposed on the basis of feeding experiments with radiolabelled tracers (Fig. 4.19) (Djerassi and Silva, 1991). The construction of the cyclopropane rings originates from a methyl or ethylcholesterol and involves a carbocationic intermediate but a SAM methylation has also been proposed in some cases (Giner, 1993).

It is believed that the sterol metabolism described for sponges is remnant of earlier metabolisms found in eukaryotes. No rational explanation has been given so far for this extreme sterol diversity.

Sterol endoperoxides are also produced by several sponge species. They were identified as chemical deterrent present in the mucus of the Caribbean sponge *A. glacialis* (Bobzin and Faulkner, 1992). Two sulphated dimeric sterols named amaroxocane A and B were isolated from the Caribbean sponge *Phorbas amaranthus*, but only amaroxocane B exhibited deterrence activity against a fish (Morinaka *et al.*, 2009).

Additional glycosylation steps lead to steroidal saponins, which were also found in some sponge species.

3.3.2. Saponins

Triterpene and steroid glycosides are widespread in the terrestrial environment and they are largely found in plants. In the marine environment, echinoderms were identified to produce a high diversity of this class of natural products but specific compounds were also isolated from some sponge species (Ivanchina *et al.*, 2011). They are characterized by structurally diverse aglycon skeletons in the sarasinosides produced by the Australian *Melophlus sarasinorum* (Kobayashi *et al.*, 1991), feroxosides from the Caribbean *Ectyoplasia ferox* (Campagnuolo *et al.*, 2001), erylosides from the Caribbean *Erylus formus* (Stead *et al.*, 2000), mycalosides from the Caribbean *Mycale laxissima* (Kalinovsky *et al.*, 2002), and more recently pandarosides (Cachet *et al.*, 2009) and acanthifoliosides from the Caribbean *P. acanthifolium* (Fig. 4.20) (Regalado *et al.*, 2011).

This family of compounds appeared as a chemotaxonomic marker of the order Astrophorida and this observation suggested that *P. acanthifolium* should be integrated into this group (Erpenbeck and van Soest, 2007).

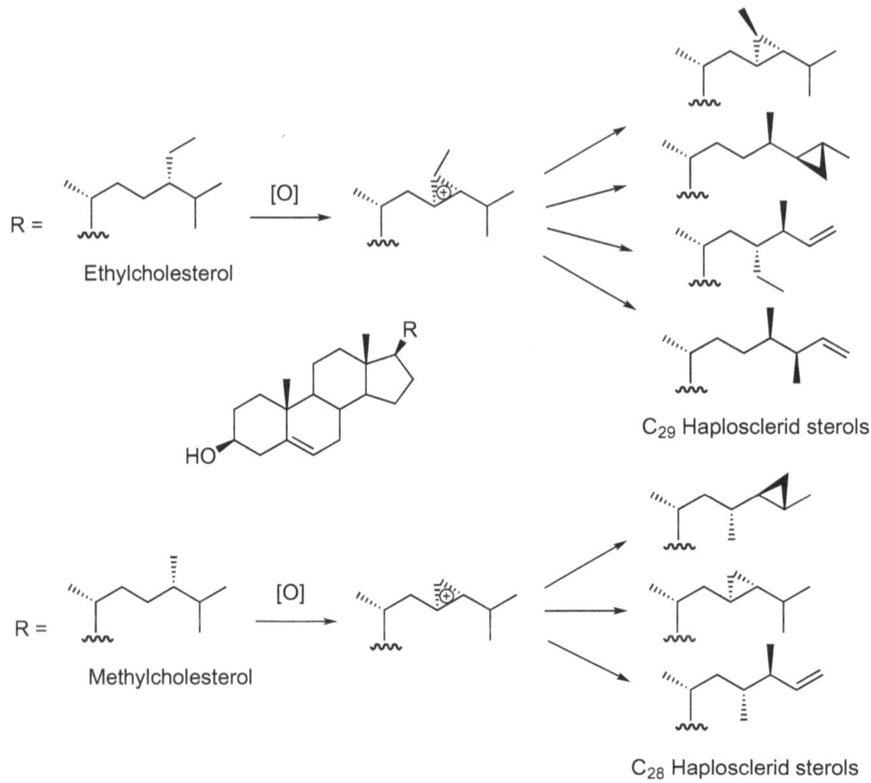

Figure 4.19 Unified biosynthetic pathways leading to Haplosclerida sponge sterols (Giner, 1993).

Triterpene glycosides from the Caribbean species *Erylus formosus* and *E. ferox* were proved to act as chemical defences (Kubanek *et al.*, 2000, 2002). They also showed an important antifouling activity.

Marine organisms are also characterized by a high incorporation of nitrogen and the following sections will focus on nitrogenated secondary metabolites some of them originating from the shikimate pathway.

4. Shikimate Pathway

4.1. Bromotyrosine derivatives

The shikimate pathway is involved in the biosynthesis of aromatic amino acids like phenylalanine and tyrosine. Although sponges are filter-feeders organisms, they possess a large variety of secondary metabolites that seem to

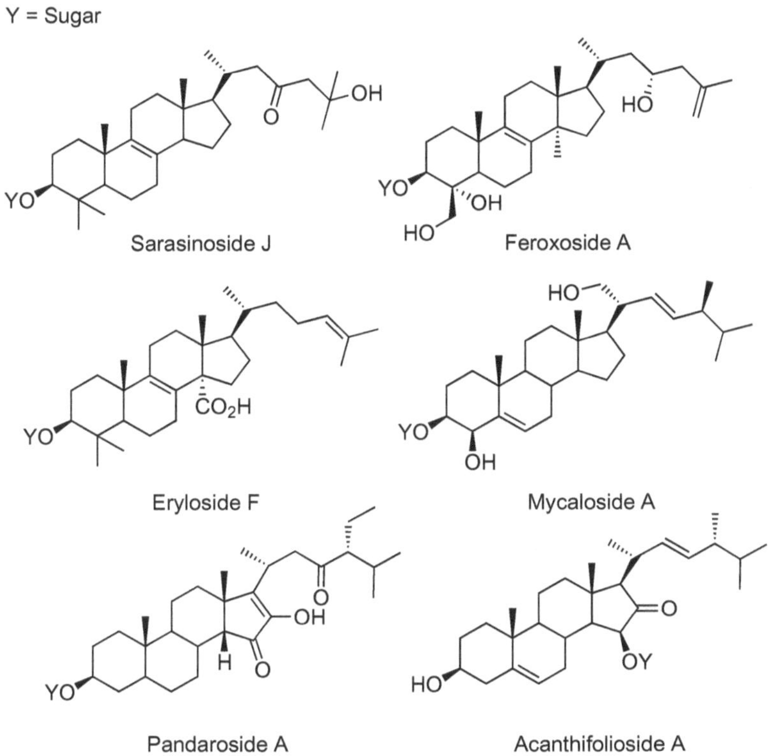

Figure 4.20 Chemodiversity of sponge saponins.

have an endogenous origin and sponges are, for example, considered the richest source of brominated natural products biogenetically derived from tyrosine (Gribble, 1998). Tymiak and Rinehart have demonstrated some steps involved in the biosynthesis of brominated secondary metabolites in *A. fistularis* (Tymiak and Rinehart, 1981). The study of the distribution of secondary metabolites within the sponge can provide important information not only on the ecological role of these compounds but also on their biosynthetic origin. Energy dispersive X-ray microanalysis was used on *A. fistularis* tissues, and brominated compounds such as aerothinonin and homoaerothionin were localized in the sperulous cells of the sponge. A more recent study came to confirm this result (Fig. 4.21; Turon *et al.*, 2000). The rate of exudation of secondary metabolites was also assessed (Thompson, 1985). The regional distribution of two brominated compounds was determined on a specimen of *Aplysina aerophoba* by a histological study and the results showed that the concentrations of the secondary metabolites changed from the surface layer (both outer and oscular region) to the centre part (Kreuter *et al.*, 1992). Studies made on *A. aerophoba*

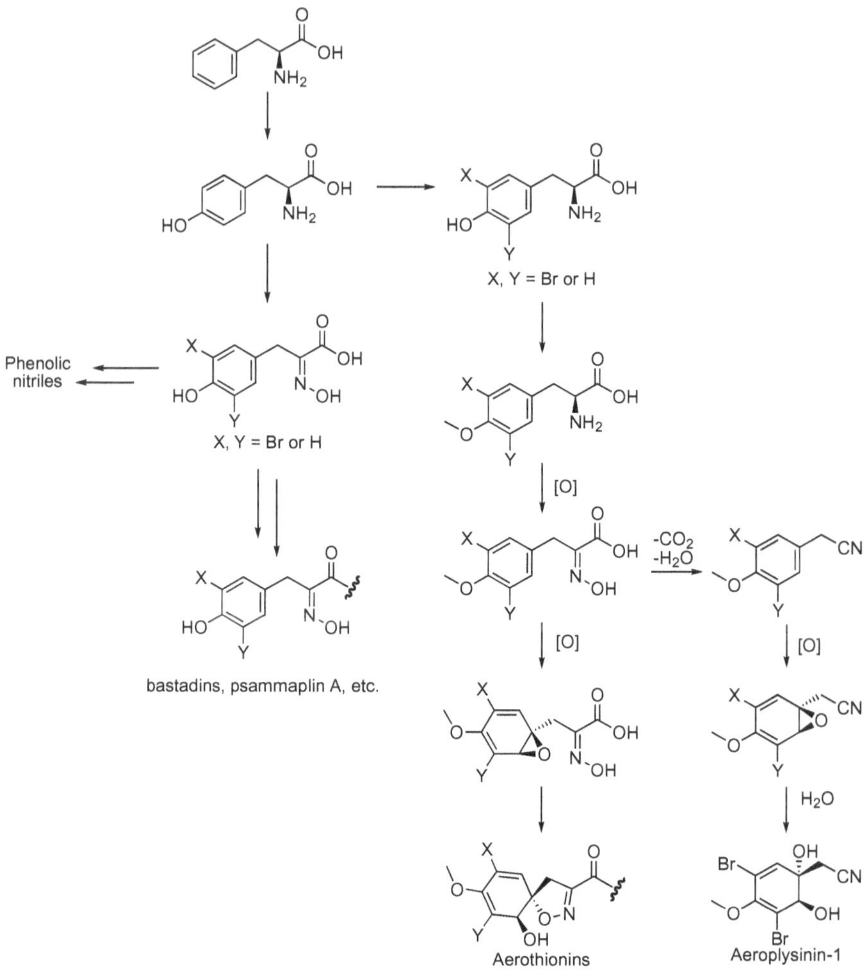

Figure 4.21 Biosynthesis hypothesis of some important bromotyrosine derivatives (Tymiak and Rinehart, 1981).

support the hypothesis that brominated secondary metabolites are produced by sponge cells. Despite this clear result, the enzymes that bind halogens to organic substrates have only been described in microorganisms, suggesting that some associated bacteria may also be involved in the biosynthesis of these compounds (Sacristan-Soriano *et al.*, 2011).

Several works were published on the biosynthesis of bromotyrosine derivatives, and an important experiment was performed by the group of Rinehart in the early 1980s (Tymiak and Rinehart, 1981). In this experiment, conducted with radiolabelled precursors, both phenylalanine and tyrosine were incorporated into homogentisamide. This result was in agreement with a previously postulated hypothesis on the skeletal rearrangement to obtain homogentisamide (Krejcarek *et al.*, 1975). The proposed mechanism should be similar to the one observed for the transformation of 4-hydroxyphenylpyruvic acid into homogentisic acid in *Streptomyces avermitilis* (Fig. 4.21; Denoya *et al.*, 1994). Different experimental results demonstrated that 3,5-dibromotyrosine and 3-bromotyrosine are incorporated into dibromoverongiaquinol and aeroplysinin-1 (Fattorusso *et al.*, 1970). Astonishingly, neither 3-bromo-4-hydroxybenzylcyanide nor 3,5-dibromo-4-hydroxybenzylcyanide were incorporated into dibromoverongiaquinol and aeroplysinin-1 by a feeding experiment (Carney and Rinehart, 1995). *O*-methylation could occur prior to the formation of the nitrile functional group, and thus the phenolic nitrile could not be the precursor of aeroplysinin-1. While simple compounds with an isoxazoline ring always possess an *O*-methyl group, complex compounds like bastadins or psammaplin A contain α-oximino amides and a free phenol group (Aoki *et al.*, 2006). Based on this observation, it has been proposed that the *O*-methyl group is required for the formation of the intermediate that leads to the dibromoverongiaquinol, aeroplysinin-1 and the isoxazoline metabolites, while metabolites with free phenols are converted into α-oximino compounds or phenolic nitriles (Carney and Rinehart, 1995). Even if the ^{14}C-labelled *O*-methyl nitriles were not incorporated into dibromoverongiaquinol and aeroplysinin-1, incorporation of tyrosine, 3-bromotyrosine and 3,5-dibromotyrosine supports the biosynthetic pathway depicted on Fig. 4.21. The absence of significant incorporation can be explained by the presence of bromine atoms interfering with the membrane permeability or a problem of sensitivity in the detection. The incorporation of the ^{14}C-labelled methyl of methionine into aeroplysinin-1 and its subsequent oxidation into a non-radioactive dienone product confirmed that methionine specifically labelled aeroplysinin-1 at the *O*-methyl group. The transfer of the methyl group from methionine to a precursor of aeroplysinin-1 is very common and implies the SAM co-substrate during the transmethylation reaction.

Like in higher organisms, it has been proposed that sponges possess active defence mechanisms (Paul and Puglisi, 2004; Paul and Ritson-Williams, 2008). Rapid wound-induced conversions of stored precursors to potent defensive compounds have been already observed. In sponges of the genus *Aplysina*, brominated isoxazoline alkaloids are transformed into more active compound such as aeroplysinin-1 or its dienone derivatives (Fig. 4.22; (Weiss *et al.*, 1996; Ebel *et al.*, 1997; Thoms *et al.*, 2006).

In studies conducted on *A. aerophoba* cell-free extracts, the unequivocal proof of the enzymatic conversion of brominated alkaloid has been observed. Indeed, the conversion of different isoxazoline alkaloids such as aerophobin-2, aplysinamisin-1 and isofistularin into aeroplysin-1 and its dienone derivatives was observed in these conditions (Weiss *et al.*, 1996; Ebel *et al.*, 1997). The ability to perform this conversion is not limited to

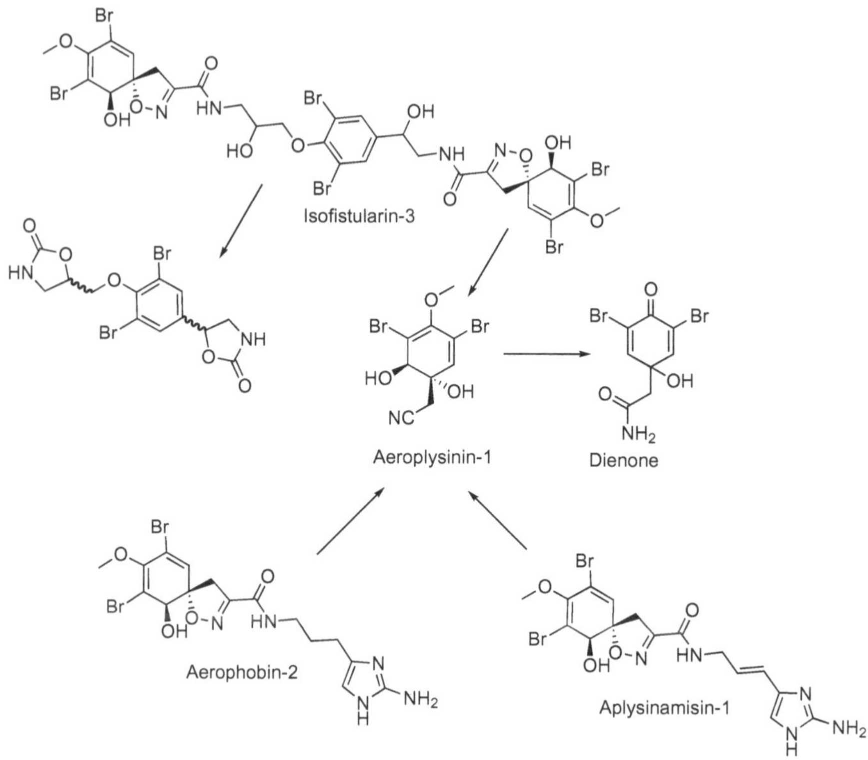

Figure 4.22 Wound induced bioconversion of the brominated isoxazoline alkaloids aerophobin-2, aplysinamisin-1 and isofistularin.

A. aerophoba, but also possible in other species belonging to this genus. Cell-free extracts of *A. cavernicola*, *A. archeri*, *A. cauliformis*, *A. fistularis*, *A. fulva* or *A. Lacunosa* have been tested as biosynthetic enzyme sources and led to the transformation of aerophobin-1 into the dienone. It is interesting to note that freeze-dried tissues of all *Aplysina* sponges contain more brominated isoxazoline alkaloids than aeroplysinin-1 or the dienone derivative. After incubation of the whole sponge in seawater, the authors observed a slight increase in the aeroplysinin-1 concentration and the dienone. After addition of freeze-dried tissues of *A. archeri*, a large increase in the concentrations of aeroplysinin-1 and the dienone was observed (Thoms *et al.*, 2006). Aeroplysinin-1 and the dienone have been tested against fish to evaluate their deterrence potential. Positive results were obtained with the polyphagous Mediterranean fish *Blennius sphinx* and with the common Caribbean wrasse *T. bifasciatum* (Ebel et al., 1997; Thoms et al., 2004). Both compounds showed additional activities in bioassays involving other marine organisms such as bacteria, microalgae or mollusks. Remarkably, the biotransformation of the protoxins aerophobin-1 and aplysinamisin-2 into the bioactive metabolites aeroplysinin-1 and the corresponding dienone occurred in less than 1 min. The chemical defence theory is corroborated by in field observations, which indicate that sponges showing signs of damage are very rare (Teeyapant and Proksch, 1993).

In the past decade, the group of Pawlik indicated that no evidence of chemical transformation was observed after tissue damage on two Caribbean *Aplysina* species (Puyana *et al.*, 2003). They highlighted that experiments on *A. aerophoba* were not performed with the living sponge, but with freeze-dried tissue samples and cell-free extracts and concluded that ecological studies should to be done preferably in field with living organism.

4.2. Discorhabdin derivatives

Another interesting group of aromatic alkaloids can be considered as chemotaxonomic maker of the genus *Latrunculia* (Hooper *et al.*, 1996). This group of natural products possesses a characteristic pyrrolo[4,3,2-de]phenantrolinone tetracyclic skeleton linked in some cases to a spiro centre at C-6 (Perry *et al.*, 1988). The first member of the discorhabdin family, discorhabdin C, was primarily isolated from a sponge of the genus *Latrunculia* (family Latrunculiidae, order Hadromerida) collected in New Zealand by Munro and co-workers in 1986 (Perry *et al.*, 1986). During a comprehensive study on *Latrunculia* sponges in New Zealand, Miller *et al.* determined the relationship between taxonomic, environmental, chemical variation within the genus (Miller *et al.*, 2001). The conclusions suggest that discorhabdins concentration variation within *Latrunculia* species, previously associated with intraspecific environmental variability, could reflect differences among species.

In 1995, Munro and co-workers presented a putative biosynthetic pathway for the discorhabdin-type compounds based on structural similarities among various groups of pyrroloiminoquinones (Fig. 4.23) (Lill *et al.*, 1995). A biosynthetic experiment was performed on slices of sponge tissue incubated with ^{14}C-phenylalanine. During the experiment, the authors were able to demonstrate that unfractionated dispersed cells of *Latrunculia* sp. retain the ability of the intact sponge and that phenylalanine is a precursor of the discorhabdins.

Tryptophan was also proposed as precursors of the discorhabdin skeleton. Tryptophan could first be decarboxylated and then hydroxylated. After an additional oxidation step, cyclization could lead to damirone. The results obtained with radiolabelled precursors were not sufficient to conclude on the nature of the tyramine incorporation that can arise after a nucleotide triphosphate-mediated amination of a damirone-like molecule or a direct Michael addition to the electronically favoured iminoquinone followed by an oxidation step. The dimer discorhabdin W has been isolated from a *Latrunculia* sponge (Lang *et al.*, 2005). Preliminary results demonstrated that

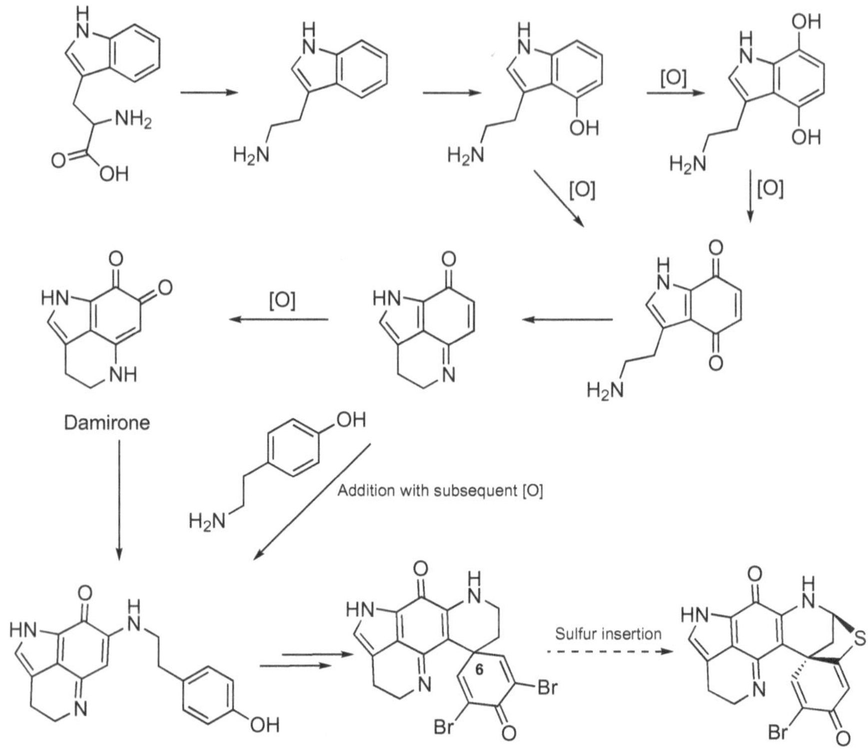

Figure 4.23 Putative biosynthetic pathways of the discorhabdins (Lill *et al.*, 1995).

discorhabdin B is involved in the biogenesis of discorhabdin W. Upon irradiation of discorhabdin B solutions with sunlight, discorhabdin B was decomposed *via* a radical reaction and discorhabdin W was detected among the products (Fig. 4.24).

Discorhabdins can act as chemical weapons and some deterrent activities have been tested (McClintock *et al.*, 1994). The authors reported that the organic extract of the dark green *Latrunculia apicalis* elicited a significant tube-foot retraction response in the major predator of Antarctic sponges, the sea star *Perknaster fuscus*. Another study showed that discorhabdin G was responsible for the feeding deterrence of *P. fuscus*. This activity was confirmed by other works using series of sea star tube-foot retraction response (Furrow *et al.*, 2003). The allocation of the bioactive compound discorhabdin G to the outermost layers of *L. apicalis* could prevent the sponge from attacks by sea stars. These results are consistent with an optimal defence theory, as *L. apicalis* sequesters is chemically feeding deterrent (discorhabdin G) in its most exposed tissues.

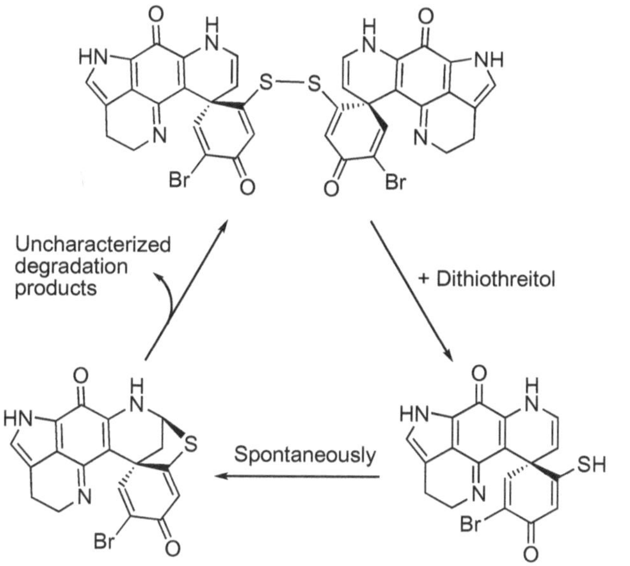

Figure 4.24 Formation and cleavage of discorhabdin W (Lang *et al.*, 2005).

5. Alkaloids

A large number of sponge secondary metabolites are characterized by a high level of nitrogens in their chemical structure. Among the alkaloids produced by sponges, peptides and especially depsipeptides have been identified early as promising bioactive compounds and they have been reviewed in 1993 (Fusetani and Matsunaga, 1993). Nevertheless, we will not focus on this important group of secondary metabolites because the biosynthetic genes leading to such compounds have been mostly identified in their associated microbes, bacteria or fungi (Dunlap *et al.*, 2007; Zhu *et al.*, 2009; Siegl and Hentschel, 2010; Zhou *et al.*, 2011). We will detail the results on the biosynthesis and the chemical ecology of three groups of sponge alkaloids: 3-alkylpiperidine, guanidine and pyrrole-imidazole alkaloids (PIAs).

5.1. 3-Alkylpiperidine alkaloids

A large number of 3-alkylpiperidine derivatives have been isolated from marine sponges (Sepcic, 2000; Sepcic and Turk, 2006; Timm *et al.*, 2008, 2010). Even if some examples are found in mollusks which can originate from a sponge diet, they seem to be restricted to sponges and especially to the order Haplosclerida (mainly from the genus *Haliclona*) and they were first accepted as chemotaxonomic markers of this order (Andersen *et al.*, 1996; Erpenbeck and van Soest, 2007). The chemical diversity in this group includes monomers like the niphatesines from *Haliclona* sp. (Kobayashi *et al.*, 1990), dimers which can be cyclic like the cyclostelletamines from *Stelletta maxima* (Fusetani *et al.*, 1994) or linear like the pachychalines from *Pachychalina* sp. (Laville *et al.*, 2008), trimers like the viscosamine from *Haliclona viscosa* (Volk and Koeck, 2003) and even the highly bioactive polymers halitoxins from sponges of the genus *Haliclona* (Fig. 4.25; Schmitz *et al.*, 1978).

More complex biosynthetic derivatives belonging to this class of natural products have also been isolated. Among others, the sarains have been isolated from the Mediterranean sponge *Haliclona sarai* (Cimino *et al.*, 1986, 1989a–c, 1990) and the manzamines from sponges of the genera *Haliclona* and *Pellina* (Fig. 4.26; Sakai et al., 1986).

Recently, some actinomycetes have been identified as producers of the manzmines (Hill *et al.*, 2004), but, with this exception, no cellular localization studies has been undertaken. It can be assumed that sponge cells should play a role in some of the biosynthetic events leading to these highly bioactive compounds due to the high concentrations of compounds usually found in the studied species. No feeding biosynthetic study has been reported so far except for the production of haminol which is produced by a Mediterranean nudibranch (Cutignano *et al.*, 2003). Nicotinic acid is

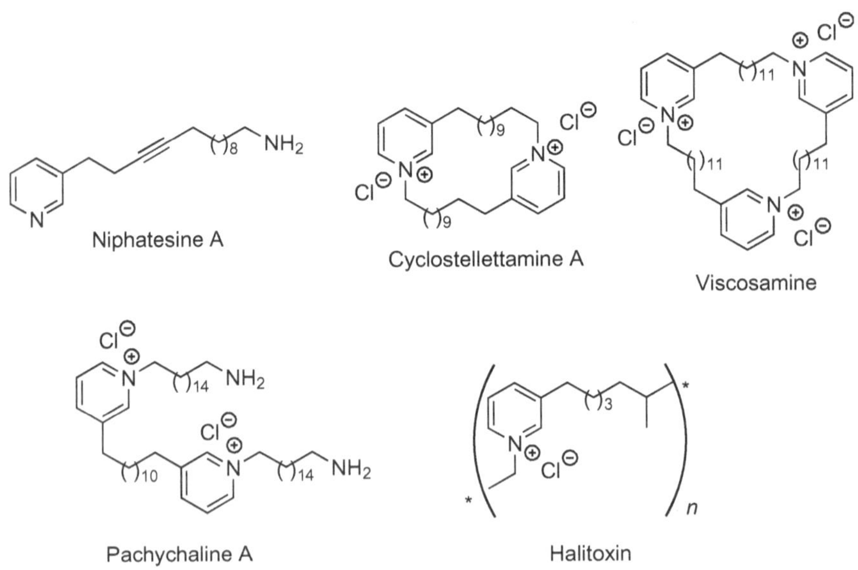

Figure 4.25 3-Alkylpyridine and 3-alkylpyridinium salts isolated from Haplosclerida sponges.

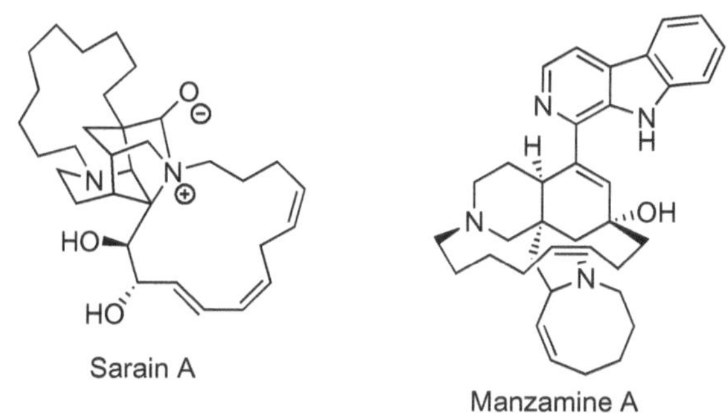

Figure 4.26 Sarains and manzamines isolated from Haplosclerida sponges.

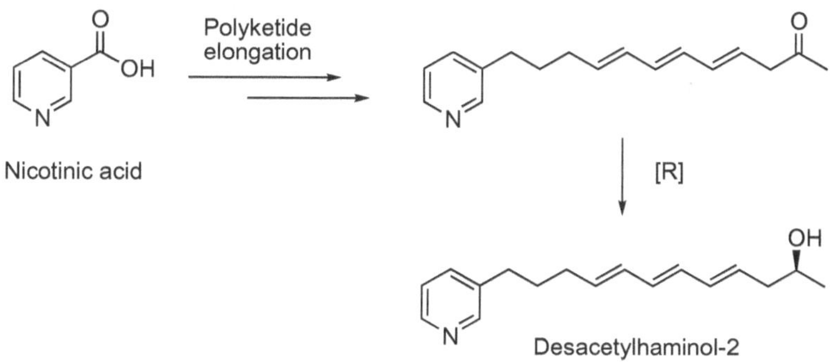

Figure 4.27 Biosynthetic pathway leading to a 3-alkylpyridine in a mollusk (Cutignano *et al.*, 2003).

proposed as the precursor of this alkaloid and could also be involved in the biosynthesis of sponge 3-alkylpiperidine in general. A polyacetate pathway was also demonstrated unambiguously for the alkyl chains (Fig. 4.27).

This result is not in accordance with the putative pathways proposed earlier by Baldwin and then Marazano using biomimetic considerations, but it remains to be verified for sponge 3-alkylpiperidine derivatives (Baldwin *et al.*, 1998; Kaiser *et al.*, 1998). Another hypothesis has been proposed recently on the basis of the structures of pachychalines involving a cyclization of terminal norspermidine moieties to yield the 3-alkylpyridines (Laville *et al.*, 2009).

Halitoxin polymers were found to be the antimicrobial constituents of the Red Sea *Amphimedon viridis* and the antifeedant compounds present in the Caribbean sponge *Amphimedon compressa* (Albrizio *et al.*, 1995; Dovi *et al.*, 2001). This class of compounds has been largely studied for their application as antifouling agents, and more precisely, the compounds produced by the Mediterranean sponge *H. sarai* by the group of Sepcic

(Sepcic *et al.*, 1999; Faimali *et al.*, 2003, 2005; Elersek *et al.*, 2008; Blihoghe *et al.*, 2011). The acyclic dimer haliclonacyclamine isolated from one *Haliclona* species was also found to exhibit interesting antifouling activity (Roper *et al.*, 2009). Applications of these compounds for their pharmaceutical or antifouling properties are ongoing.

5.2. Guanidine alkaloids

The guanidine group of alkaloids is frequently found in sponge natural products and especially in a large family of polycyclic guanidine alkaloids produced by sponges distributed worldwide (Berlinck, 1996, 1999, 2002; Berlinck and Kossuga, 2005, 2008; Berlinck *et al.*, 2010). This type of alkaloids is even recognized as chemotaxonomic markers of sponges of the family Crambeidae (order Poecilosclerida) (Erpenbeck and van Soest, 2007). Here also the structures of the compounds are extremely complex as exemplified by the bicyclic crambescine A, isolated from the Mediterranean sponge *Crambe crambe* (Berlinck *et al.*, 1990), the bicyclic/tricyclic batzelladine A isolated from the Caribbean sponge *Batzella* sp. (Patil *et al.*, 1995) and the pentacyclic ptilomycalin isolated from the Caribbean sponge *Ptilocaulis spiculifer* (Kashman *et al.*, 1989) (Fig. 4.28).

A full chemical ecological study was undertaken by the group of Turon on the guanidine alkaloids produced by the marine sponge *Crambe crambe*. Firstly, crambescines and crambescidins have been identified in the

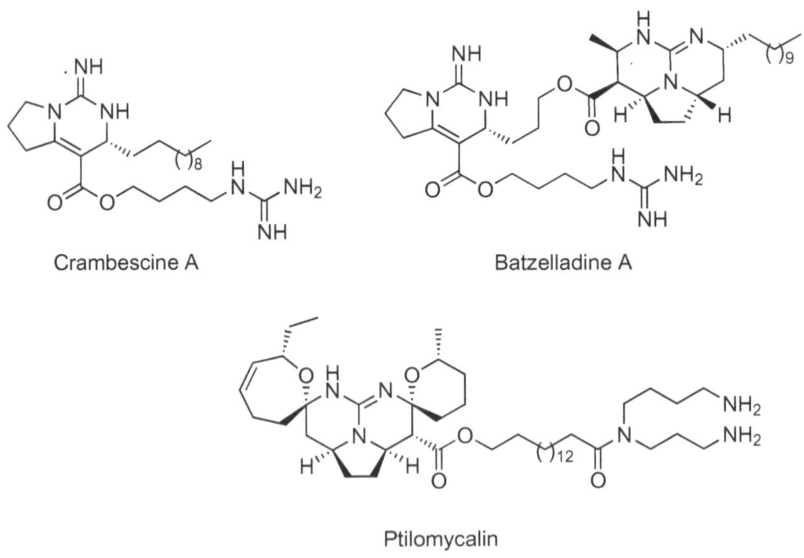

Figure 4.28 Polycyclic guanidine alkaloids isolated from sponges of the Crambeidae family.

spherulous cells of the sponge, thus suggesting at least a storage role for these cells (Uriz *et al.*, 1996). The same authors observed a variation in the production of these toxins according to the habitat or the size of the specimen, and they concluded that space competition could explain the differences in the production of the target compounds (Becerro *et al.*, 1995). A seasonal toxicity was noticed with a minimum in April and a maximum in September, and the toxicity was higher at the periphery of the sponge. They also observed that the surface of the sponge was almost axenic which must be due to an antimicrobial effect of some metabolites produced by the sponge (Becerro *et al.*, 1997b). Finally, they were able to identify a moderate antifouling activity of the extract which was in the same time devoid of any antialgal effect. The compounds were absent in the larvae until 2 weeks old and no antifeedant activity was observed before this age, while after 2 weeks, the larvae and the juveniles showed antifeedant activity. They finally underlined the importance of the methodological procedures in evaluating the right concentration of the compounds which is a requisite towards an appropriate ecological interpretation of the data (Becerro *et al.*, 1997a).

5.3. Pyrrole-imidazole alkaloids

PIAs include hundreds of secondary metabolites originating from marine sponges exclusively (Forte *et al.*, 2009). These compounds, with structure complexity ranged from the simple achiral oroidin to the extremely complex stylassidines A and B, have been isolated mainly from species of the taxons *Agelasidae*, *Axinellidae*, *Dyctionellidae* and *Hymeniacidonidae* (Assmann *et al.*, 2001). In this context, this group of compounds have been evaluated as a potential chemotaxonomic maker for the genus *Agelasidae* (Braekman *et al.*, 1992).

Studies on the cellular localization of these compounds suggested that they are located into the spherulous cells of the sponge (Richelle-Maurer *et al.*, 2003). Despite the large interest of the scientific community for this family of compounds, only few experimental works have been conducted to elucidate the biosynthesis of PIAs. The biosynthesis pathway can be separated into two categories underlining a central role for oroidin: the upstream events, that include all the biosynthetic reactions leading to oroidin, and the downstream events, including reactions conducting to more complex compounds (Fig. 4.29).

Only two biosynthetic studies have been performed for the elucidation of the upstream events so far (Andrade *et al.*, 1999; Genta-Jouve *et al.*, 2011), and four different hypotheses based on biomimetic works have been proposed since the discovery of dibromophakelin in 1982 (Foley and Büchi, 1982; Kitagawa *et al.*, 1983; Braekman *et al.*, 1992; Lindel *et al.*, 2000a; Vergne *et al.*, 2006). In 1999, Andrade *et al.* published the results obtained on a cell culture of the sponge *Cymbaxinella corrugata* (previously *Teichaxinella*

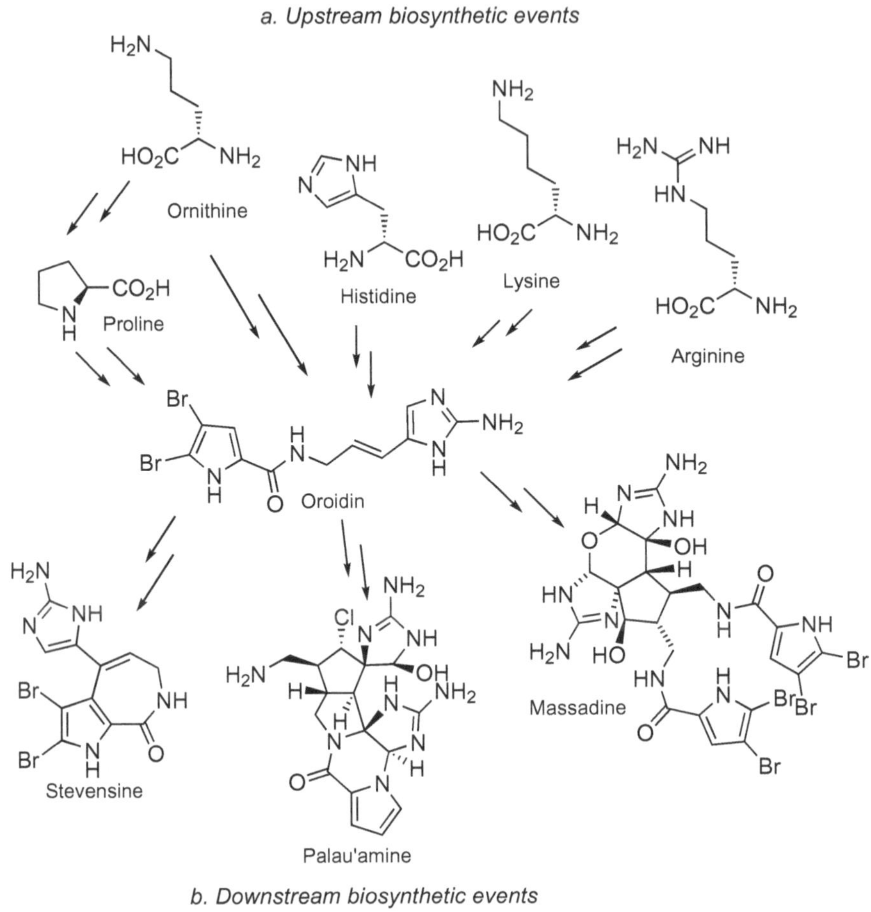

Figure 4.29 The central role of oroidin in the PIAs biosynthesis (Genta-Jouve *et al.*, 2011).

morchella) (Andrade *et al.*, 1999). Using radiolabelled precursors, they were able to incorporate ^{14}C-proline, ^{14}C-ornithine and ^{14}C-histidine into stevensine. A more recent *in vivo* experiment conducted on the Mediterranean *Cymbaxinella damicornis* using the whole organism led to a different conclusion. As previously proposed by Lindel and Köck, ^{14}C-lysine was successfully incorporated together with ^{14}C-proline into oroidin (Genta-Jouve *et al.*, 2011). According to these results, the biosynthetic pathway of oroidin can be described as below and includes the conversion of lysine into homoarginine through steps well known in the urea cycle for the guanidinylation of ornithine. Oxidation of homoarginine into γ-hydroxyhomoarginine followed by oxidation and condensation with the brominated pyrrole-2carboxylic acid lead to oroidin (Fig. 4.30)

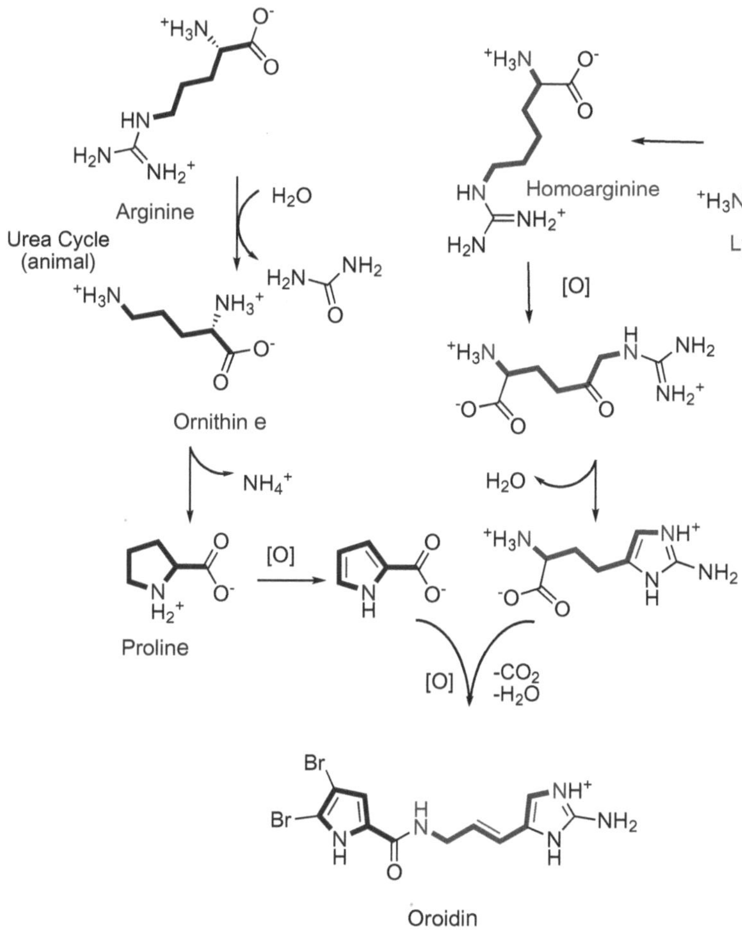

Figure 4.30 Proposed biosynthetic pathway of oroidin (adapted from Genta-Jouve *et al.*, 2011). (For colour version of this figure, the reader is referred to the online version of this chapter.)

The biosynthetic prominent role of oroidin into the biosynthesis of dimers, trimers, and tetramers is nowadays largely accepted, but only few experiments have been reported, using dichloroclathrodin as precursor (Al Mourabit et al., 2011; Seiple et al., 2011). The experience evidenced that cell-free extracts of *Agelas conifera* and *Stylissa caribica* catalyzed conversions of synthetic dichloroclathrodin to the corresponding unnatural products, tetrachlorobenzosceptrin A and tetrachloronagelamide H (Molinski *et al.*, 2011).

If the biosynthesis has been sparsely studied since the discovery of oroidin, the ecological roles of compounds belonging to this family has been largely investigated for many years. In the 1990s, the firsts experimental studies reported the eco-assay guided isolation of two feeding deterrent compounds: 4,5-dibromopyrrole-2-carboxylic acid and oroidin (Pawlik *et al.*, 1995; Chanas *et al.*, 1997). Similar antifeeding activity has also been reported for the closely related stevensine (Wilson *et al.*, 1999). A structure–activity relationship study has been performed and the imidazole moiety should not be responsible for the deterrent activity, while the bromopyrrole seems important (Lindel *et al.*, 2000b). A more recent study performed on the Mediterranean sponges *Cymbaxinella polypoides* and *C. verrucosa* suggested that these compounds may serve as a chemical defense against predators using the generalist shrimp *Palaemon elegans* in the assays (Mollo *et al.*, 2008; Haber *et al.*, 2011). The butanolic fraction of *C. verrucosa* extract showed a significant deterrent activity that the authors assigned to hymenidin, which significantly deterred shrimps feeding. These results confirmed the multiple defensive functions previously ascribed to bromopyrroles isolated from Caribbean sponges.

6. Conclusions

Research on the sponge chemical diversity has followed several important steps. During the late 1960s, the marine environment appeared as a new source of secondary metabolites and the discovery of new structural architectures from marine sponges covered the end of the twentieth century. The sea appeared as a new source of original chemical structures, often associated with interesting biological activities and applications in the therapeutic field. The main success in this research is the launching of the first antitumoral compound inspired from a sponge derived natural product (Halaven). Since the beginning of these studies, marine natural product researchers were usually chemists, but also good divers with the strong hope to find a new medicine from the Sea. The biosynthetic pathways as well as the ecological concerns on these compounds were not the main interest of the small groups of researchers working in this field. The question of the level of similarity between the terrestrial and marine environments was not fully addressed during this period even if preliminary studies were started. Today, even if some biosynthetic marine studies began in the 1970s with also some studies in chemical ecology, we still lack a global view on these topics. Difficulties may partly be attributed to a lack of funding supports for such basic researches even if industrial applications are closely associated.

Since the beginning of the twenty-first century and the awareness of the richness of our endangered marine biodiversity, several consortia have been built worldwide to deepen our knowledge on these intriguing structural architectures. The new technologies developed in the field of metabolomics should make possible a quicker assessment of the fluctuations in the metabolome. A lot of questions are still unanswered about the role of the associated micro-organisms or the factors that could be involved in the fluctuation of the compound production. Until now, most of the ecological studies have focused on the natural bioactivity of these compounds in order to assess an ecological role in the ecosystem but conclusions are usually very difficult to obtain. Geographical and temporal data are necessary to give some important clues towards a better comprehension on the underlying mechanisms of the chemical cues in the marine ecosystems. The surrounding environment may not be the main explanation to these fluctuations but physiological reasons should be taken into consideration. It must be emphasized that studies in chemical ecology is a highly interdisciplinary field where the chemist has a key role in identifying the structure of the target compounds. Precautions should be taken in the construction of the analytical chemical tools necessary to assess the concentrations of the targeted metabolites.

The chemical structures can also inspire the construction of a putative biosynthetic pathway, but experimental evidences are urgently needed. Until now, very few studies on sponge biosynthetic pathways have been reported. Indeed, very low incorporation rates of the labelled precursors are usually observed in feeding experiments due to dilution in the water and primary metabolism. With the development of highly sensitive detectors, experimental results are now expected. Finally, molecular biology allowed the identification of biosynthetic pathways for some polyketides and peptides because of a microbial origin. Other important results should appear soon for the true sponge secondary metabolites.

There remain other very important groups of sponge natural products that have been poorly studied and that deserve deep ecological and biosynthetic investigations. These studies will give important clues to assess the level of similarity with plant biosynthetic pathways. In an evolutionary perspective, this issue is also of high interest and especially in a context where microbes could have transferred their genetic material in a horizontal manner.

REFERENCES

Abdo, D., Motti, C., Battershill, C., and Harvey, E. (2007). Temperature and spatiotemporal variability of Salicylihalamide A in the sponge Haliclona. *Journal of Chemical Ecology* **33,** 1635–1645.

Al Mourabit, A., Zancanella, M. A., Tilvi, S., and Romo, D. (2011). Biosynthesis, asymmetric synthesis, and pharmacology, including cellular targets, of the pyrrole-2-aminoimidazole marine alkaloids. *Natural Product Reports* **28,** 1229–1260.

Albrizio, S., Ciminiello, P., Fattorusso, E., Magno, S., and Pawlik, J. R. (1995). Amphitoxin, a New High Molecular Weight Antifeedant Pyridinium Salt from the Caribbean Sponge Amphimedon compressa. *Journal of Natural Products* **58,** 647–652.

Andersen, R. J., van Soest, R. W. M., and Kong, F. (1996). 3-Alkylpiperidine alkaloids isolated from marine sponges in the order Haplosclerida. In "Alkaloids: Chemical and Biological Perspectives" (S. W. Pelletier, ed.), pp. 301–355. Pergamon, New York.

Andrade, P., Willoughby, R., Pomponi, S. A., and Kerr, R. G. (1999). Biosynthetic studies of the alkaloid, stevensine, in a cell culture of the marine sponge Teichaxinella morchella. *Tetrahedron Letters* **40,** 4775–4778.

Aoki, S., Cho, S.-H., Hiramatsu, A., Kotoku, N., and Kobayashi, M. (2006). Bastadins, cyclic tetramers of brominated-tyrosine derivatives, selectively inhibit the proliferation of endothelial cells. *Journal of Natural Medicines* **60,** 231–235.

Assmann, M., van Soest, R. W. M., and Köck, M. (2001). New antifeedant bromopyrrole alkaloid from the Caribbean sponge Stylissa caribica. *Journal of Natural Products* **64,** 1345–1347.

Baldwin, J. E., Claridge, T. D. W., Culshaw, A. J., Heupel, F. A., Lee, V., Spring, D. R., Whitehead, R. C., Boughtflower, R. J., Mutton, I. M., and Upton, R. J. (1998). Investigations into the manzamine alkaloid biosynthetic hypothesis. *Angewandte Chemie International Edition* **37,** 2661–2663.

Becerro, M. A., Turon, X., and Uriz, M. J. (1995). Natural variation of toxicity in encrusting sponge Crambe crambe (Schmidt) in relation to size and environment. *Journal of Chemical Ecology* **21,** 1931–1946.

Becerro, M. A., Turon, X., and Uriz, M. J. (1997a). Multiple functions for secondary metabolites in encrusting marine invertebrates. *Journal of Chemical Ecology* **23,** 1527–1547.

Becerro, M. A., Uriz, M. J., and Turon, X. (1997b). Chemically-mediated interactions in benthic organisms: the chemical ecology of Crambe crambe (Porifera, Poecilosclerida). *Hydrobiologia* **355,** 77–89.

Becerro, M. A., Paul, V. J., and Starmer, J. (1998). Intracolonial variation in chemical defenses of the sponge Cacospongia sp. and its consequences on generalist fish predators and the specialist nudibranch predator Glossodoris pallida. *Marine Ecology Progress Series* **168,** 187–196.

Bergmann, W. (1949). Marine products. XXV. Comparative biochemical studies on the lipids and sterols of marine invertebrates. *Journal of Marine Research* **8,** 137–176.

Bergmann, W., and Feeney, R. J. (1949). Marine products. XXIII. Sterols from sponges of the family Haliclonidae. *The Journal of Organic Chemistry* **14,** 1078–1084.

Bergquist, P. R., Karuso, P., Cambie, R. C., and Smith, D. J. (1991). Sterol composition and classification of the Porifera. *Biochemical Systematics and Ecology* **19,** 17–24.

Berlinck, R. G. S. (1996). Natural guanidine derivatives. *Natural Product Reports* **13,** 377–409.

Berlinck, R. G. S. (1999). Natural guanidine derivatives. *Natural Product Reports* **16,** 339–365.

Berlinck, R. G. S. (2002). Natural guanidine derivatives. *Natural Product Reports* **19,** 617–649.

Berlinck, R. G. S., and Kossuga, M. H. (2005). Natural guanidine derivatives. *Natural Product Reports* **22,** 516–550.

Berlinck, R. G. S., and Kossuga, M. H. (2008). Guanidine alkaloids from marine invertebrates. In "Modern Alkaloids" (E. Fattorusso and O. Taglialatela-Scafati, eds), pp. 305–337. Wiley-VCH Verlag GmbH & Co. KGaA, Weinheim, Germany.

Berlinck, R. G. S., Braekman, J. C., Daloze, D., Hallenga, K., Ottinger, R., Bruno, I., and Riccio, R. (1990). Two new guanidine alkaloids from the Mediterranean sponge Crambe crambe. *Tetrahedron Letters* **31,** 6531–6534.

Berlinck, R. G. S., Burtoloso, A. C. B., Trindade-Silva, A. E., Romminger, S., Morais, R. P., Bandeira, K., and Mizuno, C. M. (2010). The chemistry and biology of organic guanidine derivatives. *Natural Product Reports* **27,** 1871–1907.

Blacklock, B. J., Scheffler, B. E., Shepard, M. R., Jayasuriya, N., and Minto, R. E. (2010). Functional diversity in fungal fatty acid synthesis: The first acetylenase from the Pacific Golden Chanterelle, Cantharellus formosus. *The Journal of Biological Chemistry* **285,** 28442–28449.

Blihoghe, D., Manzo, E., Villela, A., Cutignano, A., Picariello, G., Faimali, M., and Fontana, A. (2011). Evaluation of the antifouling properties of 3-alyklpyridine compounds. *Biofouling* **27,** 99–109.

Blunt, J. W., Copp, B. R., Munro, M. H. G., Northcote, P. T., and Prinsep, M. R. (2011). Marine natural products. *Natural Product Reports* **28,** 196–268.

Bobzin, S. C., and Faulkner, D. J. (1992). Chemistry and chemical ecology of the Bahamian sponge Aplysilla glacialis. *Journal of Chemical Ecology* **18,** 309–332.

Bowling, J., Mohammed, R., Diers, J., and Hamann, M. (2010). Abundant ketone isolated from oily Plakortis sponge demonstrates antifouling properties. *Chemoecology* **20,** 207–213.

Braekman, J. C., Daloze, D., Deneubourg, F., Huysecom, J., and Vandevyver, G. (1987). I-Isocyanoaromadendrane, a new isonitrile sesquiterpene from the sponge Acanthella Acuta. *Bulletin des Societes Chimiques Belges* **96,** 539–543.

Braekman, J.-C., Daloze, D., Stoller, C., and Van Soest, R. W. M. (1992). Chemotaxonomy of Agelas (Porifera: Demospongiae). *Biochemical Systematics and Ecology* **20,** 417–431.

Cable, K. M., Herbert, R. B., Knaggs, A. R., and Mann, J. (1991). The biosynthesis of xanthocillin-X monomethyl ether in dichotomomyces cejpii. Experiments on the origin of the isocyanide carbon atoms. *Journal of the Chemical Society, Perkin Transactions* **1,** 595–599.

Cachet, N., Regalado, E. L., Genta-Jouve, G., Mehiri, M., Amade, P., and Thomas, O. P. (2009). Steroidal glycosides from the marine sponge Pandaros acanthifolium. *Steroids* **74,** 746–750.

Campagnuolo, C., Fattorusso, E., and Taglialatela-Scafati, O. (2001). Feroxosides A-B, two norlanostane tetraglycosides from the Caribbean sponge Ectyoplasia ferox. *Tetrahedron* **57,** 4049–4055.

Carballeira, N., Thompson, J. E., Ayanoglu, E., and Djerassi, C. (1986). Biosynthetic studies of marine lipids. 5. The biosynthesis of long-chain branched fatty acids in marine sponges. *The Journal of Organic Chemistry* **51,** 2751–2756.

Carney, J. R., and Rinehart, K. L. (1995). Biosynthesis of brominated tyrosine metabolites by Aplysina fistularis. *Journal of Natural Products* **58,** 971–985.

Catalan, C. A. N., Thompson, J. E., Kokke, W. C. M. C., and Djerassi, C. (1985). Biosynthetic studies of marine lipids: Experimental demonstration of the course of side chain extension in marine sterols. *Tetrahedron* **41,** 1073–1084.

Chanas, B., Pawlik, J. R., Lindel, T., and Fenical, W. (1997). Chemical defense of the Caribbean sponge Agelas clathrodes (Schmidt). *Journal of Experimental Marine Biology and Ecology* **208,** 185–196.

Cimino, G., de Rosa, S., de Stefano, S., and Sodano, G. (1982). The chemical defense of four Mediterranean nudibranchs. *Comparative Biochemistry and Physiology Part B: Comparative Biochemistry* **73,** 471–474.

Cimino, G., de Giulio, A., de Rosa, S., de Stefano, S., and Sodano, G. (1985a). Further high molecular weight polyacetylenes from the sponge Petrosia ficiformis. *Journal of Natural Products* **48,** 22–27.

Cimino, G., de Rosa, S., de Stefano, S., Morrone, R., and Sodano, G. (1985b). The chemical defense of nudibranch molluscs: Structure, biosynthetic origin and defensive properties of terpenoids from the dorid nudibranch dendrodoris grandiflora. *Tetrahedron* **41,** 1093–1100.

Cimino, G., de Rosa, S., de Stefano, S., and Sodano, G. (1986). Marine natural products: new results from Mediterranean invertebrates. *Pure and Applied Chemistry* **58,** 375–386.

Cimino, G., Mattia, C. A., Mazzarella, L., Puliti, R., Scognamiglio, G., Spinella, A., and Trivellone, E. (1989a). Unprecedented alkaloid skeleton from the Mediterranean sponge Reniera sarai: X-ray structure of an acetate derivative of saraine-A. *Tetrahedron* **45,** 3863–3872.

Cimino, G., Puliti, R., Scognamiglio, G., Spinella, A., Trivellone, E., Mattia, C. A., and Mazzarella, L. (1989b). Amazing new alkaloid skeletons from the marine sponge Reniera sarai. *Pure and Applied Chemistry* **61,** 535–538.

Cimino, G., Spinella, A., and Trivellone, E. (1989c). Isosarain-1. A new alkaloid from the Mediterranean sponge Reniera sarai. *Tetrahedron Letters* **30,** 133–136.

Cimino, G., Scognamiglio, G., Spinella, A., and Trivellone, E. (1990). Structural studies on saraine A. *Journal of Natural Products* **53,** 1519–1525.

Curran, D. P., and Sui, B. (2009). A shortcut mosher ester method to assign configurations of stereocenters in nearly symmetric environments. Fluorous mixture synthesis and structure assignment of petrocortyne A. *Journal of the American Chemical Society* **131,** 5411–5413.

Cutignano, A., Tramice, A., De, C. S., Villani, G., Cimino, G., and Fontana, A. (2003). Biogenesis of 3-alkylpyridine alkaloids in the marine mollusc Haminoea orbignyana. *Angewandte Chemie International Edition* **42,** 2633–2636.

Dai, J.-R., Hallock, Y. F., Cardellina, J. H., Gray, G. N., and Boyd, M. R. (1996). Triangulynes A-H and triangulynic acid, new cytotoxic polyacetylenes from the marine sponge Pellina triangulata. *Journal of Natural Products* **59,** 860–865.

De Rosa, S., De Giulio, A., and Iodice, C. (1994). Biological effects of prenylated hydroquinones: Structure-activity relationship studies in antimicrobial, brine shrimp, and fish lethality assays. *Journal of Natural Products* **57,** 1711–1716.

Dekas, A. E., Poretsky, R. S., and Orphan, V. J. (2009). Deep-sea Archaea fix and share nitrogen in methane-consuming microbial consortia. *Science* **326,** 422–426.

Denoya, C. D., Skinner, D. D., and Morgenstern, M. R. (1994). A Streptomyces avermitilis gene encoding a 4-hydroxyphenylpyruvic acid dioxygenase-like protein that directs the production of homogentisic acid and an ochronotic pigment in Escherichia coli. *Journal of Bacteriology* **176,** 5312–5319.

Djerassi, C., and Lam, W. K. (1991). Phospholipid studies of marine organisms. Part 25. Sponge phospholipids. *Accounts of Chemical Research* **24,** 69–75.

Djerassi, C., and Silva, C. J. (1991). Biosynthetic studies of marine lipids. 41. Sponge sterols: origin and biosynthesis. *Accounts of Chemical Research* **24,** 371–378.

Dovi, K., Yoel, K., Eugene, R., Micha, I., Ilan, I., and Yossi, L. (2001). Antimicrobial activity of the reef sponge Amphimedon viridis from the Red Sea: evidence for selective toxicity. *Aquatic Microbial Ecology* **24,** 9–16.

Dumdei, E. J., Flowers, A. E., Garson, M. J., and Moore, C. J. (1997). The biosynthesis of sesquiterpene isocyanides and isothiocyanates in the marine sponge Acanthella cavernosa (Dendy); evidence for dietary transfer to the dorid nudibranch Phyllidiella pustulosa. *Comparative Biochemistry and Physiology. Part A, Physiology* **118A,** 1385–1392.

Dunlap, W. C., Battershill, C. N., Liptrot, C. H., Cobb, R. E., Bourne, D. G., Jaspars, M., Long, P. F., and Newman, D. J. (2007). Biomedicinals from the phytosymbionts of marine invertebrates: A molecular approach. *Methods* **42,** 358–376.

Ebel, R., Brenzinger, M., Kunze, A., Gross, H. J., and Proksch, P. (1997). Wound activation of protoxins in marine sponge Aplysina aerophoba. *Journal of Chemical Ecology* **23,** 1451–1462.

Edenborough, M. S., and Herbert, R. B. (1988). Naturally occurring isocyanides. *Natural Product Reports* **5,** 229–245.

Elersek, T., Kosi, G., Turk, T., Pohleven, F., and Sepcic, K. (2008). Influence of polymeric 3-alkylpyridinium salts from the marine sponge Reniera sarai on the growth of algae and wood decay fungi. *Biofouling* **24,** 137–143.

Erpenbeck, D., and van Soest, R. W. M. (2007). Status and perspective of sponge chemosystematics. *Marine Biotechnology* **9,** 2–19.

Ettinger-Epstein, P., Tapiolas, D., Motti, C., Wright, A., Battershill, C., and de Nys, R. (2008). Production of manoalide and its analogues by the Sponge luffariella variabilis is hardwired. *Marine Biotechnology* **10,** 64–74.

Faimali, M., Sepcic, K., Turk, T., and Geraci, S. (2003). Non-toxic antifouling activity of polymeric 3-alkylpyridinium salts from the Mediterranean sponge Reniera sarai (Pulitzer-Finali). *Biofouling* **19,** 47–56.

Faimali, M., Garaventa, F., Mancini, I., Sicurelli, A., Guella, G., Piazza, V., and Greco, G. (2005). Antisettlement activity of synthetic analogues of polymeric 3-alkylpyridinium salts isolated from the sponge Reniera sarai. *Biofouling* **21,** 49–57.

Fattorusso, E., and Taglialatela-Scafati, O. (2010). Marine endoperoxides as antimalarial lead compounds. *Phytochemistry Reviews* **9,** 515–524.

Fattorusso, E., Minale, L., and Sodano, G. (1970). Aeroplysinin-I, a new bromo-compound from Aplysina aerophoba. *Journal of the Chemical Society D: Chemical Communications* 751–752.

Fisch, K. M., Gurgui, C., Heycke, N., van der Sar, S. A., Anderson, S. A., Webb, V. L., Taudien, S., Platzer, M., Rubio, B. K., Robinson, S. J., Crews, P., and Piel, J. (2009). Polyketide assembly lines of uncultivated sponge symbionts from structure-based gene targeting. *Nature Chemical Biology* **5,** 494–501.

Foley, L. H., and Büchi, G. (1982). Biomimetic synthesis of dibromophakellin. *Journal of the American Chemical Society* **104,** 1776–1777.

Fontana, A., Ciavatta, M. L., Miyamoto, T., Spinella, A., and Cimino, G. (1999). Biosynthesis of drimane terpenoids in dorid molluscs: Pivotal role of 7-deacetoxyolepupuane in two species of Dendrodoris nudibranchs. *Tetrahedron* **55,** 5937–5946.

Forte, B., Malgesini, B., Piutti, C., Quartieri, F., Scolaro, A., and Papeo, G. (2009). A submarine journey: The pyrrole-imidazole alkaloids. *Marine Drugs* **7,** 705–753.

Furrow, F. B., Amsler, C. D., McClintock, J. B., and Baker, B. J. (2003). Surface sequestration of chemical feeding deterrents in the Antarctic sponge Latrunculia apicalis as an optimal defense against sea star spongivory. *Marine Biology* **143,** 443–449.

Fusetani, N. (1997). Marine natural products influencing larval settlement and metamorphosis of benthic invertebrates. *Current Organic Chemistry* **1,** 127–152.

Fusetani, N., and Matsunaga, S. (1993). Bioactive sponge peptides. *Chemical Reviews (Washington, DC, United States)* **93,** 1793–1806.

Fusetani, N., Asai, N., Matsunaga, S., Honda, K., and Yasumuro, K. (1994). Bioactive marine metabolites. 59. Cyclostellettamines A-F, pyridine alkaloids which inhibit binding of methyl quinuclidinyl benzilate (QNB) to muscarinic acetylcholine receptors, from the marine sponge, Stelletta maxima. *Tetrahedron Letters* **35,** 3967–3970.

Garson, M. J. (1986). Biosynthesis of the novel diterpene isonitrile diisocyanoadociane by a marine sponge of the Amphimedon genus: Incorporation studies with sodium cyanide-14C and sodium acetate-2-14C. *Journal of the Chemical Society, Chemical Communications* 35–36.

Garson, M. J. (1989). Biosynthetic studies on marine natural products. *Natural Product Reports* **6,** 143–170.

Garson, M. J. (1993). The biosynthesis of marine natural products. *Chemical Reviews (Washington, DC, United States)* **93,** 1699–1733.

Garson, M. J. (2001). Ecological perspectives on marine natural product biosynthesis. In "Marine Chemical Ecology" (J. B. Mc Clinton and B. J. Baker, eds), pp. 71–114. CRC Press LLC, Boca Raton.

Garson, M. J., and Simpson, J. S. (2004). Marine isocyanides and related natural products. Structure, biosynthesis and ecology. *Natural Product Reports* **21,** 164–179.

Garson, M. J., Partali, V., Liaaen-Jensen, S., and Stoilov, I. L. (1988). Isoprenoid biosynthesis in a marine sponge of the Amphimedon genus: incorporation studies with [1-14C]-acetate, [4-14C]-cholesterol and [2-14C]-mevalonate. *Comparative Biochemistry and Physiology. Part B, Biochemistry & Molecular Biology* **91B,** 293–300.

Garson, M., Thompson, J., Larsen, R., Battershill, C., Murphy, P., and Bergquist, P. (1992). Terpenes in sponge cell membranes: Cell separation and membrane fractionation studies with the tropical marine sponge Amphimedon sp.. *Lipids* **27,** 378–388.

Genta-Jouve, G., Cachet, N., Holderith, S., Oberhänsli, F., Teyssié, J.-L., Jeffree, R., Al Mourabit, A., and Thomas, O. P. (2011). New insight into marine alkaloid metabolic pathways: Revisiting oroidin biosynthesis. *ChemBioChem* **12,** 2298–2301.

Giner, J. L. (1993). Biosynthesis of marine sterol side chains. *Chemical Reviews(Washington, DC, United States)* **93,** 1735–1752.

Gribble, G. W. (1998). Naturally occurring organohalogen compounds. *Accounts of Chemical Research* **31,** 141–152.

Haber, M., Carbone, M., Mollo, E., Gavagnin, M., and Ilan, M. (2011). Chemical defense against predators and bacterial fouling in the Mediterranean sponges Axinella polypoides and A. verrucosa. *Marine Ecology Progress Series* **422,** 113–122.

Hahn, S., Stoilov, I. L., Ha, T. B. T., Raederstorff, D., Doss, G. A., Li, H. T., and Djerassi, C. (1988). Biosynthetic studies of marine lipids. 17. The course of chain elongation and desaturation in long-chain fatty acids of marine sponges. *Journal of the American Chemical Society* **110,** 8117–8124.

Hill, R. T., Hamann, M. T., Peraud, O., and Kasanah, N. (2004). Manzamine-producing actinomycetes. 46pp. Patent: WO2004/013297.

Hitora, Y., Takada, K., Okada, S., Ise, Y., and Matsunaga, S. (2011a). (−)-Duryne and Its homologues, cytotoxic acetylenes from a marine sponge Petrosia sp.. *Journal of Natural Products* **74,** 1262–1267.

Hitora, Y., Takada, K., Okada, S., and Matsunaga, S. (2011b). Miyakosynes A-F, cytotoxic methyl branched acetylenes from a marine sponge Petrosia sp.. *Tetrahedron* **67,** 4530–4534.

Hochmuth, T., Niederkrueger, H., Gernert, C., Siegl, A., Taudien, S., Platzer, M., Crews, P., Hentschel, U., and Piel, J. (2010). Linking chemical and microbial diversity in marine sponges: Possible role for poribacteria as producers of methyl-branched fatty acids. *ChemBioChem* **11,** 2572–2578.

Hooper, G. J., Davies-Coleman, M. T., Kelly-Borges, M., and Coetzee, P. S. (1996). New alkaloids from a South African latrunculid sponge. *Tetrahedron Letters* **37,** 7135–7138.

Huyck, T. K., Gradishar, W., Manuguid, F., and Kirkpatrick, P. (2011). Eribulin mesylate. *Nature Reviews. Drug Discovery* **10,** 173–174.

Ivanchina, N. V., Kicha, A. A., and Stonik, V. A. (2011). Steroid glycosides from marine organisms. *Steroids* **76,** 425–454.

Ivanisevic, J., Perez, T., Ereskovsky, A. V., Barnathan, G., and Thomas, O. P. (2011). Lysophospholipids in the Mediterranean sponge Oscarella tuberculata: Seasonal variability and putative biological role. *Journal of Chemical Ecology* **37,** 537–545.

Jung, J. H., Im, K. S., and Bae, B. H. (2006). Bioactive polyacetylenic compounds from marine sponges. In "Biomaterials from Aquatic and Terrestrial Organisms" (M. Fingerman and R. Nagabhushanam, eds), pp. 451–511. Science Publishers, Inc., New Orleans.

Kaiser, A., Billot, X., Gateau-Olesker, A., Marazano, C., and Das, B. C. (1998). Selective entry to the dimeric or oligomeric Pyridinium sponge macrocycles via aminopentadienal derivatives. Possible biogenetic relevance with manzamine alkaloids. *Journal of the American Chemical Society* **120,** 8026–8034.

Kalinovsky, A. I., Antonov, A. S., Afiyatullov, S. S., Dmitrenok, P. S., Evtuschenko, E. V., and Stonik, V. A. (2002). Mycaloside A, a new steroid oligoglycoside with an unprecedented structure from the Caribbean sponge Mycale laxissima. *Tetrahedron Letters* **43,** 523–525.

Karuso, P., and Scheuer, P. J. (1989). Biosynthesis of isocyanoterpenes in sponges. *The Journal of Organic Chemistry* **54,** 2092–2095.

Kashman, Y., Hirsh, S., McConnell, O. J., Ohtani, I., Kusumi, T., and Kakisawa, H. (1989). Ptilomycalin A: A novel polycyclic guanidine alkaloid of marine origin. *Journal of the American Chemical Society* **111,** 8925–8926.

Keyzers, R. A., Northcote, P. T., and Davies-Coleman, M. T. (2006). Spongian diterpenoids from marine sponges. *Natural Product Reports* **23,** 321–334.

Kitagawa, I., Kobayashi, M., Kitanaka, K., Kido, M., and Kyogoku, Y. (1983). Marine natural products. XII. On the chemical constituents of the Okinawan marine sponge Hymeniacidon aldis. *Chemical and Pharmaceutical Bulletin* **31,** 2321–2328.

Kobayashi, J., Murayama, T., Kosuge, S., Kanda, F., Ishibashi, M., Kobayashi, H., Ohizumi, Y., Ohta, T., Nozoe, S., and Sasaki, T. (1990). Niphatesines A-D, new antineoplastic pyridine alkaloids from the Okinawan marine sponge Niphates sp. *Journal of the Chemical Society, Perkin Transaction* **1,** 3301–3303.

Kobayashi, M., Okamoto, Y., and Kitagawa, I. (1991). Marine natural products. XXVIII. The structures of sarasinosides A1, A2, A3, B1, B2, B3, C1, C2, and C3, nine new norlanostane-triterpenoidal oligoglycosides from the Palauan marine sponge Asteropus sarasinosum. *Chemical and Pharmaceutical Bulletin* **39,** 2867–2877.

Kornprobst, J.-M., and Barnathan, G. (2010). Demospongic acids revisited. *Marine Drugs* **8,** 2569–2577.

Krejcarek, G. E., White, R. H., Hager, L. P., McClure, W. O., Johnson, R. D., Rinehart, K. L. Jr., McMillan, J. A., Paul, I. C., Shaw, P. D., and Brusca, R. C. (1975). Rearranged dibromotyrosine metabolite from Verongia aurea. *Tetrahedron Letters* **16,** 507–510.

Kreuter, M. H., Robitzki, A., Chang, S., Steffen, R., Michaelis, M., Kljajic, Z., Bachmann, M., Schroder, H. C., and Muller, W. E. (1992). Production of the cytostatic agent aeroplysinin by the sponge Verongia aerophoba in in vitro culture. *Comparative Biochemistry and Physiology Part C: Comparative Pharmacology and Toxicology* **101,** 183–187.

Kubanek, J., Pawlik, J. R., Eve, T. M., and Fenical, W. (2000). Triterpene glycosides defend the Caribbean reef sponge Erylus formosus from predatory fishes. *Marine Ecology Progress Series* **207,** 69–77.

Kubanek, J., Whalen, K., Engel, S., Kelly, S., Henkel, T., Fenical, W., and Pawlik, J. (2002). Multiple defensive roles for triterpene glycosides from two Caribbean sponges. *Oecologia* **131,** 125–136.

Lang, G., Pinkert, A., Blunt, J. W., and Munro, M. H. G. (2005). Discorhabdin W, the first dimeric discorhabdin. *Journal of Natural Products* **68,** 1796–1798.

Laroche, M., Imperatore, C., Grozdanov, L., Costantino, V., Mangoni, A., Hentschel, U., and Fattorusso, E. (2007). Cellular localization of secondary metabolites isolated from the Caribbean sponge Plakortis simplex. *Marine Biology* **151,** 1365–1373.

Laville, R., Thomas, O. P., Berrue, F., Reyes, F., and Amade, P. (2008). Pachychalines A-C: Novel 3-alkylpyridinium salts from the marine sponge Pachychalina sp.. *European Journal of Organic Chemistry* 121–125.

Laville, R., Amade, P., and Thomas, O. P. (2009). 3-Alkylpyridinium salts from Haplosclerida marine sponges: Isolation, structure elucidations, and biosynthetic considerations. *Pure and Applied Chemistry* **81,** 1033–1040.

Lee, Y. K., Lee, J.-H., and Lee, H. K. (2001). Microbial symbiosis in marine sponges. *The Journal of Microbiology* **39,** 254–264.

Lill, R. E., Major, D. A., Blunt, J. W., Munro, M. H. G., Battershill, C. N., McLean, M. G., and Baxter, R. L. (1995). Studies on the biosynthesis of discorhabdin B in the New Zealand sponge Latrunculia sp. B. *Journal of Natural Products* **58,** 206–211.

Lindel, T., Hochgürtel, M., Assmann, M., and Köck, M. (2000a). Synthesis of the Marine Natural Product N-(4-bromopyrrolyl-2-carbonyl)-l-homoarginine, a putative biogenetic precursor of the pyrrole-imidazole alkaloids. *Journal of Natural Products* **63,** 1566–1569.
Lindel, T., Hoffmann, H., Hochgurtel, M., and Pawlik, J. R. (2000b). Structure activity relationship of inhibition of fish feeding by sponge-derived and synthetic pyrrole imidazole alkaloids. *Journal of Chemical Ecology* **26,** 1477–1496.
Litchfield, C., and Morales, R. W. (1976). Are Demospongiae membranes unique among living organisms? In "Aspects of Sponge Biology" (F. W. Harrison and R. R. Cowden, eds), pp. 183–200. Academic Press, New York.
Luibrand, R. T., Erdman, T. R., Vollmer, J. J., Scheuer, P. J., Finer, J., and Clardy, J. (1979). Ilimaquinone, a sesquiterpenoid quinone from a marine sponge. *Tetrahedron* **35,** 609–612.
Marti, R., Fontana, A., Uriz, M.-J., and Cimino, G. (2003). Quantitative assessment of natural toxicity in sponges: Toxicity bioassay versus compound quantification. *Journal of Chemical Ecology* **29,** 1307–1318.
McClintock, J. B., Baker, B. J., Slattery, M., Hamann, M., Kopitzke, R., and Heine, J. (1994). Chemotactic tube-foot responses of a spongivorous sea star Perknaster fuscus; to organic extracts from Antarctic sponges. *Journal of Chemical Ecology* **20,** 859–870.
Miller, K. M., Alvarez, B. A., Battershill, C. B., Northcote, P. N., and Parthasarathy, H. P. (2001). Genetic, morphological, and chemical divergence in the sponge genus Latrunculia (Porifera: Demospongiae) from New Zealand. *Marine Biology* **139,** 235–250.
Minale, L., Riccio, R., and Sodano, G. (1974). Avarol, a novel sesquiterpenoid hydroquinone with a rearranged drimane skeleton from the sponge Dysidea avara. *Tetrahedron Letters* **15,** 3401–3404.
Minto, R. E., and Blacklock, B. J. (2008). Biosynthesis and function of polyacetylenes and allied natural products. *Progress in Lipid Research* **47,** 233–306.
Molinski, T. F., Stout, E. P., Wang, Y.-G., and Romo, D. (2011). Sceptrin-Like Alkaloids from Oxidative Enzymatic Reactions using Cell-Free Preparations from the Marine Sponges Agelas conifera and Stylissa caribica 242nd ACS National Meeting & Exposition. American Chemical Society, Denver, CO, United States, pp. BIOL–126.
Mollo, E., Gavagnin, M., Carbone, M., Castelluccio, F., Pozone, F., Roussis, V., Templado, J., Ghiselin, M. T., and Cimino, G. (2008). Factors promoting marine invasions: A chemoecological approach. *Proceedings of the National Academy of Sciences USA* **105,** 4582–4586.
Moore, B. S. (2006). Biosynthesis of marine natural products: macroorganisms (Part B). *Natural Product Reports* **23,** 615–629.
Morales, R. W., and Litchfield, C. (1977). Incorporation of 1-14C-acetate into C26 fatty acids of the marine sponge Microciona prolifera. *Lipids* **12,** 570–576.
Morinaka, B. I., Pawlik, J. R., and Molinski, T. F. (2009). Amaroxocanes A and B: Sulfated dimeric sterols defend the Caribbean Coral Reef Sponge Phorbas amaranthus from Fish Predators. *Journal of Natural Products* **72,** 259–264.
Müller, W. E., Diehl-Seifert, B., Sobel, C., Bechtold, A., Kljajic, Z., and Dorn, A. (1986). Sponge secondary metabolites: Biochemical and ultrastructural localization of the antimitotic agent avarol in Dysidea avara. *The Journal of Histochemistry and Cytochemistry* **34,** 1687–1690.
Müller, W. E. G., Böhm, M., Batel, R., De Rosa, S., Tommonaro, G., Müller, I. M., and Schröder, H. C. (2000). Application of cell culture for the production of bioactive compounds from sponges: Synthesis of avarol by primmorphs from Dysidea avara. *Journal of Natural Products* **63,** 1077–1081.
Müller, W. E. G., Klemt, M., Thakur, N. L., Schröder, H. C., Aiello, A., D'Esposito, M., Menna, M., and Fattorusso, E. (2004). Molecular/chemical ecology in sponges: evidence for an adaptive antibacterial response in Suberites domuncula. *Marine Biology* **144,** 19–29.

Nguyen, T., Ishida, K., Jenke-Kodama, H., Dittmann, E., Gurgui, C., Hochmuth, T., Taudien, S., Platzer, M., Hertweck, C., and Piel, J. (2008). Exploiting the mosaic structure of trans-acyltransferase polyketide synthases for natural product discovery and pathway dissection. *Nature Biotechnology* **26,** 225–233.

Noyer, C., Thomas, O. P., and Becerro, M. A. (2011). Patterns of chemical diversity in the Mediterranean sponge Spongia lamella. *PLoS One* **6,** e20844.

Page, M., West, L., Northcote, P., Battershill, C., and Kelly, M. (2005). Spatial and temporal variability of cytotoxic metabolites in populations of the New Zealand sponge Mycale hentscheli. *Journal of Chemical Ecology* **31,** 1161–1174.

Parrill, A. L. (2008). Lysophospholipid interactions with protein targets. *Biochimica et Biophysica Acta (BBA)-Molecular and Cell Biology of Lipids* **1781,** 540–546.

Patil, A. D., Kumar, N. V., Kokke, W. C., Bean, M. F., Freyer, A. J., de Brosse, C., Mai, S., Truneh, A., Carte, B., Breen, A. L., Hertzberg, R. P., Johnson, R. K., Westley, J. W., and Potts, B. C. M. (1995). Novel alkaloids from the Sponge Batzella sp.: Inhibitors of HIV gp120-Human CD4 binding. *The Journal of Organic Chemistry* **60,** 1182–1188.

Paul, V. J., and Puglisi, M. P. (2004). Chemical mediation of interactions among marine organisms. *Natural Product Reports* **21,** 189–209.

Paul, V. J., and Ritson-Williams, R. (2008). Marine chemical ecology. *Natural Product Reports* **25,** 662–695.

Paul, V. J., Puglisi, M. P., and Ritson-Williams, R. (2006). Marine chemical ecology. *Natural Product Reports* **23,** 153–180.

Paul, V. J., Ritson-Williams, R., and Sharp, K. (2011). Marine chemical ecology in benthic environments. *Natural Product Reports* **28,** 345–387.

Pawlik, J. R. (1993). Marine invertebrate chemical defenses. *Chemical Reviews (Washington, DC, United States)* **93,** 1911–1922.

Pawlik, J. R., Chanas, B., Toonen, R. J., and Fenical, W. (1995). Defenses of Caribbean sponges against predatory reef fish. I. Chemical deterrency. *Marine Ecology Progress Series* **127,** 183–194.

Pawlik, J. R., McFall, G., and Zea, S. (2002). Does the odor from sponges of the genus Ircinia Protect them from fish predators? *Journal of Chemical Ecology* **28,** 1103–1115.

Perry, N. B., Blunt, J. W., McCombs, J. D., and Munro, M. H. G. (1986). Discorhabdin C, a highly cytotoxic pigment from a sponge of the genus Latrunculia. *The Journal of Organic Chemistry* **51,** 5476–5478.

Perry, N. B., Blunt, J. W., and Munro, M. H. G. (1988). Cytotoxic pigments from new zealand sponges of the genus latrunculia: Discorhabdins a, b and c. *Tetrahedron* **44,** 1727–1734.

Piel, J. (2009). Metabolites from symbiotic bacteria. *Natural Product Reports* **26,** 338–362.

Piel, J., Hui, D., Wen, G., Butzke, D., Platzer, M., Fusetani, N., and Matsunaga, S. (2004). Antitumor polyketide biosynthesis by an uncultivated bacterial symbiont of the marine sponge Theonella swinhoei. *Proceedings of the National Academy of Sciences of the United States of America* **101,** 16222–16227.

Porter, J. W., and Targett, N. M. (1988). Allelochemical interactions between sponges and corals. *The Biological Bulletin* **175,** 230–239.

Puyana, M., Fenical, W., and Pawlik, J. R. (2003). Are there activated chemical defenses in sponges of the genus Aplysina from the Caribbean? *Marine Ecology Progress Series* **246,** 127–135.

Raederstorff, D., Shu, A. Y. L., Thompson, J. E., and Djerassi, C. (1987). Biosynthetic studies of marine lipids. 11. Synthesis, biosynthesis, and absolute configuration of the internally branched demospongic acid, 22-methyl-5,9-octacosadienoic acid. *The Journal of Organic Chemistry* **52,** 2337–2346.

Regalado, E. L., Jimenez-Romero, C., Genta-Jouve, G., Tasdemir, D., Amade, P., Nogueiras, C., and Thomas, O. P. (2011). Acanthifoliosides, minor steroidal saponins from the Caribbean sponge Pandaros acanthifolium. *Tetrahedron* **67,** 1011–1018.

Richelle-Maurer, E., de Kluijver, M. J., Feio, S., Gaudêncio, S., Gaspar, H., Gomez, R., Tavares, R., Van de Vyver, G., and Van Soest, R. W. M. (2003). Localization and ecological significance of oroidin and sceptrin in the Caribbean sponge Agelas conifera. *Biochemical Systematics and Ecology* **31,** 1073–1091.

Rogers, S. D., and Paul, V. J. (1991). Chemical defenses of three Glossodoris nudibranchs and their dietary Hyrtios sponges. *Marine Ecology Progress Series* **77,** 221–232.

Rohmer, M. (1999). The discovery of a mevalonate-independent pathway for isoprenoid biosynthesis in bacteria, algae and higher plants[dagger]. *Natural Product Reports* **16,** 565–574.

Roper, K., Beamish, H., Garson, M., Skilleter, G., and Degnan, B. (2009). Convergent antifouling activities of structurally distinct bioactive compounds synthesized within two sympatric Haliclona demosponges. *Marine Biotechnology* **11,** 188–198.

Sacristan-Soriano, O., Banaigs, B., Casamayor, E. O., and Becerro, M. A. (2011). Exploring the links between natural products and bacterial assemblages in the sponge Aplysina aerophoba. *Applied and Environmental Microbiology* **77,** 862–870.

Sakai, R., Higa, T., Jefford, C. W., and Bernardinelli, G. (1986). Manzamine A, a novel antitumor alkaloid from a sponge. *Journal of the American Chemical Society* **108,** 6404–6405.

Sarma, N. S., Krishna, M. S. R., and Rao, S. R. (2005). Sterol ring system oxidation pattern in marine sponges. *Marine Drugs* **3,** 84–111.

Scheuer, P. J. (1992). Isocyanides and cyanides as natural products. *Accounts of Chemical Research* **25,** 433–439.

Schirmer, A., Gadkari, R., Reeves, C. D., Ibrahim, F., DeLong, E. F., and Hutchinson, C. R. (2005). Metagenomic analysis reveals diverse polyketide synthase gene clusters in microorganisms associated with the marine sponge Discodermia dissoluta. *Applied and Environmental Microbiology* **71,** 4840–4849.

Schmitz, F. J., Hollenbeak, K. H., and Campbell, D. C. (1978). Marine natural products: Halitoxin, toxic complex of several marine sponges of the genus Haliclona. *The Journal of Organic Chemistry* **43,** 3916–3922.

Seiple, I. B., Su, S., Young, I. S., Nakamura, A., Yamaguchi, J., Jorgensen, L., Rodriguez, R. A., O'Malley, D. P., Gaich, T., Köck, M., and Baran, P. S. (2011). Enantioselective total syntheses of (−)-palau'amine, (−)-axinellamines, and (−)-massadines. *Journal of the American Chemical Society* **133,** 14710–14726.

Seo, Y., Cho, K. W., Rho, J.-R., Shin, J., and Sim, C. J. (1998). Petrocortynes and petrosiacetylenes, novel polyacetylenes from a sponge of the genus Petrosia. *Tetrahedron* **54,** 447–462.

Sepcic, K. (2000). Bioactive alkylpyridinium compounds from marine sponges. *Journal of Toxicology: Toxin Reviews* **19,** 139–160.

Sepcic, K., and Turk, T. (2006). 3-Alkylpyridinium compounds as potential non-toxic antifouling agents. *Progress in Molecular and Subcellular Biology* **42,** 105–124.

Sepcic, K., Musevic, I., Lahajnar, G., Turk, T., and Macek, P. (1999). AFM imaging of surface adsorbed polymeric 3-alkylpyridinium salts from the marine sponge Reniera sarai. *International Journal of Biological Macromolecules* **26,** 353–356.

Shigemori, H., Madono, T., Sasaki, T., Mikami, Y., and Kobayashi, J. (1994). Nakijiquinones A and B, new antifungal sesquiterpenoid quinones with an amino acid residue from an Okinawan marine sponge. *Tetrahedron* **50,** 8347–8354.

Shin, J., Seo, Y., Cho, K. W., Rho, J.-R., and Paul, V. J. (1998). Osirisynes A-F, highly oxygenated polyacetylenes from the sponge Haliclona osiris. *Tetrahedron* **54,** 8711–8720.

Siegl, A., and Hentschel, U. (2010). PKS and NRPS gene clusters from microbial symbiont cells of marine sponges by whole genome amplification. *Environmental Microbioliogy Reports* **2,** 507–513.

Silva, C. J., Wuensche, L., and Djerassi, C. (1991). Biosynthetic studies of marine lipids. 35. The demonstration of de novo sterol biosynthesis in sponges using radiolabeled isoprenoid precursors. *Comparative Biochemistry and Physiology. B: Biochemistry and Molecular Biology* **99B,** 763–773.

Simpson, J. S., and Garson, M. J. (1998). Thiocyanate biosynthesis in the tropical marine sponge Axinyssa n.sp.. *Tetrahedron Letters* **39,** 5819–5822.

Simpson, J. S., and Garson, M. J. (2004). Biosynthetic pathways to isocyanides and isothiocyanates; precursor incorporation studies on terpene metabolites in the tropical marine sponges Amphimedon terpenensis and Axinyssa n.sp.. *Organic and Biomolecular Chemistry* **2,** 939–948.

Simpson, J. S., Raniga, P., and Garson, M. J. (1997). Biosynthesis of dichloroimines in the tropical marine sponge Stylotella aurantium. *Tetrahedron Letters* **38,** 7947–7950.

Sjögren, M., Dahlström, M., Hedner, E., Jonsson, P. R., Vik, A., Gundersen, L.-L., and Bohlin, L. (2008). Antifouling activity of the sponge metabolite agelasine D and synthesised analogs on Balanus improvisus. *Biofouling* **24,** 251–258.

Sladic, D., and Gasic, M. (2006). Reactivity and biological activity of the marine sesquiterpene hydroquinone avarol and related compounds from sponges of the order Dictyoceratida. *Molecules* **11,** 1–33.

Srivastava, M., Simakov, O., Chapman, J., Fahey, B., Gauthier, M. E. A., Mitros, T., Richards, G. S., Conaco, C., Dacre, M., Hellsten, U., Larroux, C.Putnam, N. H., *et al.* (2010). The Amphimedon queenslandica genome and the evolution of animal complexity. *Nature* **466,** 720–726.

Stead, P., Hiscox, S., Robinson, P. S., Pike, N. B., Sidebottom, P. J., Roberts, A. D., Taylor, N. L., Wright, A. E., Pomponi, S. A., and Langley, D. (2000). Eryloside F, a novel penasterol disaccharide possessing potent thrombin receptor antagonist activity. *Bioorganic & Medicinal Chemistry Letters* **10,** 661–664.

Teeyapant, R., and Proksch, P. (1993). Biotransformation of brominated compounds in the marine sponge Verongia aerophoba—Evidence for an induced chemical defense? *Naturwissenschaften* **80,** 369–370.

Thompson, J. E. (1985). Exudation of biologically-active metabolites in the sponge Aplysina fistularis. I. Biological evidence. *Marine Biology* **88,** 23–26.

Thompson, J. E., Walker, R. P., Wratten, S. J., and Faulkner, D. J. (1982). A chemical defense mechanism for the nudibranch cadlina luteomarginata. *Tetrahedron* **38,** 1865–1873.

Thompson, J. E., Walker, R. P., and Faulkner, D. J. (1985). Screening and bioassays for biologically-active substances from forty marine sponge species from San Diego, California, USA. *Marine Biology* **88,** 11–21.

Thompson, J. E., Murphy, P. T., Berquist, P. R., and Evans, E. A. (1987). Environmentally induced variation in diterpene composition of the marine sponge Rhopaloeides odorabile. *Biochemical Systematics and Ecology* **15,** 595–606.

Thoms, C., Wolff, M., Padmakumar, K., Ebel, R., and Proksch, P. (2004). Chemical defense of Mediterranean sponges Aplysina cavernicola and Aplysina aerophoba. *Zeitschrift für Naturforschung, C: A Journal of Biosciences* **59,** 113–122.

Thoms, C., Ebel, R., and Proksch, P. (2006). Activated chemical defense in Aplysina sponges revisited. *Journal of Chemical Ecology* **32,** 97–123.

Timm, C., Volk, C., Sasse, F., and Köck, M. (2008). The first cyclic monomeric 3-alkylpyridinium alkaloid from natural sources: Identification, synthesis, and biological activity. *Organic and Biomolecular Chemistry* **6,** 4036–4040.

Timm, C., Mordhorst, T., and Köck, M. (2010). Synthesis of 3-alkylpyridinium alkaloids from the arctic sponge Haliclona viscosa. *Marine Drugs* **8,** 483–497.

Tsoukatou, M., Hellio, C., Vagias, C., Harvala, C., and Roussis, V. (2002). Chemical defense and antifouling activity of three Mediterranean sponges of the genus Ircinia. *Zeitschrift für Naturforschung, C: A Journal of Biosciences* **57,** 161–171.

Tsoukatou, M., Marechal, J.-P., Hellio, C., Novakovic, I., Tufegdzic, S., Sladic, D., Gasic, M., Clare, A. S., Vagias, C., and Roussis, V. (2007). Evaluation of the activity of the sponge metabolites avarol and avarone and their synthetic derivatives against fouling micro- and macroorganisms. *Molecules* **12,** 1022–1034.

Turon, X., Becerro, M. A., and Uriz, M. J. (2000). Distribution of brominated compounds within the sponge Aplysina aerophoba: Coupling of X-ray microanalysis with cryofixation techniques. *Cell and Tissue Research* **301,** 311–322.

Tymiak, A. A., and Rinehart, K. L. Jr. (1981). Biosynthesis of dibromotyrosine-derived antimicrobial compounds by the marine sponge Aplysina fistularis (Verongia aurea). *Journal of the American Chemical Society* **103,** 6763–6765.

Uriz, M. J., Becerro, M. A., Tur, J. M., and Turon, X. (1996). Location of toxicity within the Mediterranean sponge Crambe crambe (Demospongiae: Poecilosclerida). *Marine Biology* **124,** 583–590.

Vergne, C., Boury-Esnault, N., Perez, T., Martin, M.-T., Adeline, M.-T., Tran Huu Dau, E., and Al-Mourabit, A. (2006). Verpacamides A-D, a Sequence of C11N5 Diketopiperazines Relating Cyclo(Pro-Pro) to Cyclo(Pro-Arg), from the Marine Sponge Axinella vaceleti: Possible biogenetic precursors of pyrrole-2-aminoimidazole alkaloids. *Organic Letters* **8,** 2421–2424.

Volk, C. A., and Koeck, M. (2003). Viscosamine: The first naturally occurring trimeric 3-alkyl pyridinium alkaloid. *Organic Letters* **5,** 3567–3569.

Watanabe, K., Tsuda, Y., Yamane, Y., Takahashi, H., Iguchi, K., Naoki, H., Fujita, T., and Van, S. R. W. M. (2000). Strongylodiols A, B and C, new cytotoxic acetylenic alcohols isolated from the Okinawan marine sponge of the genus Strongylophora as each enantiomeric mixture with a different ratio. *Tetrahedron Letters* **41,** 9271–9276.

Weiss, B., Ebel, R., Elbraechter, M., Kirchner, M., and Proksch, P. (1996). Defense metabolites from the marine sponge Verongia aerophoba. *Biochemical Systematics and Ecology* **24,** 1–12.

Wilson, D. M., Puyana, M., Fenical, W., and Pawlik, J. R. (1999). Chemical defense of the caribbean reef sponge Axinella corrugata against predatory fishes. *Journal of Chemical Ecology* **25,** 2811–2823.

Yong, K. W. L., De Voss, J. J., Hooper, J. N. A., and Garson, M. J. (2011). Configurational assignment of cyclic peroxy metabolites provides an insight into their biosynthesis: Isolation of plakortolides, seco-plakortolides, and plakortones from the Australian marine sponge Plakinastrella clathrata. *Journal of Natural Products* **74,** 194–207.

Zhou, K., Zhang, X., Zhang, F., and Li, Z. (2011). Phylogenetically diverse cultivable fungal community and polyketide synthase (PKS), non-ribosomal peptide synthase (NRPS) Genes associated with the South China Sea sponges. *Microbial Ecology* **62,** 644–654.

Zhu, P., Zheng, Y., You, Y., Yan, X., and Shao, J. (2009). Molecular phylogeny and modular structure of hybrid NRPS/PKS gene fragment of Pseudoalteromonas sp. NJ6-3-2 isolated from marine sponge hymeniacidon perleve. *Journal of Microbiology and Biotechnology* **19,** 229–237.

CHAPTER FIVE

Biosilica: Molecular Biology, Biochemistry and Function in Demosponges as well as its Applied Aspects for Tissue Engineering[1]

Xiaohong Wang*,†, Heinz C. Schröder†, Matthias Wiens†, Ute Schloßmacher† and Werner E. G. Müller†,2

Contents

* National Research Center for Geoanalysis, Chinese Academy of Geological Sciences, Beijing 100037, China
† ERC Advanced Investigator Grant Research Group, Institute for Physiological Chemistry, University Medical Center of the Johannes Gutenberg University Mainz, Duesbergweg 6, 55128 Mainz, Germany

[1] This contribution is dedicated to Professor Jean Vacelet (Université de la Méditerranée Aix-Marseille II, Marseille, France) for his innovative contributions in sponge systematics and cell biology.

[2] Corresponding author: Email: wmueller@uni-mainz.de

Advances in Marine Biology, Volume 62
ISSN 0065-2881, DOI: 10.1016/B978-0-12-394283-8.00005-9

Abstract

Biomineralization, biosilicification in particular (i.e. the formation of biogenic silica, SiO_2), has become an exciting source of inspiration for the development of novel bionic approaches following 'nature as model'. Siliceous sponges are unique among silica-forming organisms in their ability to catalyze silica formation using a specific enzyme termed silicatein. In this study, we review the present state of knowledge on silicatein-mediated 'biosilica' formation in marine demosponges, the involvement of further molecules in silica metabolism and their potential applications in nano-biotechnology and bio-medicine.

While most forms of multicellular life have developed a calcium-based skeleton, a few specialized organisms complement their body plan with silica. Only sponges (phylum Porifera) are able to polymerize silica enzymatically mediated in order to generate massive siliceous skeletal elements (spicules) during a unique reaction, at ambient temperature and pressure. During this biomineralization process (i.e. biosilicification), hydrated, amorphous silica is deposited within highly specialized sponge cells, ultimately resulting in structures that range in size from micrometres to metres. This peculiar phenomenon has been comprehensively studied in recent years, and in several approaches, the molecular background was explored to create tools that might be employed for novel bioinspired biotechnological and biomedical applications. Thus, it was discovered that spiculogenesis is mediated by the enzyme silicatein and starts intracellularly. The resulting silica nanoparticles fuse and subsequently form concentric lamellar layers around a central protein filament, consisting of silicatein and the scaffold protein silintaphin-1. Once the growing spicule is extruded into the extracellular space, it obtains final size and shape. Again, this process is mediated by silicatein and silintaphin-1/silintaphin-1, in combination with other molecules such as galectin and collagen. The molecular toolbox generated so far allows the fabrication of novel micro- and nano-structured composites, contributing to the economical and sustainable synthesis of biomaterials with unique characteristics. In this context, first bioinspired approaches implement recombinant silicatein and silintaphin-1 for applications in the field of biomedicine (biosilica-mediated regeneration of tooth and bone defects) with promising results.

Key Words: Biosilica; silicatein; silintaphin-1; biomaterials; Porifera; sponges; biomedicine; biotechnology

1. Introduction

Since the beginning of modern biology in the eighteenth century, the elucidation of the evolutionary origin of sponges (phylum Porifera) was problematic. It remained enigmatic whether the sponges should be grouped to plants or animals. A reflection of this uncertainty is the classification of sponges as 'Zoophyta' (Donati, 1753), a view that was also accepted by Pallas (1787), who grouped these animals to the 'Thierpflanzen' especially because of their apparent lack of a nervous or a muscle system. No further data were given by Spencer (1867) who defined sponges as aggregates which, according to his knowledge, are composed of cells that are physically but not physiologically integrated. Again, in line with this assumption was the classification of Haeckel (1868) who placed the sponges to the Coelenterata/Pflanzenthiere, a progress, since 2 years earlier he attributed the sponges to the Protista and separated them from the Animalia (Haeckel, 1866). With Gesner (1558), with whom also the illustrative descriptions of sponges started, and one sponge was especially in the focus of the scientists, *Tethya lyncurium*. The morphological, physical, and later also the chemical characteristics of spicules were analyzed. He gave the first illustration of a sponge, of *T. lyncurium* (Fig. 5.1A and E). Later, Donati (1753) provided the first illustration of spicule bundles that radiate from the centre of the body (medulla) to its surface (cortex) (Fig. 5.1B and F); then Pallas (1787) (Fig. 5.1C and G) and Johnston (1842) (Fig. 5.1D and H) followed. It is interesting that this sponge was described to undergo a fast evolution in the Northern Adriatic Sea (Müller and Zahn, 1968) in the 'Limski Kanal' (Fig. 5.1J). There, within the fiord-like canal, the red coloured species, termed *Tethya limski*, lives, while the yellow related species, *T. lyncurium*, is abundantly present in the open sea. (Fig. 5.1I and K).

Only after the introduction of molecular cloning technologies, genes encoding for structural and/or functional proteins, a rapid increase of our knowledge on the evolution of the Porifera took pace (Pfeifer *et al.*, 1993). The data also facilitated the subdivision of this Porifera phylum into the three classes: the Hexactinellida, the Demospongiae and the Calcarea. In the following, a number of subsequent publications (see Müller, 1998a–c), all published by our group, clearly demonstrated that the Porifera have to be grouped together with the other metazoan phyla into one monophyletic unit (Müller *et al.*, 1994; Müller, 1995). The hypothetical ancestor of all metazoan phyla was finally named Urmetazoa (Müller, 2001; Müller *et al.*, 2001). Having laid the grounds, the next fascinating question about of the origin of the metazoan 'Bauplan' could be addressed. This view is still valid today (Harcet *et al.*, 2010).

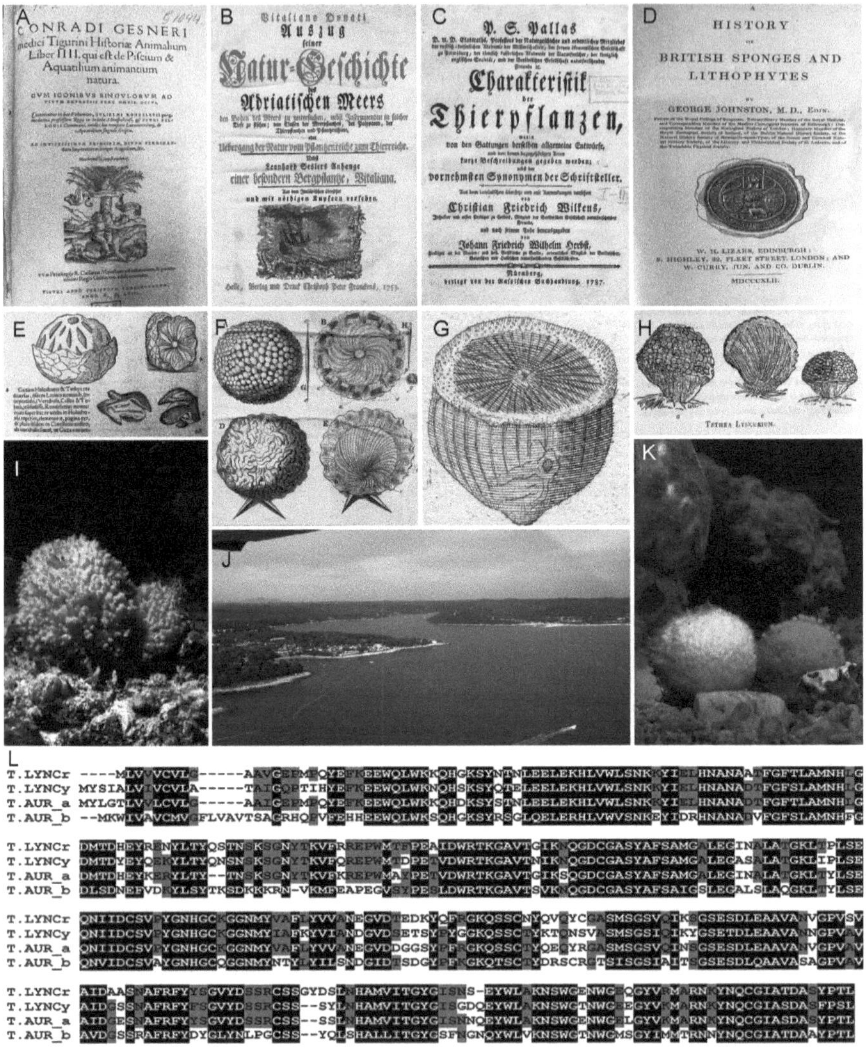

Figure 5.1 First illustration of sponge species by Gesner (1558) (A and E), and later by Donati (1753) (B and F), Pallas (1787) (C and G) and Johnston (1842) (D and H). Especially, the species *Tethya lyncurium* had been portrayed in those oeuvres (E–H). This species is very abundant in the 'Limski Kanal (Northern Adriatic Sea)' (J) where two species had been described, *T. lyncurium* (yellow coloured) (I) living in the open sea and the red, related species *T. limski* (K) (left: *T. lyncurium;* right: *T. limski*). (L) From those species, the *silicatein* genes have been isolated and the deduced amino acid sequences (T.LYNCr [red], and T.LYNCy [yellow]) were compared with the sequences from *T. aurantium*, the presumed next relative of *T. lyncurium* (T.AUR_a [alpha isoform of silicatein] and T.AUR_b [beta isoform of silicatein]). The identical and similar amino acids in all sequences are white on black and those in at least three are in black on grey. (For interpretation of the references to colour in this figure legend, the reader is referred to the online version of this chapter.)

2. Siliceous Sponge Skeleton

The skeletal framework of some sponges is highly ordered. In two examples, the demosponge *Lubomirskia baicalensis* (Müller *et al.*, 2006a) and the hexactinellid *Euplectella aspergillum* (Müller, 2005), it can strikingly be seen that the growth of the sponges proceeds in a radiate accretive manner, meaning that growth zones, which are highly ordered, are delimited by growth lines. These growth centres are constructed by an ordered arrangement of the spicules within the body. Most siliceous sponges are composed of larger megascleres (>100 μm) and smaller microscleres (<100 μm). Some microscleres have a bizarre shape, while the monaxonal megascleres usually have a long central rod with an end that in some cases is decorated with a spine. Interestingly, the spicules have the enzyme silicatein both in the central hollow canal as well as around the outer concentric silica layers. The central core rod of the spicule is synthesized first by a silicatein fibre, which exists in the central hole around which a 1- to 3-μm thick siliceous hollow fibre is formed. This layer becomes surrounded by an organic silicatein layer, which synthesizes the second siliceous layer and so on. This process of growth is very reminiscent of the process of bone formation.

2.1. Morphogenesis and axis formation

Morphogenetic molecules can be grouped into soluble morphogens or transcription factors; they are crucial for the establishment of a multicellular organism which is built of differentiated cells. Both solute molecules (morphogens), for example, the endothelial monocyte-activating polypeptide also termed myotrophin (Schröder *et al.*, 2000), and transcription factors, allowing and controlling differential gene expression, are active in sponges during embryogenesis, larval development and adult tissue formation. Examples are the metazoan Homeobox molecules (axis formation during embryogenesis), T-box factors (generation of cell lineage diversity in the developing embryo), Forkhead (organ development and differentiation), or LIM/homeobox protein (polarity of an organ) (Adell *et al.*, 2003; Müller *et al.*, 2004). Furthermore, sponges possess molecules that are found in tight junctions. These molecules were key innovations during the course of metazoan evolution since they set the basis for the formation of tight epithelial layers. A example is the scaffold protein MAGI, a membrane-associated guanylate kinase, or the tetraspan receptor (Adell *et al.*, 2004; Müller *et al.*, 2004).

2.2. Anabolic enzyme for the synthesis of silica: Silicatein, one key innovation during metazoan evolution

One major breakthrough in the understanding of spicule formation came after the discovery of the key enzyme involved in spiculogenesis. The group of Morse (Shimizu *et al.*, 1998; Cha *et al.*, 1999) discovered that the organic filament in the central canal of spicules is composed of a cathepsin L (Krasko *et al.*, 1997)-related enzyme, which they termed silicatein. They cloned two of the proposed three isoforms of silicateins, the α- and β-forms from the marine demosponge *Tethya aurantium* (Cha *et al.*, 1999; Müller *et al.*, 2005). In subsequent years, these molecules were cloned also from other sponges, among them there are the marine sponge *Suberites domuncula* (Krasko *et al.*, 2000, 2002) and the freshwater sponge *L. baicalensis* (Wiens *et al.*, 2006).

Silicatein exists in the axial canal (diameter of around 1 μm) of spicules as axial filament which can be freed from the silica mantel with hydrofluoric acid (HF) and subsequently visualized by staining with Coomassie Brilliant Blue (Müller *et al.*, 2006b). The *S. domuncula* cDNA encoding silicatein was isolated and characterized (Krasko *et al.*, 2000). The 1169-bp long sequence has an open reading frame of 993 nucleotides (nt); the predicted translation product of 330 amino acids (aa) (Fig. 5.2) has a calculated size of M_r 36,306. As outlined before (Shimizu *et al.*, 1998; Krasko *et al.*, 2000; Müller *et al.*, 2003), silicatein is a new member of the cathepsin subfamily. The three aa Cys, His and Asn, which form the catalytic triad of cysteine proteases, are present in the sponge cathepsin at the characteristic sites: Cys_{125}, His_{164} and Asn_{184}. Furthermore, the silicatein comprises one cluster of characteristic hydroxy amino acids (serine) (Fig. 5.2B).

Phylogenetic analysis was performed with the silicateins from the demosponges *L. baicalensis*, *T. aurantium*, *S. domuncula*, *Spongilla lacustris* and *Ephydatia fluviatilis*, the silicateins from the hexactinellids (*E. aspergillum*; *Crateromorpha meyeri* and *Monorhaphis chuni*), the cathepsins L from these sponges and the glass sponge *Aphrocallistes vastus* and as well as the papain cysteine peptidase from the plant *Arabidopsis thaliana* (Fig. 5.2A). The polypeptides were aligned and a non-rooted tree was calculated. This tree shows that the cathepsin L sequences form one branch which separated from the silicateins. In addition, the silicateins form a single branch that emerges from origin of the demosponge cluster (Fig. 5.2A).

We could demonstrate that silicatein occurs in the axial filament in several isoforms (Schröder *et al.*, 2003). By two-dimensional gel electrophoretic analysis of the axial filament (=silicatein), we showed that in the axial filament silicatein undergoes stepwise phosphorylation (Müller *et al.*, 2005). Five phospho-isoforms with p*I* values of 5.5, 4.8, 4.6, 4.5 and 4.3 have been identified (Müller *et al.*, 2005). The sizes of these phosphorylated proteins were around 25 kDa, compatible with the predicted mature form

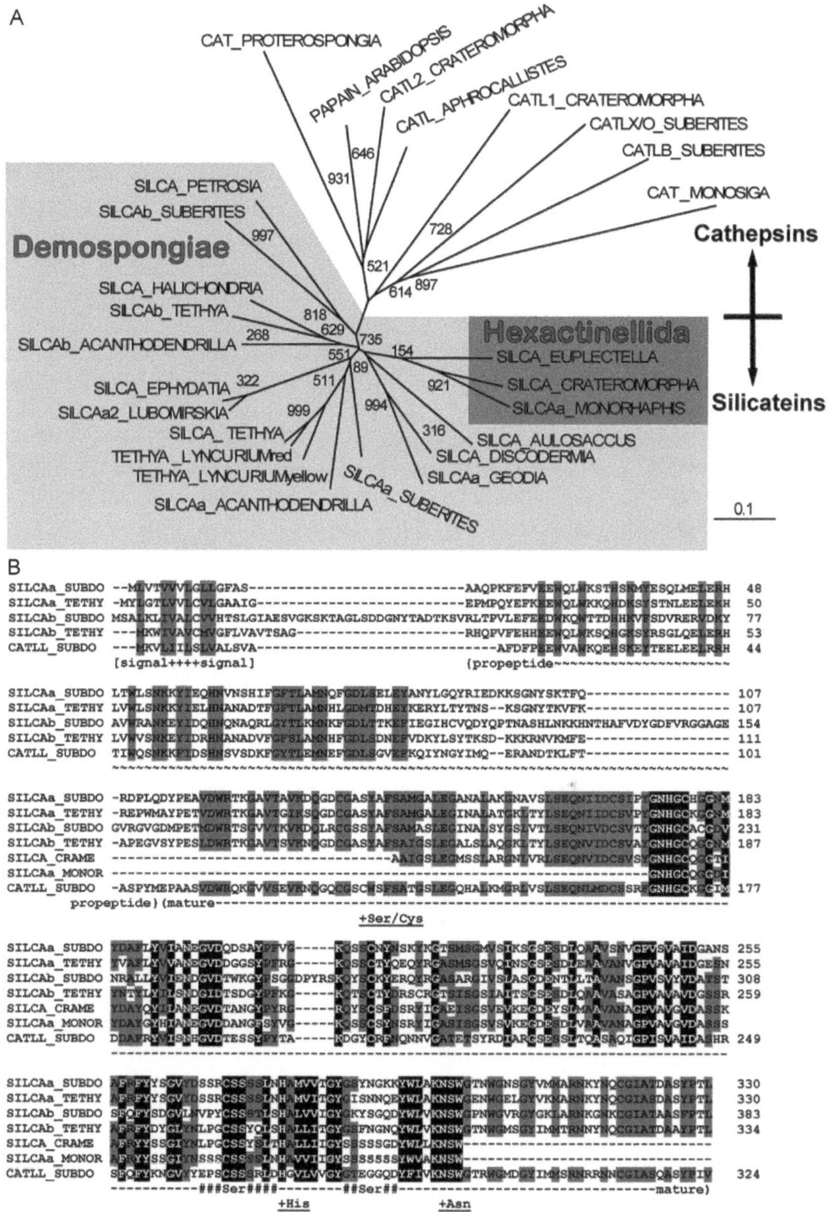

Figure 5.2 Phylogeny based on the biosilica skeleton. (A) Phylogenetic analysis of silicatein within the cathepsin family. The deduced proteins were aligned and the phylogenetic tree was constructed. The hitherto-known four hexactinellid sequences were included; silicatein from *E. aspergillum* (SILCA_EUPLECTELLA; FR748156), *C. meyeri* (SILCA_CRATEROMORPHA; AM920776), *M. chuni* (SILCAa_MONORHAPHIS; FN394978) and the

of silicatein. Interestingly, the silicatein proteins (around 25 kDa) appeared as two (perhaps also three) spots.

2.3. Proposed reaction mechanism of silicatein

Recently, a revised model of the originally proposed mechanism for the polycondensation reaction of silicatein (Cha *et al.*, 1999) has been presented (Schröder *et al.*, 2012). This reaction mechanism bases on the predicted three-dimensional structure of the mature silicatein-α molecule from the

---

silicatein-like protein *Aulosaccus* sp. (SILCA_AULOSACCUS; ACU86976.1). The bulk of silicatein sequences have been identified in demosponges. Firstly, the silicateins-α sequences from *Suberites domuncula* (SILCAa_SUBERITES; CAC03737.1), *Tethya aurantium* (SILCAa_TETHYA; AAC23951.1), *Tethya lycurium* red variation (TETHYA_LYCURIUMred; FR748154) and *Tethya lycurium* yellow variation (TETHYA_LYCURIUMyellow; FR748155), *Discodermia japonica* (SILCA_DISCODERMiA; FR748157), *Geodia cydonium* (SILCAa_GEODIA; CAM57981.1) and *Acanthodendrilla* sp. *Vietnam* (SILCAa_ACANTHODENDRILLA; ACH92669.1), as well as from *Lubomirskia baicalensis* (SILCAa2_LUBOMIRSKIA; AJ968945) and *E. fluviatilis* (SILCA_EPHYDATIA; BAE54434.1). Secondly, the silicatein-β sequences from *Suberites domuncula* (SILCAb_SUBERITES; CAH04635.1), *Tethya aurantium* (SILCAb_TETHYA; AF098670_1) and *Acanthodendrilla* sp. *Vietnam* (SILCAb_ACANTHODENDRILLA; FJ013043.1). Thirdly, the silicateins that had been identified in marine sponges from which only one isoform had been obtained; silicatein from *Petrosia ficiformis* (SILCA_PETROSIA; AAO23671.1) and *Halichondria okadai* (SILCA_HALICHONDRIA; BAB86343.1). As reflected in the rooted tree, these silicateins derived from the cathepsins among which in this tree the following sequences have been included; cathepsin-like protein 2 *C. meyeri* (CATL2_CRATEROMORPHA; CAP17585.1), cathepsin-like protein 1 *C. meyeri* (CATL1_CRATEROMORPHA; CAP17584.1), mRNA for cathepsin L (catl gene) *A. vastus* (CATL_APHROCALLISTES); AJ968951cathepsin B *Suberites domuncula* (CATLB_SUBERITES; CAH04630.1), cathepsin X/O *Suberites domuncula* (CATLX/O_SUBERITES; |CAH04633.1). The resulting tree was rooted with the sequence from the papain-like cysteine peptidase XBCP3 *A. thaliana* (PAPAIN_ARABIDOPSIS; AF388175_1). In addition, the cathepsins from choanoflagellates had been included to show that they derived, according to this tree, from the sponge cathepsins; the cysteine protease from *Proterospongia* sp. (PRL00000040) and the cathepsin from *Monosiga ovata* MNL00000103. (B) A selected set of silicateins, the silicatein-α from *S. domuncula* (SILICAa_SUBDO) and *T. aurantium* (SILICAa_TETHY), as well as the silicatein-β from *S. domuncula* (SILICAb_SUBDO), and *T. aurantium* (SILICAb_TETHY) had been included together with cathepsin L from *S. domuncula* (CATLL_SUBDO). Residues conserved (similar or related to their physico-chemical properties) in all sequences are shown in white on black, and those in at least five sequences in black on grey. The characteristic sites in the sequences are marked; the catalytic triad amino acids, Ser in silicateins, Cys in cathepsin, as well as His and Asn, are highlighted. The borders of the signal peptide (signal), the propeptide (propeptide) and the mature silicatein (mature) are given. (For interpretation of the references to colour in this figure legend, the reader is referred to the online version of this chapter.)

demosponge *S. domuncula* that has been obtained using the X-ray structure of the human procathepsin S as template for the modelling process (Fig. 5.3A). The model revealed that the synthetic substrate (Cha *et al.*, 1999), TEOS (tetraethyl orthosilicate), fits into the substrate pocket of the silicatein-α molecule (Fig. 5.3A). In the catalytic centre, the calculated distance between the C-α atom of Ser to the C-α of His is 7.8 Å; between Asn and Ser, and between Asn and His, the distance is 9.6 Å and 9.77 Å, respectively. In order to outline the optimal position of TEOS within the catalytic centre of the enzyme, the diameter of the TEOS sphere (5.76 Å)

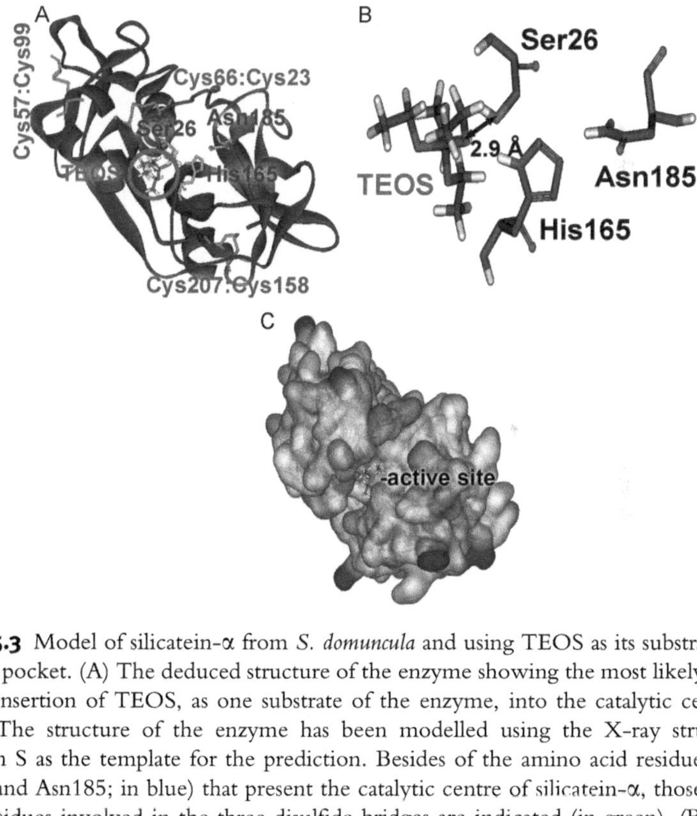

Figure 5.3 Model of silicatein-α from *S. domuncula* and using TEOS as its substrate in the catalytic pocket. (A) The deduced structure of the enzyme showing the most likely position for the insertion of TEOS, as one substrate of the enzyme, into the catalytic centre (red circle). The structure of the enzyme has been modelled using the X-ray structure of cathepsin S as the template for the prediction. Besides of the amino acid residues (Ser26, His165 and Asn185; in blue) that present the catalytic centre of silicatein-α, those cystcine (Cys) residues involved in the three disulfide bridges are indicated (in green). (B) Sterical proximity of the three amino acids of the catalytic centre with respect to the TEOS molecule. The distance between the oxygen atom of the Ser hydroxyl group and the substrate (TEOS) is indicated. (C) A proposed three-dimensional model of the electrostatic charge distribution of silicatein-α is given showing the inserted TEOS molecule in the catalytic centre [active site] (positive charges: red; negative charges: blue; hydrophobic areas: white). From Schröder *et al.* (2012) with permission. (For interpretation of the references to colour in this figure legend, the reader is referred to the online version of this chapter.)

has been calculated (Fig. 5.3B). Assuming an effective diameter of 8 Å for that molecule including its aqueous environment, the predicted three-dimensional structure of silicatein-α was found to allow this substrate to enter into the catalytic pocket and to interact with the oxygen atom of the Ser residue of the catalytic centre of the molecule (Fig. 5.3C). The calculated distance between the Ser hydroxyl group involved in the nucleophilic attack and the silicon atom of the TEOS molecule was found to be 2.93 Å (Fig. 5.3B), below that for hydrogen-bond interaction (3.6 Å), thus permitting the enzymatic cleavage (hydrolysis) of the substrate molecule. Even more space is available for the natural substrate, orthosilicic acid, which is much smaller than the synthetic silica precursor (TEOS).

The mechanism of silicatein reaction proposed by us bases on the assumption that the presence of both the serine and histidine residue in the catalytic centre of the silicatein molecule is essential for the interaction with the substrate orthosilicate and the subsequent reactions. The catalytic mechanism can be described in several phases as outlined (Schröder *et al.*, 2012). In the first step, a nucleophilic attack (S_N2 type) takes place from the (electronegative) oxygen atom of the hydroxyl group of the serine residue to the (electropositive) silicon atom of a silicic acid molecule, a process that proceeds in the catalytic pocket of the silicatein molecule (Fig. 5.4; Step 1). This mechanism is facilitated by the hydrogen bond between Ser and His, which increases the nucleophilicity of the serine OH group. Then, the proton transfers from the histidine nitrogen (Ser–His hydrogen bond) to one of the silicon OH ligands and proceeds to the pentavalent intermediate resulting in a release of a water molecule (Fig. 5.4; Step 2). This reaction results in a transition state during which the silicon is covalently bound to the serine residue positioned in the catalytic centre of silicatein. Then, this serine-bound silicic acid molecule undergoes a nucleophilic attack at the silicon atom of a second orthosilicic acid molecule entering the substrate pocket of the enzyme. During this process, the nucleophilicity of the attacking oxygen atom of the OH ligand of the first silicic acid molecule is increased by formation of a hydrogen bond to the nitrogen imidazole of the His residue of the catalytic centre of the enzyme (Step 3). The concomitant loss of a water molecule, which is facilitated by proton transfer from the nitrogen imidazole of the His residue, generates a disilicic acid molecule (Step 4) that is bound to silicatein through an ester-like linkage. Next, rotation of the ester bond is required to allow an interaction of a second OH ligand of the enzyme-bound silicic acid unit with the nitrogen imidazole of the catalytic centre His residue, facilitating further growth of the disilicic acid by nucleophilic attack to a third orthosilicic acid molecule (Step 5). Repeating this reaction cycle (nucleophilic attack, proton transfer, loss of water, rotation) will result in generation of higher membered silicic acid oligomers. It should be mentioned that the release of silicatein-bound trisicilic acid (or perhaps also the tetrasilicic acid) is proposed to be initiated

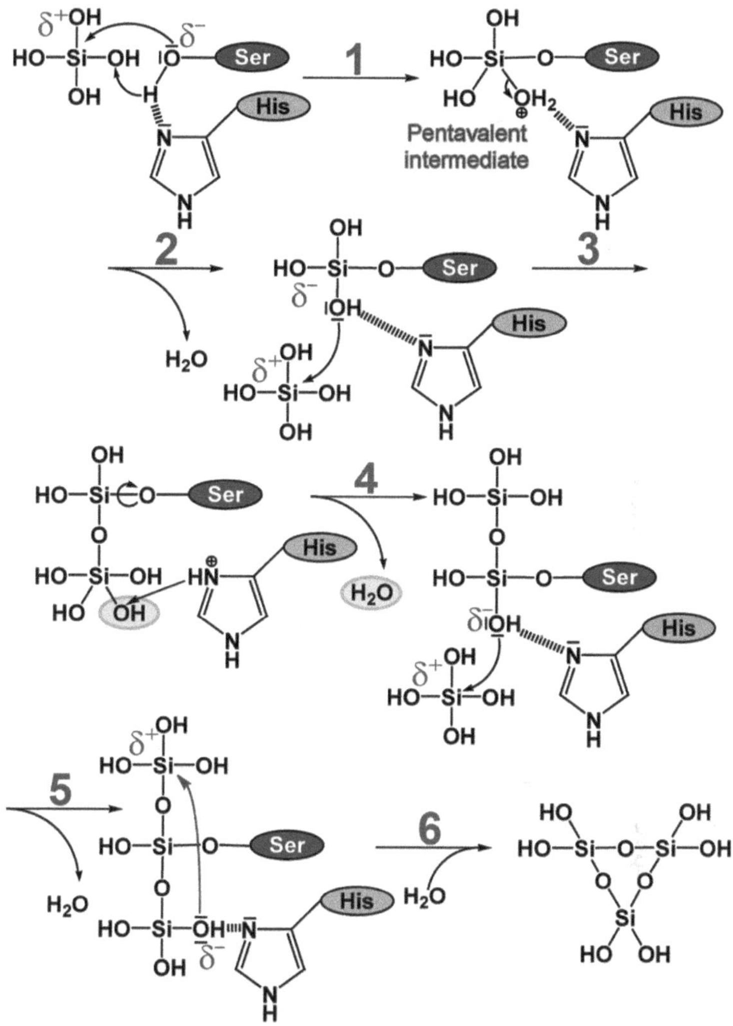

Figure 5.4 The proposed mechanism of silicatein-catalyzed polymerization of orthosilicic acid (Schröder *et al.*, 2012), involving six steps. Step 1: Nucleophilic attack of the (negatively charged [δ^-]) Ser oxygen atom at the (positively charged [δ^+]) silicon atom of the orthosilicic acid substrate, followed by a transfer of a proton (originating from the Ser–His hydrogen bridge) from the imidazole nitrogen of the His residue to an OH ligand of the silicic acid molecule. Step 2: Release of a water molecule from the formed pentavalent intermediate. Step 3: Nucleophilic attack of the oxygen atom of one of the OH ligands of the covalently bound silicic acid molecule at the silicon atom of a second orthosilicic acid molecule, facilitated by hydrogen bridge formation of the bound silicic acid to the imidazole nitrogen [N^+]. Step 4: Loss of water after proton transfer from the imidazole group. Step 5: Nucleophilic attack of the oxygen atom of a further OH ligand of the first silicic acid molecule [δ^-], after rotation of the Si—O—C bond between this molecule and the enzyme

by cyclization of the oligomers (Step 6). Cyclization reactions of siloxane polymers and their mechanisms have long been studied. The strain in the enzyme-bound three-membered (and higher membered) cyclic siloxane oligomers may facilitate this reaction. The covalent linkage between the enzyme and the silicic acid oligomer (ring) is hydrolyzed by water (Fig. 5.4, Step 6). Cyclization reactions of siloxane polymers and their mechanisms have been studied since a long time (Flory and Semlyen, 1966; Hench and West, 1995). The tension in the enzyme-bound three-membered (and higher membered) cyclic siloxane oligomers may facilitate the polycondensation reaction.

2.4. Catabolic enzyme: Silicase

In the course to further elucidate the metabolism of siliceous spicules, another enzyme, silicase, was identified in the marine sponge *S. domuncula*. Silicase is able to depolymerize amorphous silica. The cDNA was isolated and the deduced polypeptide disclosed its relationship to be related to carbonic anhydrases. Recombinant silicase displays besides of a carbonic anhydrase activity, the ability to dissolve amorphous silica under formation of free silicic acid (Schröder *et al.*, 2003). This enzyme will surely have a wide application in nano-biotechnology, for example, in lithography (Schröder *et al.*, 2008).

2.5. Silicatein-associated proteins: Silintaphin-1

The molecular biological technique of the two-hybrid system has been applied to identify molecules that interact specifically with the silicatein molecules. The first strong silicatein interactor was identified and subsequently termed silintaphin-1 (Wiens *et al.*, 2009a). The existence of this 42.5-kDa silicatein interactor (386 aa) is restricted to sponges. Since it shows no sequence similarities to any other known proteins, silintaphin-1 probably represents the prototypic member of a new family. However, the protein reveals significant similarities along a stretch of ca. 110 aa to the pleckstrin homology (PH) domain, a common protein interaction domain

---

Ser residue [δ^+], at the silicon atom of a third orthosilicic acid species. This reaction is supposed to be facilitated again by hydrogen bridge formation to the imidazole nitrogen. Step 6: Cyclization of the resulting (enzyme-bound) trisilicic acid by nucleophilic attack of the (negatively charged) oxygen of an OH ligand of the lastly incorporated silicic acid species at the (positively charged) silicon of the second silicic acid species and release of the formed reactive trisiloxane ring after hydrolysis of the Si—O—C bond. From Schröder *et al.* (2012) with permission. (For colour version of this figure, the reader is referred to the online version of this chapter.)

that occurs in many unrelated proteins. PH domains usually form a structure that can serve as scaffolds for presenting different types of binding sites. In silintaphin-1, the PH domain likely facilitates binding to silicatein and consequently assists assembly of the axial filament (Wiens *et al.*, 2009a). The PH domain interacts in eukaryotic cells with a series of membrane-associated molecules, for example, inositol lipids, receiving and responding to upstream signals resulting in a targeting of host proteins to the distinct cellular sites (reviewed in Blomberg *et al.*, 1999). Other PH domains recognize protein ligands through their phosphotyrosine site(s) and are involved in the architecture of the cytoskeleton and by that to construct modules of intracellular signalling systems. A binding of silintaphin-1 to one phosphotyrosine site(s) at silicatein-α can be structurally proposed (Schlossmacher *et al.*, 2011). In turn, it is likely that each of the tetrameric silicatein-α molecules, forming the platform for one silicatein-β molecule bridging them to the adjacent silicatein-α tetramer, binds one silintaphin-1 molecule. Besides of this phosphotyrosine-controlled protein:protein interaction of the PH domain, structural evidence has been proposed by Blomberg *et al.* (2000) that this domain interacts with clusters of charged aa via electrostatic interactions. These authors showed that the PH domain of UNC-89, a giant muscle protein in *Caenorhabditis elegans*, has a strong negative overall electrostatic potential on the surface of this molecule. Modelling studies revealed that this region interacts with positively charged residues. More recently, it has been shown that binding of Ca^{2+} to the EF-hand motifs can modulate binding of proteins to PtdIns(4,5)P_2 via their PH domain (Yamamoto *et al.*, 2001).

The PH domain is flanked on both sides by several nearly identical aa sequence repeats rich in Pro and charged/polar residues (Glu, Asp, Thr). This combination of repeated sequence composition and ligand-binding domain is also found in titin (connectin), a filamentous protein of sarcomers with unique elastic properties and distinct bending rigidities (Trombitas *et al.*, 1998). In analogy, silintaphin-1 might hence convey a similar elasticity and flexibility required during spiculogenesis. Silintaphin-1 expression was observed to be significantly up-regulated in regenerating *S. domuncula* tissue that requires profound synthesis of spicules and/or reorganization of the skeletal elements. Furthermore, the expression was also susceptible to the silicon concentration in the surrounding aquatic medium, indicating that silintaphin-1 is tightly interwoven in the pathways of poriferan silicon metabolism and spiculogenesis.

With the discovery of silintaphin-1, the first member of the silicatein-interacting protein (Wiens *et al.*, 2009a), it became overt that the stabilization of the silicatein proteins within a silicatein-based axial filament is supported and likely also controlled by additional polypeptides. As mentioned, silintaphin-1 has been shown to control the process of silicatein assembly to filaments *in vitro* and very likely also *in vivo* (Wiens *et al.*, 2009a).

Silintaphin-1 was found to accelerate axial filament formation from silicatein units and causes a change of the morphology of the filaments in dependence of the presence of metal additions. For the isolation of silintaphin-1, the yeast two-hybrid library screening system had been applied, a system that allows a comprehensive and reliable identification of biologically relevant interacting proteins, provided that all available controls have been conducted.

2.6. Silicatein-associated proteins: Silintaphin-2

In a further, more recent study (Wiens *et al.*, 2011), we described a new solid-phase pull-down assay that allowed the identification of a further silicatein-interacting protein, of silintaphin-2. In order to confirm the reliability of this technique, we applied in parallel the yeast two-hybrid library screening system as well. Again, for these studies, we have used the demosponge *S. domuncula* since its skeleton is only composed of megascleres (Arndt, 1935) that are classified as oxeas/styles or tylostyles (90–320 μm × 5–8 μm). Using this approach, the 15-kDa polypeptide (silintaphin-2) has been identified and cloned; sequence analysis and also binding studies revealed that silintaphin-2 represents a sponge-specific Ca^{2+}-binding protein that functions *in situ* in concert with silicatein(s).

With silintaphin-2, the second interactor molecule for the sponge silicatein(s) has been identified and cloned. It is, like silintaphin-1, a direct interactor of the sponge enzyme that crucially involved in the synthesis and formation of the siliceous matrix of spicular skeletal elements. As identified (Wiens *et al.*, 2011), the sponge silintaphin-2 comprises within the polypeptide helical molecule three (potential) EF-hand motifs that have been attributed to those sites that bind $^{45}CaCl_2$ under the experimental conditions used. Following this line of argument, silintaphin-2 would bind to silicatein-α indirectly via silintaphin-1 and Ca^{2+}. In a study, the recombinant and purified silicatein-α has been used as the solid-phase in the pull-down assay and the expressed *silicatein-α* gene protein in the yeast two-hybrid library system for the identification of silintaphin-2. Therefore, a direct interaction of silintaphin-2 with silicatein-α has to be proposed.

Based on the studies with the two interactors (silintaphins), we can conclude that the silicateins are the crucial molecules involved in the enzymatic polycondensation reaction of poly(silicate) which represents the inorganic scaffold of the siliceous spicules. Hence, this system in sponges allows an intriguing and unique way to study and understand the formation of complex skeletal structures starting from the genetic toolkit. Even though the silicatein(s) molecules provide in the initial phase the polarity and the orientation of the growing spicules through their fractal and subsequently filamentous assembly (Müller *et al.*, 2007a), additional polypeptides are required to understand the formation of the species-specific size and morphology of the spicules. The first silicatein interactor silintaphin-1 stabilizes and facilitates the axial orientation of the silicateins, allowing the formation of 100- to 500-nm thick rods like axial filaments existing in the axial canal of the spicules (Wiens *et al.*, 2009a). Silintaphin-2 as a Ca^{2+}-binding protein could function through a binding to silintaphin-1, or more likely as a direct interactor to silicatein, as a modulator of enzymes involved in the post-translational modification processing of the silicateins.

2.7. Extent of enzymatic biosilica formation

In order to determine the enzymatic polycondensation *in vitro* under defined conditions, the recombinant silicatein-α was used with pre-hydrolyzed TEOS (orthosilicate) as substrate. After an incubation period of 1 h at 22 °C, the biosilica product was hydrolyzed by NaOH and subsequently the concentration of soluble silicic acid was determined colorimetrically. It was found that 6 μg of enzyme catalyzes 77 μg of biosilica per assay (Fig. 5.5A). The biosilica-forming activity of recombinant silintaphin-1 was significantly lower (28 μg per assay). Conversely, if both proteins were allowed to interact prior to the addition of TEOS, the extent of biosilica formation was substantially higher. At a silicatein-α/silintaphin-1 in stoichiometric molar ratio of 1:1 (4:1), both proteins produced 201 (415) μg per assay. A further increase in the ratio resulted in lowered activity, with 165 μg per assay. In control assays, which contained pre-hydrolyzed TEOS but no protein, the poly(silicate)-forming activity was negligible, at 15 μg per assay (Fig. 5.5A). In parallel, the activity of isolated spicule proteins (spicular extract [composed primarily of silicateins]), containing the native silicateins, was assayed as well. Following incubation with pre-hydrolyzed TEOS in assays supplemented with 6 μg of protein produced a comparably high activity of 321 μg biosilica per assay (Fig. 5.5B). The activity could be significantly reduced after blocking of the proteins with antibodies raised against silicatein-α, silicatein-β and silintaphin-1.

We have determined earlier that the silicateins associate in a stoichiometric ratio to each other. For these experiments, native silicateins were isolated by a Tris/glycerol extraction procedure from *S. domuncula* (Müller *et al.*, 2007a). It was found that silicateins isolated by this procedure were monomeric (24 kDa) but readily formed dimers through non-covalent linkages; they showed a considerable proteolytic activity that increased during the polymerization phase of the protein. The filament/aggregate formation from disassembled silicatein could be visualized by light microscopy and also by TEM analyses. Since in *S. domuncula* silicatein-α is four times more abundant in the axial filament than silicatein-β, we propose that four silicateins form a platform with serine clusters directed to the centre. These serines of the con-axially arranged silicateins interact with silicatein-β. We conclude that initially the silicateins reassemble chaotically, and in the

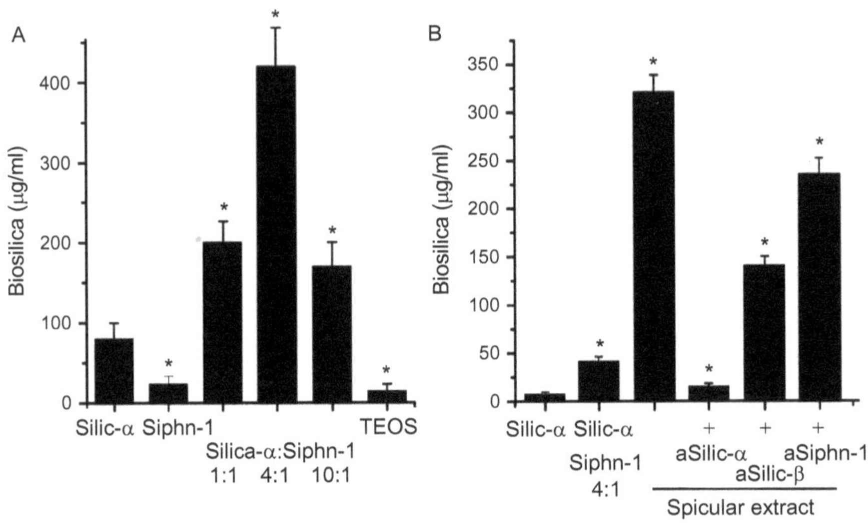

Figure 5.5 Silicatein-mediated synthesis of biosilica *in vitro*. For the *in vitro* assays, recombinant silicatein and pre-hydrolyzed TEOS [orthosilicate] as substrate was used. (A) With pre-hydrolyzed TEOS, the activity of recombinant silicatein-α, silintaphin-1 or silicatein-α/silintaphin-1 (in stoichiometric ratios of 1:1, 4:1 or 10:1) was assessed by using the molybdenum blue colorimetric method. In parallel, control experiments, the deposition of poly (silicate) was determined in the absence of protein and in the presence of TEOS only. (B) In addition, the activity of spicular protein extracts in a glycerol-based Tris buffer system was determined. The activity of the extracts (spicular extract) was analyzed in the absence of antibodies or after preincubation of the extract with aSilic-α, aSilic-β, or aSiphn-1-01 antibodies. For comparison, the activities of silicatein-α and silicatein-α/silintaphin-1 (4:1) are given in parallel. The results are expressed as means (10 experiments each) ± standard error of the mean; $^{*}P<0.01$. From Schlossmacher *et al.* (2011) with permission.

second phase order themselves to fractal-like structures, which subsequently form the filaments. Based on these two reports, we can propose that four silicatein-β molecules form one tetramer that associates with another silicatein-β tetramer together with silintaphin-1 to more stable complexes.

It is well established that the silica-forming activity, mediated by silicateins, is generally high in sponge tissue and causes a considerable spicule growth rate. It had been calculated that the growth rate of spicules *in vivo* is 1–10 μm/h^{-1} for spicules of the demosponge *E. fluviatilis*, which have an average length of 200–350 μm and a thickness of 15 μm (Müller *et al.*, 2008a). In turn, the total weight of one spicule amounts to approximately 88.3×10^{-9} g, which is equivalent to roughly 4.4×10^{13} molecules of biosilica being formed per hour in one spicule (Schlossmacher *et al.*, 2011). Adopting that 5% of protein exists in a spicule, as measured for *M. chuni* spicules (Müller *et al.*, 2008a), it follows that 2.2×10^{-10} g of protein (mainly of silicatein-α) catalyzes the synthesis of 4.4×10^{-9} g of biosilica per hour. As the molecular mass of mature silicatein is 25 kDa, it follows that 8.8×10^{-15} mol of silicatein (5.3×10^{9} molecules) synthesize 4.4×10^{-9} g [7.33×10^{-11} mol (4.4×10^{13} molecules)] of biosilica per hour. With regard to native silicatein, the turnover number of silicate per silicatein on a molar basis is 8×10^{3} molecules of converted substrate per silicatein molecule per hour in spicules. Turning back to recombinant silicatein-α, it has been determined that the enzyme catalyzes the conversion of the synthetic substrate bis(*p*-aminophenoxy)-dimethylsilane with a turnover rate of 18.7×10^{3}/h (Müller *et al.*, 2008b). This value shows that an almost as high activity as that of recombinant silicatein-α, with a turnover rate of 5.4×10^{3} molecules (on the basis of 77 μg of biosilicate formed per 6 μg of protein) per molecule of silicatein-α and per hour, was determined in the more recently performed analysis (Schlossmacher *et al.*, 2011).

2.8. Structure formation by association of silicateins with silintaphin-1 *in vitro*

The phenomenon that silica had been formed during the sol-gel process forming highly structural morphologies is known since Bütschli (1909). He published that silica which had been treated/heated in the presence of Na-carbonate forms nicely arranged micro-structures that vary from radial to concentric. While those structures had been obtained by very high silica concentrations, the biosilica polycondensation process proceeds at concentrations only of around 100 μM with respect to orthosilicate, a reaction that is catalyzed enzymatically. The potency of silicatein to form structural complexes with silintaphin-1, that is the self-assembly of recombinant silicatein-α with silintaphin-1, was studied *in vitro* (Fig. 5.6). In a first set of studies, self-assembly of recombinant silicatein-α was investigated in Tris buffer (lacking glycerol) by SEM analysis. When silicatein-α was added to the buffer at a concentration of 50 μg/ml (2 μM), no structures could be resolved by SEM after 1 min (Fig. 5.6A). However, after 30 min, randomly organized aggregates became visible (Fig. 5.6B), which showed increases in size and compactness of packing after 30 min (Fig. 5.6C). After extending the incubation time [60 min (Fig. 5.6D) and 120 min (Fig. 5.6E)], the aggregates continued to grow. In comparison, co-incubation of recombinant silicatein-α and silintaphin-1 at a stoichiometric ratio of 4:1 elicited drastically different self-assembly (Fig. 5.6G–J), with fractal-like patterns. Even though in this case, no aggregation was seen after 1 min of incubation, 30-nm spherical assemblies were formed after an incubation period of >30 min and continue to grow to >100-nm large aggregates (Fig. 5.6I). After total incubation time of 30 (Fig. 5.6G) and 60 min (Fig. 5.6H), the size and compactness of the assemblies increased until, finally, filament-like structures were formed

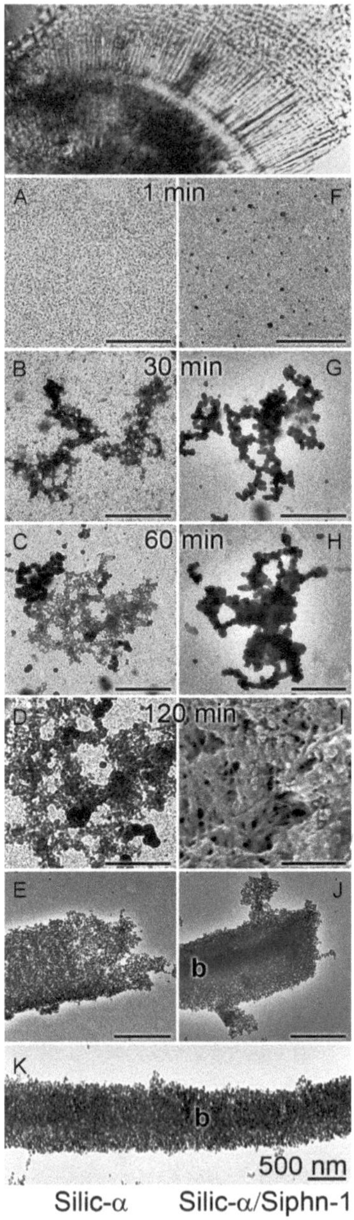

Figure 5.6 (Upper image) Silica gel prepared together with 5% Na-carbonate forms, after heating for 4 h in a platinum jar, nicely arranged structures (Bütschli, 1909); magnification: 2000×. Self-assembly of silicatein-α/silintaphin-1 *in vitro*. (A–E) TEM micrographs of recombinant silicatein-α alone and (G–M) of recombinant silicatein-α incubated together with silintaphin-1 in a molar ratio of 4:1 for the indicated times. Silicatein-α was allowed to

after >120 min (Fig. 5.6I). The lengths of the filaments exceeded 5 μm, but the diameters remained almost constant with 20 nm (Fig. 5.6I).

The process of self-assembly was followed also during the biosilica formation. For those experiments, aggregates of assembled recombinant silicatein-α (after 240 min of incubation) were exposed to pre-hydrolyzed TEOS for 24 h (22 °C) and then inspected by SEM (Fig. 5.6K). The micrographs showed that particles (nanospheres), with an average size of about 30 nm, were densely packed into a rod-shaped form. A similar arrangement of particles was observed following co-incubation of silicatein-α/silintaphin-1. Under these conditions, a filamentous backbone structure could be visualized (Fig. 5.6J); furthermore, rod-shaped structures had been formed in the silicatein-α/silintaphin-1 assays (Fig. 5.6K) that had been longer than those which assembled with silicatein-α alone (4 μm vs. <2 μm). For those experiments with silicatein-α/silintaphin-1, a stoichiometric ratio of 4:1 had been chosen. Other ratios, for example, 10:1 or 1:10 did not result in differently shaped aggregates.

The data show that silicatein-α and its interactor, the scaffold protein silintaphin-1, are fundamental prerequisites for the synthesis of biosilica. Silicatein-α forms multimers (also including silicatein-β) that assemble with silintaphin-1 in a stoichiometric manner to form fractal-like structures. These aggregates reveal a greatly enhanced capacity for biosilica synthesis, suggesting that the proteinaceous polymeric matrix facilitates silica polycondensation (perhaps in a processive manner). Moreover, these data also might contribute to the morphogenesis of the biosiliceous product.

2.9. Two-step reactions during biosilica formation in spicules: Polycondensation is followed by syneresis

Based on these data, it is obvious that in demosponges the formation of the siliceous spicules is initiated by the enzymatic polycondensation of orthosilicate to biosilica, mediated by silicateins. While this process has been elucidated to a large extent, the subsequent step of maturation/aging of

assemble in a Tris/HCl buffer, lacking glycerol, for 1 min (A), 30 min (B), 60 min (C) and 120 min (D). Then, the samples were inspected by TEM. In a parallel series of experiments, silicatein-α and silintaphin-1 were incubated for the same time periods: 1 min (F), 30 min (G), 60 min (H) and 120 min [SEM] (I). Finally, after a total incubation period of 240 min, protein assemblies were supplemented with pre-hydrolyzed TEOS for an additional 24 h (22 °C). During that time, clusters of silica particles were formed. In the presence of silicatein-a (Silic-a), the clusters appeared homogeneous (E), but the clusters formed in the presence of silicatein-α/silintaphin-1 (Silica/Siphn-1; J) frequently showed a backbone-like structure (b). (K) The length of the silica deposits in the assays with silicatein-α/silintaphin-1 often exceeded 4 μm. Size bars in all images: 500 nm; from Schlossmacher *et al.* (2011) with permission.

the biosilica (gel network) formed had only be studied recently (Wang *et al.*, 2011). As known from sol-gel polycondensation processes, silica gel formation involves the phases of polymerization of silica, coarsening and finally transformation (Iler, 1979; Brinker and Scherer, 1990). During this process, silanol groups (Si—OH) are linked together to siloxane bonds (Si—O—Si) under water release (see Iler, 1979). Auto-polycondensation and Ostwald ripening, a process during which small silica oligomers fuse with larger silica particles, progress and involve depolymerization and re-polymerization reactions during which the most 'sticking' arrangement, the most stable configuration of the silica units and smaller particles, are established (Iler, 1979; Brinker and Scherer, 1990). Also hardening/aging reactions proceed that are paralleled by syneresis and shrinkage of the silica network (Brinker and Scherer, 1990).

Studies on the elucidate and the importance of the syneresis, expulsion of water, from the gel network in the biological system during spicule formation became possible after the recent observation that primmorphs, 3D-cell cultures, exposed to Mn-sulphate down-regulate *aquaporin* expression (Wang *et al.*, 2011). It had been found that aquaporin(s), integral cell membrane pore protein, are embedded in the cell membrane also of sponge cells. Aquaporins allow water to pass but block the passage of ions in order to prevent dissipation (Engel *et al.*, 2000). These pores prevent formation of water gradients within and between cells. In *S. domuncula*, it is the *aquaporin-8 pore* that is expressed; in the presence of Mn-sulphate, the synthesis of this pore protein is suppressed. Aquaporin-8 molecules control absorption of water from the intestine and, in turn, regulate the intracellular osmohomeostasis and mucosal fluid fluxes (Calamita *et al.*, 2001).

In our studies, primmorphs, a special form of 3D-cell aggregates that are composed of proliferating and differentiating cells (Müller *et al.*, 1999), were co-incubated with spicules. Those primmorphs had been incubated with Mn-sulphate in order to suppress the expression of aquaporins. In the presence of this salt (Mn-sulphate), the cells produced silica deposits onto the spicules through which those spicules subsequently underwent a process of bio-sintering. The newly formed bridges and deposits had irregular shapes and forms and comprised by far no defined structures that are characteristic for the spicules, as analyzed by SEM. If positioned adjacent to another spicule, irrespective of whether homologous from *S. domuncula*, or even heterologous, for example, from *Baikalospongia intermedia*, the siliceous spicules fused/sintered together. The newly formed deposits on the spicules have been identified to be rich in the elements silicon and oxygen as verified by EDX analyses as well FT-IR spectroscopy. In addition, it is shown that those newly formed deposits onto the spicules contain a higher percentage of water, compared to the aged/mature spicules. While the mature spicules comprise approximately 12.8% water, the covering deposits contained 14.5% water; silica gel contained a slightly higher amount of water with 14.9%. These values are in the range of the published data, with 13.7% (Sandford, 2003). These findings, obtained thermogravimetrically and also FT-IR spectroscopically, show that the newly formed deposits comprise freshly polycondensated biosilica that has not yet extensively passed the phase of syneresis.

Even though recognized only recently (Müller *et al.*, 2009a), the process of bio-sintering in sponge spicules is a frequently occurring process in the two phyla of the siliceous sponges: the demosponges and hexactinellids. While an intra-spicular bio-sintering is seen in every spicule formed in demosponges, an inter-spicular bio-sintering can abundantly be found in hexactinellids. In hexactinellids, the bio-sintering process, occurring between spicules, is a secondary process that occurs after completion of the synthesis of individual spicules (Leys *et al.*, 2007) and starts from the surfaces of existing spicules under formation of silica bridges via simple spot soldering. Hence, this secondary silicification is a processive reaction and starts from distinct spots until complete secondary silica lamellae between spicules are extended and formed. In view of the data reported here, it is reasonable to assume that also under physiological conditions in the animals the secondary silica deposits are formed and fused/bio-sintered together immediately after their synthesis and prior to the phases of syneresis and shrinking. The proposed two step process, polycondensation followed by syneresis, is schematically shown in Fig. 5.7.

2.10. Synthesis of spicules in *S. domuncula*

The skeleton of *S. domuncula* is composed of only two types of megascleres, monactinal tylostyles and a smaller fraction of diactinal oxeas. The formation of those spicules has been described in details (Müller *et al.*, 2005; reviewed in Müller *et al.*, 2006b). Growth of the spicules is a rapid process that surely occurs more frequently in embryos or in primmorphs, than in adult specimens. Hence, we applied the primmorph system and could establish that the initial steps of spicule formation that occur intracellularly in the sclerocytes (Müller *et al.*, 2005). The sclerocytes in primmorphs produce after a cultivation period of 5 days readily spicules. As shown in Fig. 5.8A, the 15-μm large sclerocyte produces one spicule of a length of 6 μm; the lengths of the spicules in those cells vary between 0.7 and 8 μm with diameters of up to 0.9 μm. After the spicules reach a size of >5 μm, the sclerocytes contract and expand to surround the entire spicule. At higher magnification, it could be observed that the blunt end of one spicule, within a sclerocyte, is tightly associated with fibrils, which are very likely involved in the extrusion of the spicules out of the cells (Perović-Ottstadt *et al.*, 2005). Spicule formation starts within vesicles of the cells (Fig. 5.8A) with the formation of the first silica layer(s) around the axial filament. Then this spicule is extruded into the extracellular space. There, during growth in the

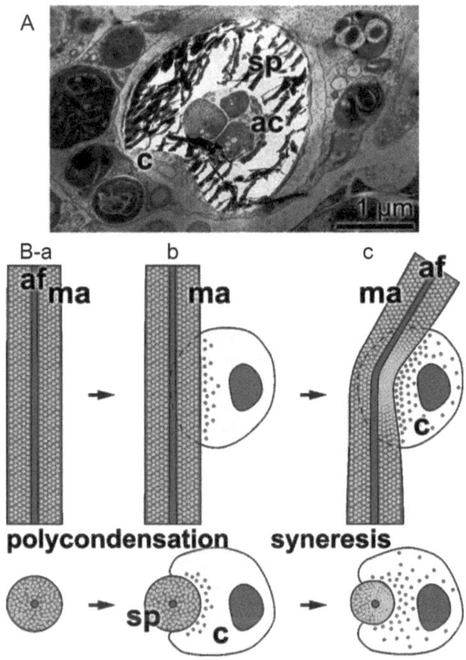

Figure 5.7 The proposed two-step reactions during spicule formation: polycondensation and syneresis. (A) Thin section through sponge tissue (*S. domuncula*). One cell (c) is shown that is closely associated with a growing sponge spicule. It is striking that the spicule, with its axial canal (ac), is punched in at that region where this cell is closely attached to the spicule. (B) Schematic sequence process during maturation of spicules. In the first step, biosilica is formed around the axial filament (af) (built by silicatein) by enzymatic polycondensation. Then a cell attaches and removes the reaction water (syneresis), forming during the sol gel process. As a consequence, the biosilica material of the growing spicule shrinks and becomes bent. In the lower panel, the uptake of the reaction water, formed during polycondensation, through the aquaporin water pores is sketched.

extracellular space, the diameter of the axial filament decreases from 1.5 to 0.8 μm, while simultaneously the size of the spicule increases up to 450 μm in length and to 5 μm in diameter (Fig. 5.8B). In the next growth stage, two types of structures can be distinguished within the axial canal. The 1.6-μm-wide axial canal is filled primarily with membrane structures (Fig. 5.8B) or with small cell remains in addition to membrane structures (Fig. 5.8C–E). The 0.8- to 0.2-mm-thick axial filament has the same electron density as in the initial stage of spicule formation. Among the cellular structures there are many 10- to 15-nm length round fibrils of not yet defined nature. The axial filament in the canal increases again in size and becomes higher in electron density (Fig. 5.8E). In the final stage, the axial canal changes from nearly circular to triangular (Fig. 5.8F).

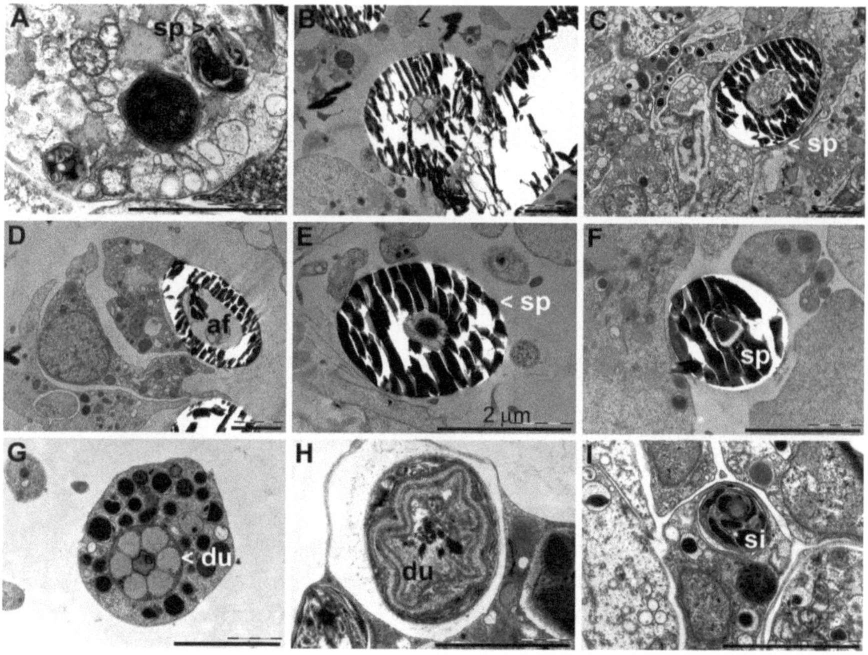

Figure 5.8 Developmental stages of the axial filament within the growing spicules in primmorphs; TEM analysis of the different phases of spicule formation in primmorphs. (A) In sclerocytes, the spicule forming cells, the formation of the skeletal elements starts within vesicles. (B) The axial canal of newly growing spicule comprises hollow fibrillar structures of three (shown here) to six tube-like ducts. (C) Subsequently, the axial filament comprises membrane structures, which in a later stage (D) remain as thin, electron dense axial filament. (E) The electron dense axial filament in the canal increases and shows the membrane structures. (F) Finally, the axial canal changes from nearly circular to triangular and comprises the electron dense axial filament. (G) The cross-section of one hollow fibrillar structure, in the initial phase of spicule formation in the sclerocyte, is shown. Six ducts (du) are seen. (H) All ducts fuse together. Both in the centre and around the hollow fibrillar structures, very dense silica deposits (de) are seen. (I) Finally, the hollow ducts disappear and larger silica (si) deposits are seen in the primordial spicule. Scale bars: 2 µm. From Müller *et al.* (2006b) with permission.

Interesting are the hollow fibrillar structures which are seen in sections through spicules in the transition phase (Fig. 5.8B). Those spicules comprise complete membrane structures in the axial canal (Fig. 5.8C) and show in the canal a more compact axial filament (Fig. 5.8F). In the transition phase (Fig. 5.8B), the hollow fibrillar structure fills almost the complete space in the canal. These hollow fibrillar structures show six tube-like ducts, which are initially regularly arranged around a central filled canal of spicules inside the cells (Fig. 5.8G). The diameter of those hollow fibrillar structures measures approximately 1.3 mm. In the following stage, the regular arrangement of these structures changes and the hollow ducts fuse with each other and also with the central canal (Fig. 5.8H). Then, first silica depositions are seen within the structure and also outside it. Finally, the fused hollow ducts disappear and larger silica deposits are observed (Fig. 5.8I); in this phase, the growing spicule is still located within the cell. At present, we assume that the function of the hollow fibrillar structure is to provide the guidance for the formation of the silicatein axial filament and to allow also the deposition of the first silica clods.

The extracellularly located spicules are surrounded by sclerocytes which are filled with both electron-dense and electron-poor vesicles. EDX/SEM studies revealed that the electron-dense vesicles are filled of silicon/silica; therefore, we termed them silicasomes (Schröder *et al.*, 2007a). It could be shown that the cells are filled with 0.2–0.8 μm large electro-dense granules. Between the cells, collagen fibres are abundantly present. Especially in close neighbourhood to the spicules, the silicasomes are released into the extracellular mesohyl compartment; this release process is shown in Fig. 5.9. Moreover, after contrasting of sections through tissue followed with labelled antibodies against silicatein-α, it could be demonstrated by TEM analysis that the electro-dense silicasomes are surrounded by molecules/structures reacting with the labelled antibodies. Frequently, the fibres oriented in parallel to the surface of the cells contain vesicles, surrounded

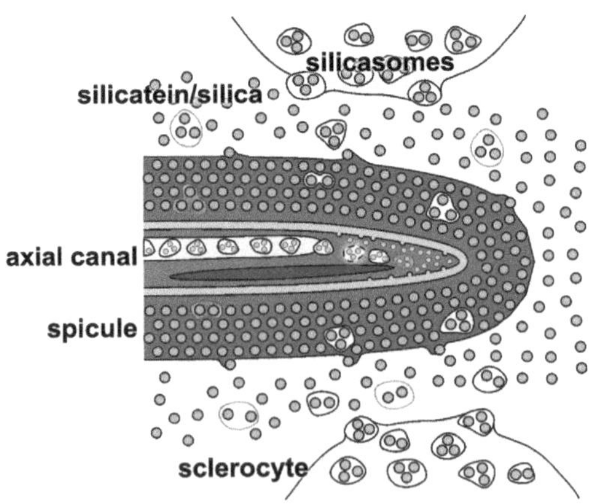

Figure 5.9 Schematic outline of the extracellular growth of the sponge spicules (*S. domuncula*). The spicules are surrounded by cells, the sclerocytes. From these cells, silicasomes are release into the extracellular space discharging there both silicatein and silica. Subsequently, the enzyme is depositing on the surface of the existing spicule additional biosilica. (For colour version of this figure, the reader is referred to the online version of this chapter.)

by structures/molecules that react with the antibodies. In order to obtain a evaluation of silicon/silica deposition in the granules, EDX-micro-analyses on sections were performed (Schröder *et al.*, 2007a). The result of those scans through an electron-poor vacuole shows two prominent peaks of carbon and oxygen but not of silicon. In contrast, if the analyses of the two selected electron-dense particles were performed, the silicon peaks were prominently identified.

The extracellular deposition of biosilica onto the surface of growing spicules had been analyzed by immuno-TEM analyses (Schröder *et al.*, 2006; reviewed in Müller *et al.*, 2006b). Again, the development of the spicules had been studied in primmorphs. The fine structure analysis is summarized in Fig. 5.10 (Müller *et al.*, 2006b). Firstly, concentric rings are seen around the forming spicules which are 0.2–0.5 μm apart (Fig. 5.10B). Subsequently, the inner rings fuse and electron dense linear clods become visible (Fig. 5.10C). Later, during maturation, the number of the concentric rings increases from 2–3 rings with a diameter of 2–3 μm (Fig. 5.10B) to 10 rings and a total diameter 4–6 μm (Fig. 5.10D). At higher magnification, the clods appear round shaped (Fig. 5.10E and F) and explained as the first biosilica deposits. Finally, the first layer of silica is formed, showing again around these silica structures concentric rings of gold particles, reflecting silicatein antibodies (Fig. 5.10G). The number of gold/antibody rings around the primordial spicules, composed of a first silica layer, is larger than three (Fig. 5.10H). Under the experimental conditions used here, also the axial filament (Z silicatein) reacts with the antibodies (Fig. 5.10H and I). It should be mentioned that around the axial filament and the first layer of the silica deposition, a space filled with non-reacting material is seen (Fig. 5.10I). Probably, there the substrate for the silicatein is stored.

Based on these studies, a comprehensive description of the spicule growth can be outlined.

2.11. Phases of silica deposition during spicule formation

The process of spicule formation can be divided into two phases (Müller *et al.*, 2009b): the initial intracellular steps as well as the extracellular final and shaping phases.

(*i*) *The intracellular phase in the sclerocytes:* Silica is taken up actively by a $Na^+/HCO_3^-[Si(OH)_4]$ cotransporter (Schröder *et al.*, 2004). In the first steps, silicatein is synthesized as a pro-enzyme (signal peptide-propeptide-mature enzyme: 36.3 kDa) and processed via the 34.7 kDa form (propeptide-mature enzyme) to the 23 kDa mature enzyme. Very likely during the transportation through the endoplasmic reticulum and the Golgi complex, silicatein undergoes phosphorylation and is transported into vesicles where forms rods, the axial filaments. After assembly to filaments, the first layer(s) of silica is (are) formed. Silica deposition occurs in two directions; first from

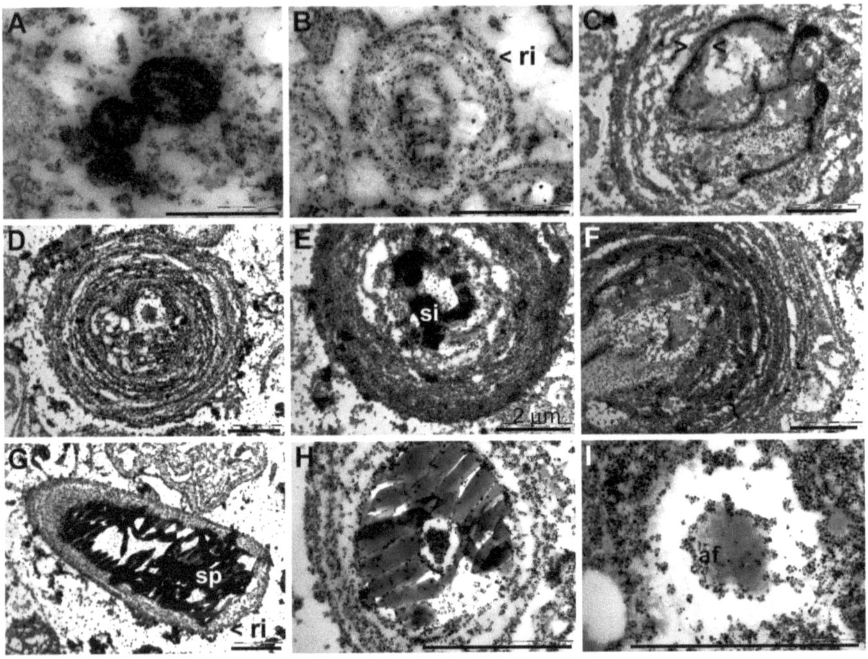

Figure 5.10 Immunogold electron microscopy of cross-sections through growing spicules in primmorphs; the immune complexes among the polyclonal anti-silicatein antibodies had been visualized with nanogold anti-rabbit IgG; TEM images. (A) Sclerocytes with a strong accumulation of gold granules, reflecting the condensation of silicatein molecules. (B) At a higher magnification concentric rings (ri) are seen that surround a centre with granules of a higher density. (C) In a later stage, first electron dense linear clods (><) become visible around which the rings are arranged. (D) The inner rings fuse and electron dense linear clods are formed. (E) The electron dense material (silica [Si]) accumulates in the centre of the multi-lamellar silicatein-reacting material. (F) Higher magnification of (E). (G) The first layer of silica at a growing spicule (sp) is formed around which the concentric ring(s) (ri) remains present. (H) Finally, the silica layer grows and the number of rings increases. (I) The silicatein antibodies react with the axial filament (af) and the deposits of the silica rings, leaving space for non-reacting material. Scale bars: 2 μm. From Müller *et al.* (2006b) with permission.

the axial canal to the surface (centrifugal orientation) and second from the mesohyl to the surface of the spicule (centripedal). Finally, the spicules are released into the extracellular space where they grow in length and diameter by appositional growth.

(*ii*) *Extracellular phase (appositional growth):* Silicatein is present also in the extracellular space. It came surprising that also there the silicatein molecules are organized to larger entities. The immunogold electron microscopical analysis showed that the silicatein molecules are arranged along strings, which are organized in parallel to the surfaces of the spicules (Schröder *et al.*, 2006). In the presence of Ca^{2+}, silicatein associates with galectin and allows the appositional growth of the spicules. Since the surface of a new siliceous spicule is also covered with silicatein, the appositional growth/ thickening of a spicule hence proceeds from two directions (centrifugal and centripetal).

(*iii*) *Extracellular phase* (shaping): In the next step, the galectin-containing strings are organized by collagen fibres to net-like structures. It is very likely that collagen, which is released by the specialized cells, the collencytes, provides the organized platform for the morphogenesis of the spicules. The longitudinal growth of the spicules can be explained by the assumption that at the tips of the spicules, the galectin/silicatein complexes are incorporated into deposited biosilica under formation and elongation of the axial canal.

2.12. Formation of the morphology of the spicules

In siliceous sponges (phylum Porifera; class Demospongiae), the silicatein is located inside the spicules as axial filament in a distinct canal, termed axial canal (see Uriz, 2006). There, several isoforms of silicatein have been identified, with silicatein-α, -β and -γ as the main forms in marine sponges (Shimizu *et al.*, 1998), while in the freshwater species, several silicatein-α isoforms exist in the axial filament (Wiens *et al.*, 2009b). As mentioned above, the individual silicatein molecules undergo self-assembly to oligomers (Murr and Morse, 2005), a process that proceeds and results in the formation of filaments via fractal intermediates (Murr and Morse, 2005; Müller *et al.*, 2007a). The association of the different isoforms is stoichiometrically controlled; four silicatein-α molecules associate and allow one silicatein-β molecule to stick inside this pentamer (Müller *et al.*, 2007a) in a way that the active centres of the enzyme are exposed to the environment. This association sequence, starting from a chaotic aggregation, leads to fractal-like structures that result in the building of filaments of higher-ordered pattern. Based on diffraction experiments, using a small-angle X-ray scattering approach, it had been found that those patterns are not identical in the different sponge species (Croce *et al.*, 2003). The protein units within the filaments are highly organized and show two different orientations of the lattices formed from silicatein molecules that are spaced by distances of 5.8 nm. The orientations of the lattices are either regular in 2D hexagonal arrangements of protein units in perpendicular direction along the spicule axis (like in *Geodia cydonium*), or two different 2D lattices exist in which the repeating protein units are inclined by a defined angle (e.g. in the hexactinellid *Scolymastra joubini*) (Croce *et al.*, 2003). This view is supported by the fact that the morphologies of the spicules are manifold and filigree and they are characteristic in the appearance for each sponge species (see Uriz, 2006).

While initially it had been proposed that the pattern formation of poly (silicate) structures in biological systems is determined by physico-chemical laws only, for example, by electromagnetic vibrations (see Thompson, 1942), a paradigm change emerged with the discovery of the silicateins. Since the discovery of this class of enzymes (Shimizu *et al.*, 1998) and the subsequent finding that the initial form of the siliceous spicules, the axial orientation, is determined by the rod-like axial filament assembled inside the sponge cells (Müller *et al.*, 2005). It is assumed that the control of the sizes and shapes of spicules, their morphology, is exclusively genetically controlled (Müller, 2005). This view is also supported by the finding that morphologically differently formed and sized spicules in a given species, for example, in *G. cydonium*, comprise different sets of silicateins (Müller *et al.*, 2007b). While in this species, the larger spicules, the megascleres, contain three silicatein isoforms [-α, -β and -γ], the smaller microscleres contain only one form of silicatein, which has been termed silicatein-α/β. Besides of different sets of silicatein isoforms, also different extents of post-translation modifications, for example, phosphorylation (Müller *et al.*, 2005) or methylations (Armirotti *et al.*, 2009), have been proposed to be involved in the determination of the morphology of the axial filaments and the biological activity derived from them.

It might be mentioned that especially for the Class of Hexactinellida, a different biosilicification mechanism has been proposed, which is based on collagen (reviewed in Ehrlich, 2010). However, the presented experiments had been performed at ortho-silica concentrations that are higher than those used for the silicatein-driven reaction.

3. Biomedical/Biotechnological Application of Silicatein and Silintaphin-1

The recombinant silica enzymes (protected by patents hold by us; silicatein, EP1320624; silicase, DE10246186) are of high interest and potential importance for a variety of medical and technical applications, for example, surface modification of glasses and other materials including biomaterials, the preparation of resins, insulators, mesoporous molecular sieves, and catalysts as reviewed (reviewed in Schröder *et al.*, 2007b, 2008). First strategies have been designed for the application of these enzymes for surface modifications (coatings) of biomaterials for use in biomedical fields. The advantage of such bio-based approaches compared to conventional technical procedures is that the enzymatic silicatein reaction occurs under mild physiological conditions, whereas physical–chemical methods require high temperatures or pressures, and the use of caustic chemicals which possibly damage the organic (bio)materials used. Such biomaterials may include collagen used in tissue engineering and in bone replacement materials. The use of biocatalytically (silicatein) formed silica may also be a suitable approach for coating metal implants to increase biocompatibility or to attach bioactive substances to the relatively inert metal surface. The same strategy could be employed for the encapsulation of drugs, hormones and other bioactive molecules and the controlled release of these compounds.

In addition, recombinant silica enzymes may also be applied for the synthesis of the nanostructures of amorphous silica. In industry, micro- and mesoporous silicas are used as reinforcing fillers in plastics, paints, sealants and rubber materials, as adsorbents and catalyst supports, as desiccant agent, and as filters in separation technologies. Such applications require silicas with specific mechanical strength, pore volume, surface area and pore-size distribution. Finally, the application of techniques in lithography based on enzyme (silicatein)-mediated biosilicification may represent an innovative approach in the field of fabrication of microelectronics. Two fields are highlighted in the following section.

3.1. Biotechnological approach

The discovery of silintaphin-1 established the basis to control *in vitro* biosilicification, since during bioinspired approaches co-incubation of refolded, recombinant silicatein and silintaphin-1 (equimolar concentrations) resulted in the formation of biomimetic filamentous protein structures with a diameter of 20–40 nm, and a length of several micrometres, that strongly resemble natural axial filaments (Wiens *et al.*, 2009a). This observation confirmed that silintaphin-1 represents a scaffold protein that is required for the filamentous organization of silicatein also *in vivo*. In poriferan cells, spiculogenesis is initiated by the synthesis of the axial filament around which silica nanoparticles assemble to form a compact concentric layer. In a further biomimetic approach, this process was reconstructed *in vitro*. For that purpose, amorphous silica nanoparticles were co-incubated with refolded recombinant silicatein. Fourier transform spectroscopy with attenuated total reflectance and immunoblot analyses confirmed the adsorption of the enzyme on the surfaces of nanoparticles even after several washing steps, indicating the significant binding capacity of silicatein to amorphous silica. Subsequent incubation of the silica/protein complexes with native silintaphin-1 resulted in the formation of structures with spicular to rod-like morphology, thus revealing the cross-linking capacities of silintaphin-1.

After the initial intracellular stage of spiculogenesis, the developing spicule is transported into the extracellular space, where it grows through apposition of lamellar silica layers. There, spicules obtain their final sizes, between 450 μm (Demospongiae) and 2.5 m (Hexactinellida), and morphology. Based on the processes observed during the extracellular growth of spicules, the natural principle was applied during biomimetic and (nano)biotechnological approaches ("Nature as Model"; Müller *et al.*, 2009b). As outlined above, growth of spicules is guided by a proteinaceous cylinder containing silicatein that facilitates the formation of biosilica

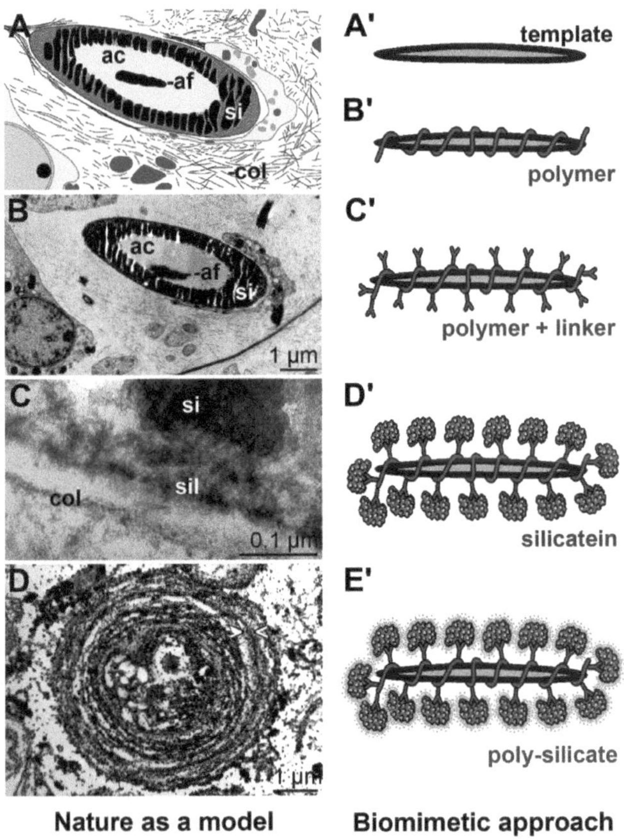

Figure 5.11 "Nature as model"—a biomimetic approach. Left panel: (A–D) Synthesis of poriferan spicules in the extracellular space. (A) (scheme) and (B) (TEM) Cross-section through a growing spicule with the silica coat (si) that surrounds axial canal (ac) and axial filament (af). In the extracellular space collagen (col), fibres are indicated. (C) High-magnification TEM image showing the silicatein molecules (sil), associated with galectin, surrounding the silica material (si). Collagen fibres (col) have no contact to the silicatein/silica complex. (D) Immunogold-labelled silicatein visualizing the organic cylinders which guide and mediate the synthesis of the silica layers in an appositional manner (marked ><). Right panel: (A′–E′) The biomimetic approach. The template (A′) is successively functionalized with a reactive ester polymer (B′) and the NTA linker (C′). (D′) recombinant silicatein is bound via His-tag and Ni^{2+} to the NTA-polymer and subsequently mediates formation and assembly of polysilica formation (E′). From Müller *et al.* (2009b) with permission. (For colour version of this figure, the reader is referred to the online version of this chapter.)

lamellae (Fig. 5.11; left panel [A–D]). In a bioinspired approach, biosilica was synthesized on "inert" surfaces (matrices) from monomeric precursors (Tahir *et al.*, 2004). However, prior to the biosilica formation, the matrices had to be functionalized with a reactive polymer that is subsequently able to chemisorb nitrilotriacetic acid (NTA), which in turn is required to bind His-tagged recombinant silicatein (Fig. 5.11; right panel [A′–E′]). Silicatein that had been immobilized on this matrix has the capacity to synthesize nanoparticulate biosilica, biotitania and biozirconia from monomeric precursors. This striking example shows that nature could be used as a biological blueprint for biomedical and biotechnological applications.

Natural sponge spicules can act as optical glass fibres that transmit light with high efficiency (Müller *et al*, 2006c), as initially described by Cattaneo-Vietti *et al.* (1996). In addition, transmission of light is very selective since only wavelengths between 615 and 1310 nm can pass through the spicules (sharp high- and low-pass filters). Spicules comprise a high refractive index core, a surrounding lower refractive-index cylindrical tube, and an outer portion with a progressively increasing refractive index. These optical characteristics of spicules are complemented by surprising mechanical properties based on the composite structure, lamellar architecture, and presence of dopants. Thus, spicules revealed enhanced fracture toughness (Sundar *et al*, 2003; Aizenberg *et al.*, 2004). Accordingly, it is tempting to employ the directed assembly of silica nanoparticles by the combined action of silicatein and silintaphin-1 to fabricate light-guiding micro- and nanostructured composites that represent an economical alternative to industrial glass fibres currently used in micro-optical approaches.

3.2. Biomedical approach: Teeth

Biosilica is an essential nutrient for the natural ecosystem in general (Struyf and Conley, 2008), for humans and other vertebrates in particular (Carlisle, 1986; Dyck *et al.*, 2000), where silicon deprivation results in severe skeletal malformations (Carlisle, 1972). Thus, in avian connective tissue, the highest silicon concentrations were determined as compared to heart or muscle tissue. Moreover, a spatial correlation could be established between the areas of bone formation within animal tissue and the accumulation of silicon (Fig. 5.12A). Thus, a burst of silicon accumulation was seen around the osteoid and osteoid–bone interfaces, suggesting that this inorganic component is essential for bone formation. Consequently, the effect of biosilica on the activity of osteoblasts was investigated in depth. Indeed, the cell model used (human osteogenic sarcoma cells; SaOS-2) displayed an increased mineralization activity, when cultivated on biosilica surfaces in the presence of β-glycerophosphate, an organic phosphate donor (Schröder *et al.*, 2005).

In particular, concurrent coating of the substratum with biosilica and type 1 collagen not only increased the cellular Ca-phosphate deposition but also stimulated cell proliferation. In subsequent studies, the effect of biosilica and silica-based components on the expression of key genes involved in formation of tooth enamel, amelogenin, ameloblastin and enamelin, was investigated. These studies revealed that the combination of β-glycerophosphate and silica-based components increased the expression of these marker genes (Fig. 5.12B). This finding was further supported by HRSEM, visually demonstrating the increased deposition of hydroxyapatite crystallites in treated cells (Müller *et al.*, 2007c). First attempts to evaluate the biomedical application of silicatein/biosilica for treatment of bone/tooth defects and dental care are promising. Thus, during a reaction of silicatein with the substrate sodium metasilicate, nanoscale biosilica layers (50–150 nm) were formed on dental (pig teeth; Fig. 5.12C and D) and bone hydroxyapatite (rat femurs; Fig. 5.12E and F; HR-SEM). This data pave the way for biomedical approaches that aim, for example, to generate protective biosilica layers on teeth (reducing the risk of bacterial induced caries/cavities) or to regenerate bone tissue via biosilica-stimulated activity of mineralizing cells (resulting in an increased deposition of hydroxyapatite).

3.3. Biomedical approach: Bone-osteoporosis

Bone formation depends on the interplay of multiple factors within a functionally defined spatio-temporal context. Among the molecules involved, proteinaceous [e.g. bone morphogenetic protein/alkaline phosphatase (Kamegai *et al.*, 1990; Aldinger *et al.*, 1991)], organic chemical [e.g. organophosphates/β-glycerophosphate (Chung *et al.*, 1992)], inorganic chemical [e.g. silica (Carlisle, 1972; Schwarz and Milne, 1972)], but also physical factors [e.g. physical stress or ultrasound (Borsje *et al.*, 2010)] play a role. Hence, bone reconstitution studies must consider an array of well-tuned (molecular) interactions and pathways.

It is not yet possible to denominate the one master transcription factor that regulates differentiation of osteoblast and osteoclast precursor cell lineages. The discovery of the osteoclastogenesis inhibitory factor osteoprotegerin [OPG] (Simonet *et al.*, 1997), a cytokine that is expressed in osteoblasts, has significantly contributed to understanding the mechanisms that control bone mineral density and volume. Since then, many efforts have been directed towards the screening for factors that maintain OPG synthesis on a physiological level, forestalling pathophysiological disorders, such as osteosclerosis (Wang *et al.*, 2004) or osteoporosis (Lane and Yao, 2009). In our more recent study (Wiens *et al.*, 2010), we investigated the effect of silica, which fundamentally affects osteogenesis, on gene/protein expression. As model system to evaluate the effect of biosilica on OPG

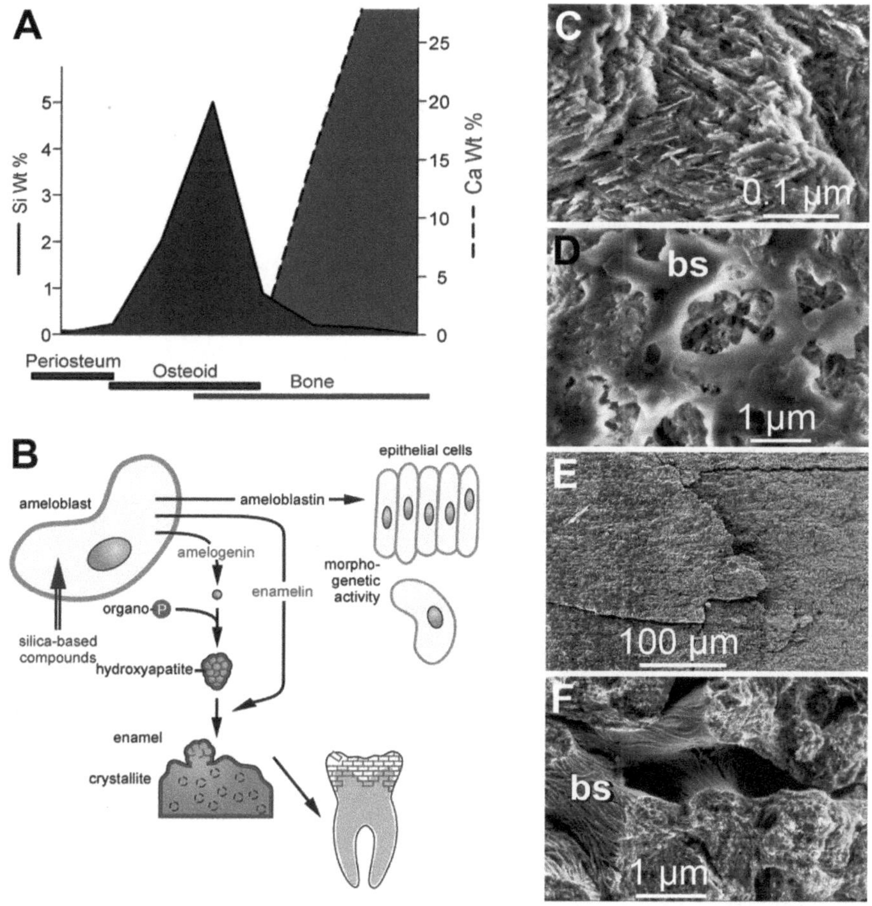

Figure 5.12 Biomedical application of biosilica and silicatein. (A) Spatial relationship between silicon accumulation and calcium composition during early stages of bone formation in rats (modified according to Carlisle, 1986). (B) Schematic representation of the effect of silica-based components on the expression of the three marker genes (amelogenin, ameloblastin and enamelin) in ameloblasts. The silica-based components stimulate the expression of amelogenin, resulting in the formation of nanospherical hydroxyapatite around which hydroxyapatite crystals are deposited. (C and D) Formation of biosilica layers on pig molars. After tissue removal, the teeth were treated with phosphate buffered saline, supplemented with protease and phosphatase inhibitors, according to Aoba *et al.* (1987). Subsequently, dental hydroxyapatite was incubated with sodium metasilicate (100 μM) in the absence (C) or presence (D) of recombinant silicatein (4 μg/ml PBS) for 12 h at 20 °C. Then, the samples were examined by HR-SEM. (E and F) In parallel, biosilica formation on femur bone samples was examined: untreated control (E) or silicatein-treated (F). The biosilica layers are marked (bs). From Müller *et al.* (2009b) with permission. (For colour version of this figure, the reader is referred to the online version of this chapter.)

expression and cell mineralization, an osteosarcoma cell line was chosen, SaOS-2 (Wiens *et al.*, 2010). This cell line possesses osteoblastic features and generates a mineralization-competent extracellular matrix. It was found that biosilica positively affects growth and mineralization activity of SaOS-2 cells. Biosilica was supplied enzymatically through recombinant silicatein. We found that biosilica stimulates mineralization activity of SaOS-2 cells. Furthermore, and likewise important we found that biosilica significantly up-regulated OPG-expression, but the cytokine was also increasingly released into the culture medium, while RANKL (receptor activator for NF-κB ligand) expression remained unchanged. Accordingly, Fig. 5.13 depicts the putative *in vivo* regulatory role of silica/biosilica on OPG expression/release in osteoblasts. Consequential secondary effects negatively affect pre-osteoclast maturation and osteoclast activation since by binding to OPG, RANKL is sequestered and unavailable to bind to its (pre)ostecloastic receptor, RANK. Hence, biosilica reveals a considerable biomedical potential with therapeutical and prophylactic implementations, for example, concerning osteoporosis.

4. Conclusion

The elucidation of the different phases of spiculogenesis, and the molecules involved in it, allows a wide application of silicatein in the field of nano-biotechnology. In sponges, silica deposition is a biologically controlled process. The key enzyme is silicatein which is non-covalently linked to galectin (Schröder *et al.*, 2006). The organized process of biosilica deposition is guided by an organic matrix, collagen. This principle has been utilized by the group of Tremel (Tahir *et al.*, 2004, 2005) to synthesize on a gold surface (matrix) biosilica from monomeric precursors. The matrix used had been activated by cysteamine to allow binding to a reactive polymer; in turn, this polymer is able to chemisorb the nitrilotriacetic acid ligand. Finally, this architecture allowed the binding of histidine-tagged silicatein. Surprisingly enough, silicatein immobilized onto this matrix has the capacity to catalyze bio-titania and bio-zirconia from the monomeric metal oxide precursors. This is a further striking example to show that nature can be used as a biological blueprint for nanobiotechnological applications. In addition, this technology introduces a new concept, the synthesis of inorganic polymers by an organic molecule (silicatein). Just in opposite, in 1828, Wöhler succeeded with his epochal experiments on the synthesis of urea, an organic compound, from inorganic basic materials ["die künstliche Erzeugung eines organischen, und zwar animalischen, Stoffes aus unorganischen Stoffen"] (Wöhler, 1828).

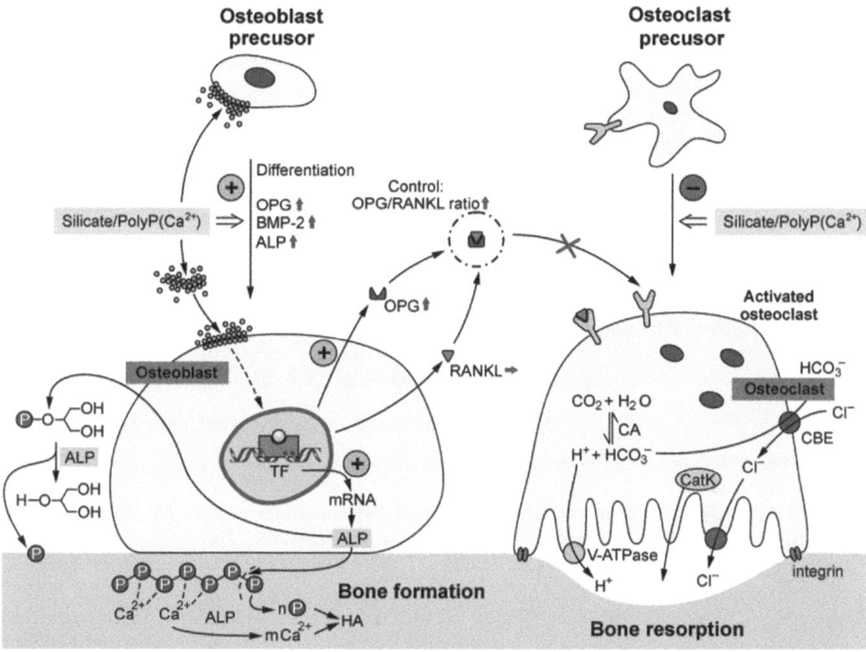

Figure 5.13 Proposed effects of biosilica on osteoblasts, osteoclasts and their progenitors. Biosilica enhances expression of OPG in osteoblasts. Osteoblasts have the potential to differentiate to hydroxyapatite-forming osteocytes and lining cells. OPG counteracts various effects of RANKL, a cytokine that induces preosteoclast maturation and osteoclast activation. In concert with biosilica, polyphosphate (PolyP) was found to contribute to the anabolic bone metabolism. Scheme. Modified according to Müller *et al.* (2011). (For colour version of this figure, the reader is referred to the online version of this chapter.)

ACKNOWLEDGEMENTS

W. E. G. M. is a holder of an ERC Advanced Investigator Grant (No. 268476 BIOSILICA). This work was supported by grants from the German Bundesministerium für Bildung und Forschung (project "Center of Excellence *BIOTEC*marin"), the Deutsche Forschungsgemeinschaft (Schr 277/10-1), the European Commission/EUREKA (EUROSTARS, No. 4289—SILIBACTS), the International Human Frontier Science Programme, the European Commission (project No. 244967—Mem-S Project), the Public Welfare Project of Ministry of Land and Resources of the People's Republic of China (Grant No. 201011005-06) and the International S & T Cooperation Programme of China (Grant No. 2008DFA00980).

REFERENCES

Adell, T., Grebenjuk, V. A., Wiens, M., and Müller, W. E. G. (2003). Isolation and characterization of two T-box genes from sponges, the phylogenetically oldest metazoan taxon. *Development Genes and Evolution* **213,** 421–434.

Adell, T., Gamulin, V., Perović-Ottstadt, S., Wiens, M., Korzhev, M., Müller, I. M., and Müller, W. E. G. (2004). Evolution of metazoan cell junction proteins: The scaffold protein MAGI and the transmembrane receptor tetraspanin in the demosponge *Suberites domuncula*. *Journal of Molecular Evolution* **59,** 41–50.

Aizenberg, J., Sundar, V. C., Yablon, A. D., Weaver, J. C., and Chen, G. (2004). Biological glass fibers: Correlation between optical and structural properties. *Proceedings of the National Academy of Sciences of the United States of America* **101,** 3358–3363.

Aldinger, G., Herr, G., Kiisswetter, W., Reis, H. J., Thielemann, F. W., and Holz, U. (1991). Bone, morphogenetic protein: A review. *International Orthopaedics* **15,** 169–177.

Aoba, T., Fukae, M., Tanabe, T., Shimizu, M., and Moreno, E. C. (1987). Selective adsorption of porcine amelogenins onto hydroxyapatite and their inhibitory activity on seeded crystal growth of hydroxyapatite. *Calcified Tissue International* **41,** 281–289.

Armirotti, A., Damonte, G., Pozzolini, M., Mussino, F., Cerrano, C., Salis, A., Benatti, U., and Giovine, M. (2009). Primary structure and post-translational modifications of silicatein beta from the marine sponge *Petrosia ficiformis* (Poiret, 1789). *Journal of Proteome Research* **8,** 3995–4004.

Arndt, W. (1935). "Die Tierwelt der Nord- und Ostsee" vol. IIIa. Porifera. Akademische Verlagsgesellschaft, Leipzig, pp. IIIa2–IIIa2140.

Blomberg, N., Baraldi, E., Nilges, M., and Saraste, M. (1999). The PH superfold: A structural scaffold for multiple functions. *Trends in Biochemical Sciences* **24,** 441–445.

Blomberg, N., Baraldi, E., Sattler, M., Saraste, M., and Nilges, M. (2000). Structure of a PH domain from the *C. elegans* muscle protein UNC-89 suggests a novel function. *Structure* **8,** 1079–1087.

Borsje, M. A., Ren, Y., de Haan-Visser, H. W., and Kuijer, R. (2010). Comparison of low-intensity pulsed ultrasound and pulsed electromagnetic field treatments on OPG and RANKL expression in human osteoblast-like cells. *The Angle Orthodontist* **80,** 498–503.

Brinker, C. J., and Scherer, G. W. (1990). Sol-Gel Science: The Physics and Chemistry of Sol-Gel Processing: The Physics and Chemistry of Sol-Gel Processing. Academic Press, Boston.

Bütschli, O. (1909). Untersuchungen über die Mikrostruktur künstlicher und natürlicher Kieselsäuregallerten. Winter, Heidelberg.

Calamita, G., Mazzone, A., Bizzoca, A., Cavalier, A., Cassano, G., Thomas, D., and Svelto, M. (2001). Expression and immunolocalization of the aquaporin-8 water channel in rat gastrointestinal tract. *European Journal of Cell Biology* **80,** 711–719.

Carlisle, E. M. (1972). Silicon: An essential element for the chick. *Science* **178,** 619–621.

Carlisle, E. M. (1986). Silicon as an essential trace element in animal nutrition. In "Ciba Foundation symposium 121." pp. 123–139. Wiley, Chichester, UK.

Cattaneo-Vietti, R., Bavestrello, G., Cerrano, C., Sarà, A., Benatti, U., Giovine, M., and Gaino, E. (1996). Optical fibres in an Antarctic sponge. *Nature* **383,** 397–398.

Cha, J. N., Shimizu, K., Zhou, Y., Christianssen, S. C., Chmelka, B. F., Stucky, G. D., and Morse, D. E. (1999). Silicatein filaments and subunits from a marine sponge direct the polymerization of silica and silicones in vitro. *Proceedings of the National Academy of Sciences of the United States of America* **96,** 361–365.

Chung, C. H., Golub, E. E., Forbes, E., Tokuoka, T., and Shapiro, I. M. (1992). Mechanism of action of beta-glycerophosphate on bone cell mineralization. *Calcified Tissue International* **51,** 305–311.

Croce, G., Frache, A., Milanesio, M., Viterbo, D., Bavestrello, G., Benatti, U., Giovine, M., and Amenitsch, H. (2003). Fiber diffraction study of spicules from marine sponges. *Microscopy Research and Technique* **62,** 378–381.

Donati, V. (1753). Auszug seiner Natur-Geschichte des Adriatischen Meeres. Francken, Halle, Germany.

Dyck, K. V., Robberecht, H., Cauwenbergh, R. V., Vlaslaer, V. V., and Deelstra, H. (2000). Indication of silicon essentiality in humans. *Biological Trace Element Research* **77,** 25–32.
Ehrlich, H. (2010). Biological Materials of Marine Origin. Springer Science, Dordrecht.
Engel, A., Fujiyoshi, Y., and Agre, P. (2000). The importance of aquaporin water channel protein structures. *The EMBO Journal* **19,** 800–806.
Flory, P. J., and Semlyen, J. A. (1966). Macrocyclization equilibrium constants and the statistical configuration of poly(dimethylsiloxane) chains. *Journal of the Chemical Society* **88,** 3209–3212.
Gesner, C. (1558). Historiae Animalium. Vol. 20. C Forer, Zürich.
Haeckel, E. (1866). Generelle Morphologie der Organismen. Verlag von Georg Reimer, Berlin.
Haeckel, E. (1868). Natürliche Schöpfungsschichte. Verlag von Georg Reimer, Berlin.
Harcet, M., Roller, M., Cetkovic, H., Perina, D., Wiens, M., Müller, W. E. G., and Vlahovicek, K. (2010). Demosponge EST sequencing reveals a complex genetic toolkit of the simplest metazoans. *Molecular Biology and Evolution* **27,** 2747–2756.
Hench, L. L., and West, J. K. (1995). Molecular orbital models of silica. *Annual Review of Materials Science* **25,** 37–68.
Iler, R. K. (1979). The Chemistry of Silica: Solubility, Polymerization, Colloid and Surface Properties and Biochemistry of Silica. Wiley-Interscience Publication, New York.
Johnston, G. (1842). A History of British Sponges and Lithophytes. Lizars, Edinburgh.
Kamegai, A., Tanabe, T., Nagahara, K., Kumasa, S., and Mori, M. (1990). Pathologic and enzyme histochemical studies on bone formation induced by bone morphogenetic protein in mouse muscle tissue. *Acta Histochemica* **89,** 25–35.
Krasko, A., Gamulin, V., Seack, J., Steffen, R., Schröder, H. C., and Müller, W. E. G. (1997). Cathepsin, a major protease of the marine sponge Geodia cydonium: purification of the enzyme and molecular cloning of cDNA. *Molecular Marine Biology and Biotechnology* **6,** 296–307.
Krasko, A., Batel, R., Schröder, H. C., Müller, I. M., and Müller, W. E. G. (2000). Expression of silicatein and collagen genes in the marine sponge *Suberites domuncula* is controlled by silicate and myotrophin. *European Journal of Biochemistry* **267,** 4878–4887.
Krasko, A., Schröder, H. C., Batel, R., Grebenjuk, V. A., Steffen, R., Müller, I. M., and Müller, W. E. G. (2002). Iron induces proliferation and morphogenesis in primmorphs from the marine sponge *Suberites domuncula*. *DNA & Cell Biololgy* **21,** 67–80.
Lane, N. E., and Yao, W. (2009). Developments in the scientific understanding of osteoporosis. *Arthritis Research & Therapy* **11,** 228.
Leys, S. P., Mackie, G. O., and Reiswig, H. M. (2007). The biology of glass sponges. *Advances in Marine Biology* **52,** 1–145.
Müller, W. E. G. (1995). Molecular phylogeny of Metazoa [Animals] (1995) Monophyletic origin. *Naturwissenschaften* **82,** 321–329.
Müller, W. E. G., (ed.) (1998a). Progress in Molecular and Subcellular Biology. "In Molecular Evolution: Evidence for Monophyly of Metazoa," Vol. 19. Springer Verlag, Berlin, Heidelberg, New York.
Müller, W. E. G., (ed.) (1998b). Progress in Molecular and Subcellular Biology. "In Molecular Evolution: Towards the Origin of Metazoa," Vol. 21. Springer Verlag, Berlin, Heidelberg, New York.
Müller, W. E. G. (1998c). Origin of Metazoa: sponges as living fossils. *Naturwissenschaften* **85,** 11–25.
Müller, W. E. G. (2001). How was metazoan threshold crossed: the hypothetical Urmetazoa. *Comparative Physiology and Biochemistry [A]* **129,** 433–460.
Müller, W. E. G. (2005). Spatial and temporal expression patterns in animals. (R. A. Meyers, ed.), Encyclopedia of Molecular Cell Biology and Molecular Medicine, **13,** pp. 269–309. Wiley-VCH Press, Weinheim.

Müller, W., and Zahn, R. K. (1968). *Tethya limski* n.sp., eine Tethyide aus der Adria (Porifera: Homosclerophorida: *Tethyidae*). *Senckenbergiana Biologica* **49,** 469–478.
Müller, W. E. G., Müller, I. M., and Gamulin, V. (1994). On the monophyletic evolution of the Metazoa. *Brazilian Journal of Medical and Biological Research* **27,** 2083–2096.
Müller, W. E. G., Wiens, M., Batel, R., Steffen, R., Borojevic, R., and Custodio, M. R. (1999). Establishment of a primary cell culture from a sponge: primmorphs from *Suberites domuncula. Marine Ecology Progress Series* **178,** 205–219.
Müller, W. E. G., Schröder, H. C., Skorokhod, A., Bünz, C., Müller, I. M., and Grebenjuk, V. A. (2001). Contribution of sponge genes to unravel the genome of the hypothetical ancestor of Metazoa (Urmetazoa). *Gene* **276,** 161–173.
Müller, W. E. G., Krasko, A., Le Pennec, G., Steffen, R., Ammar, M. S. A., Wiens, M., Müller, I. M., and Schröder, H. C. (2003). Molecular mechanism of spicule formation in the demosponge *Suberites domuncula*: silicatein–collagen–myotrophin. *Progress in Molecular and Subcellular Biology* **33,** 195–221.
Müller, W. E. G., Wiens, M., Adell, T., Gamulin, V., Schröder, H. C., and Müller, I. M. (2004). Bauplan of urmetazoa: basis for genetic complexity of metazoa. *International Review of Cytology* **235,** 53–92.
Müller, W. E. G., Rothenberger, M., Boreiko, A., Tremel, W., Reiber, A., and Schröder, H. C. (2005). Formation of siliceous spicules in the marine demosponge *Suberites domuncula. Cell and Tissue Research* **321,** 285–297.
Müller, W. E. G., Kaluzhnaya, O. V., Belikov, S. I., Rothenberger, M., Schröder, H. C., Reiber, A., Kaandorp, J. A., Manz, B., Mietchen, D., and Volke, F. (2006a). Magnetic resonance imaging of the siliceous skeleton of the demosponge *Lubomirskia baicalensis. Journal of Structural Biology* **153,** 31–41.
Müller, W. E. G., Belikov, S. I., Tremel, W., Perry, C. C., Gieskes, W. W. C., Boreiko, A., and Schröder, H. C. (2006b). Siliceous spicules in marine demosponges (example *Suberites domuncula*). *Micron* **37,** 107–120.
Müller, W. E. G., Wendt, K., Geppert, C., Wiens, M., Reiber, A., and Schröder, H. C. (2006c). Novel photoreception system in sponges? Unique transmission properties of the stalk spicules from the hexactinellid Hyalonema sieboldi. *Biosensors and Bioelectronics* **21,** 1149–1155.
Müller, W. E. G., Boreiko, A., Schloßmacher, U., Wang, X. H., Tahir, M. N., Tremel, W., Brandt, D., Kaandorp, J. A., and Schröder, H. C. (2007a). Fractal-related assembly of the axial filament in the demosponge *Suberites domuncula*: relevance to biomineralization and the formation of biogenic silica. *Biomaterials* **28,** 4501–4511.
Müller, W. E. G., Schloßmacher, U., Eckert, C., Krasko, A., Boreiko, A., Ushijima, H., Wolf, S. E., Tremel, W., and Schröder, H. C. (2007b). Analysis of the axial filament in spicules of the demosponge Geodia cydonium: different silicatein composition in microscleres [asters] and megascleres [oxeas and triaenes]. *European Journal of Cell Biology* **86,** 473–487.
Müller, W. E. G., Boreiko, A., Wang, X. H., Krasko, A., Geurtsen, W., Custódio, M. R., Winkler, T., Lukić-Bilela, L., Link, T., and Schröder, H. C. (2007c). Morphogenetic activity of silica and bio-silica on the expression of genes, controlling biomineralization using SaOS-2 cells. *Calcified Tissue International* **81,** 382–393.
Müller, W. E. G., Jochum, K., Stoll, B., and Wang, X. H. (2008a). Formation of giant spicule from quartz glass by the deep sea sponge Monorhaphis. *Chemistry of Materials* **20,** 4703–4711.
Müller, W. E. G., Schloßmacher, U., Wang, X. H., Boreiko, A., Brandt, D., Wolf, S. E., Tremel, W., and Schröder, H. C. (2008b). Poly(silicate)-metabolizing silicatein in siliceous spicules and silicasomes of demosponges comprises dual enzymatic activities (silica-polymerase and silica-esterase). *The FEBS Journal* **275,** 362–370.

Müller, W. E. G., Wang, X. H., Cui, F. Z., Jochum, K. P., Tremel, W., Bill, J., Schröder, H. C., Natalio, F., Schloßmacher, U., and Wiens, M. (2009a). Sponge spicules as blueprints for the biofabrication of inorganic-organic composites and biomaterials. *Applied Microbiology and Biotechnology* **83,** 397–413.

Müller, W. E. G., Belikov, S. I., Kaluzhnaya, O. V., Chernogor, L., Krasko, A., and Schröder, H. C. (2009b). Symbiotic interaction between dinoflagellates and the demosponge *Lubomirskia baicalensis*: aquaporin-mediated glycerol transport. In "Biosilica in Evolution, Morphogenesis, and Nanobiotechnology" (W. E. G. Müller and M. A. Grachev, eds), Marine Molecular Biotechnology, **47,** pp. 145–170. Springer-Verlag, Berlin Heidelberg.

Müller, W. E. G., Wang, X. H., Diehl-Seifert, B., Kropf, K., Schloßmacher, U., Lieberwirth, I., Glasser, G., Wiens, M., and Schröder, H. C. (2011). Inorganic polymeric phosphate/polyphosphate as an inducer of alkaline phosphatase and a modulator of intracellular Ca^{2+} level in osteoblasts (SaOS-2 cells) *in vitro*. *Acta Biomaterialia* **7,** 2661–2671.

Murr, M. M., and Morse, D. E. (2005). Fractal intermediates in the self-assembly of silicatein filaments. *Proceedings of the National Academy of Sciences of the United States of America* **102,** 11657–11662.

Pallas, P. S. (1787). Charakteristik der Thierpflanzen. Raspiche Buchhandlung, Nürnberg.

Perović-Ottstadt, S., Schröder, H. C., Batel, R., Giovine, M., Wiens, M., Krasko, A., Müller, I. M., and Müller, W. E. G. (2005). Arginine kinase in the demosponge *Suberites domuncula*: regulation of its expression and catalytic activity by silicic acid. *The Journal of Experimental Biology* **208,** 637–646.

Pfeifer, K., Haasemann, M., Gamulin, V., Bretting, H., Fahrenholz, F., and Müller, W. E. G. (1993). S-type lectins occur also in invertebrates: high conservation of the carbohydrate recognition domain in the lectin genes from the marine sponge Geodia cydonium. *Glycobiology* **3,** 179–184.

Sandford, F. (2003). Physical and chemical analysis of the siliceous skeleton in six sponges of two groups (Demospongiae and Hexactinellida). *Microscopical Research Technology* **62,** 336–355.

Schlossmacher, U., Wiens, M., Schröder, H. C., Wang, X. H., Jochum, K. P., and Müller, W. E. G. (2011). Silintaphin-1: interaction with silicatein during structureguiding biosilica formation. *The FEBS Journal* **278,** 1145–1155.

Schröder, H. C., Krasko, A., Batel, R., Skorokhod, A., Pahler, S., Kruse, M., Müller, I. M., and Müller, W. E. G. (2000). Stimulation of protein (collagen) synthesis in sponge cells by a cardiac myotrophin-related molecule from *Suberites domuncula*. *The FASEB Journal* **14,** 2022–2031.

Schröder, H. C., Krasko, A., Le Pennec, G., Adell, T., Hassanein, H., Müller, I. M., and Müller, W. E. G. (2003). Silicase, an enzyme which degrades biogenous amorphous silica: contribution to the metabolism of silica deposition in the demosponge *Suberites domuncula*. *Progress in Molecular and Subcellular Biology* **33,** 249–268.

Schröder, H. C., Perović-Ottstadt, S., Rothenberger, M., Wiens, M., Schwertner, H., Batel, R., Korzhev, M., Müller, I. M., and Müller, W. E. G. (2004). Silica transport in the demosponge *Suberites domuncula*: fluorescence emission analysis using the PDMPO probe and cloning of a potential transporter. *The Biochemical Journal* **381,** 665–673.

Schröder, H. C., Borejko, A., Krasko, A., Reiber, A., Schwertner, H., and Müller, W. E. G. (2005). Mineralization of SaOS-2 cells on enzymatically (Silicatein) modified bioactive osteoblast-stimulating surfaces. *Journal of Biomedical Materials Research. Part B, Applied Biomaterials* **75B,** 387–392.

Schröder, H. C., Boreiko, A., Korzhev, M., Tahir, M. N., Tremel, W., Eckert, C., Ushijima, H., Müller, I. M., and Müller, W. E. G. (2006). Co-expression and functional interaction of silicatein with galectin: matrix-guided formation of siliceous spicules in the marine demosponge *Suberites domuncula*. *The Journal of Biological Chemistry* **281,** 12001–12009.

Schröder, H. C., Natalio, F., Shukoor, I., Tremel, W., Schloßmacher, U., Wang, X. H., and Müller, W. E. G. (2007a). Apposition of silica lamellae during growth of spicules in the demosponge *Suberites domuncula*: biological/biochemical studies and chemical/biomimetical confirmation. *Journal of Structural Biology* **159,** 325–334.

Schröder, H. C., Brandt, D., Schloßmacher, U., Wang, X. H., Tahir, M. N., Tremel, W., Belikov, S. I., and Müller, W. E. G. (2007b). Enzymatic production of biosilica-glass using enzymes from sponges: basic aspects and application in nanobiotechnology (material sciences and medicine). *Naturwissenschaften* **94,** 339–359.

Schröder, H. C., Wang, X. H., Tremel, W., Ushijima, H., and Müller, W. E. G. (2008). Biofabrication of biosilica-glass by living organisms. *Natural Product Report* **25,** 455–474.

Schröder, H. C., Wiens, M., Schloßmacher, U., Brandt, D., and Müller, W. E. G. (2012). Silicatein-mediated polycondensation of orthosilicic acid: modeling of catalytic mechanism involving ring formation. *Silicon* **4,** 33–38.

Schwarz, K., and Milne, D. B. (1972). Growth promoting effects of silicon in rats. *Nature* **239,** 333–334.

Shimizu, K., Cha, J., Stucky, G. D., and Morse, D. E. (1998). Silicatein alpha: cathepsin L-like protein in sponge biosilica. *Proceedings of the National Academy of Sciences of the United States of America* **95,** 6234–6238.

Simonet, W. S., Lacey, D. L., Dunstan, C. R., Kelley, M., Chang, M. S., Lüthy, R., Nguyen, H. Q., Wooden, S., Bennett, L., Boone, T., Shimamoto, G., DeRose, M., Elliott, R., Colombero, A., Tan, H. L., Trail, G., Sullivan, J., Davy, E., Bucay, N., Renshaw-Gegg, L., Hughes, T. M., Hill, D., Pattison, W., Campbell, P., Sander, S., Van, G., Tarpley, J., Derby, P., Lee, R., and Boyle, W. J. (1997). Osteoprotegerin: a novel secreted protein involved in the regulation of bone density. *Cell* **89,** 309–319.

Spencer, H. (1867). The Principles of Biology. Williams and Norgate, London.

Struyf, E., and Conley, D. J. (2008). Silica: an essential nutrient in wetland biogeochemistry. *Frontiers in Ecology and the Environment* **7,** 88–9410.1890/070126.

Sundar, V. C., Yablon, A. D., Grazul, J. L., Ilan, M., and Aizenberg, J. (2003). Fibre-optical features of a glass sponge. *Nature* **424,** 899–900.

Tahir, M. N., Théato, P., Müller, W. E. G., Schröder, H. C., Janshoff, A., Zhang, J., Huth, J., and Tremel, W. (2004). Monitoring the formation of biosilica catalysed by histidin-tagged silicatein. *Chemical Communications* **24,** 2848–2849.

Tahir, M. N., Théato, P., Müller, W. E. G., Schröder, H. C., Borejko, A., Faiss, S., Janshoff, A., Huth, J., and Tremel, W. (2005). Formation of layered titania and zirconia catalysed by surface-bound silicatein. *Chemical Communications* **44,** 5533–5535.

Thompson, D'Ar y W. (1942). On Growth and Form. University Press, Cambridge.

Trombitas, K., Greaser, M., Labeit, S., Jin, J. P., Kellermayer, M., Helmes, M., and Granzier, H. (1998). Titin extensibility in situ: entropic elasticity of permanently folded and permanently unfolded molecular segments. *The Journal of Cell Biology* **140,** 853–859.

Uriz, M. J. (2006). Mineral spiculogenesis in sponges. *Canadian Journal of Zoology* **84,** 322–356.

Wang, J. C., Hemavathy, K., Charles, W., Zhang, H., Dua, P. K., Novetsky, A. D., Chang, T., Wong, C., and Jabara, M. (2004). Osteosclerosis in idiopathic myelofibrosis is related to the overproduction of osteoprotegerin (OPG). *Experimental Hematology* **32,** 905–910.

Wang, X. H., Schröder, H. C., Brandt, D., Wiens, M., Lieberwirth, I., Glasser, G., Schloßmacher, U., Wang, S., and Müller, W. E. G. (2011). Sponge bio-silica formation involves syneresis following polycondensation in vivo. *Chembiochem* **12,** 2316–2324.

Wiens, M., Belikov, S. I., Kaluzhnaya, O. V., Krasko, A., Schröder, H. C., Perovic-Ottstadt, S., and Müller, W. E. G. (2006). Molecular control of serial module formation along the apical-basal axis in the sponge *Lubomirskia baicalensis*: silicateins, mannose-binding lectin and Mago Nashi. *Development Genes and Evolution* **216,** 229–242.

Wiens, M., Bausen, M., Natalio, F., Link, T., Schlossmacher, U., and Müller, W. E. G. (2009a). The role of the silicatein-a interactor silintaphin-1 in biomimetic biomineralization. *Biomaterials* **30,** 1648–1656.

Wiens, M., Wrede, P., Grebenjuk, V. A., Kaluzhnaya, O. V., Belikov, S. I., Schröder, H. C., and Müller, W. E. G. (2009b). Towards a molecular systematics of the Lake Baikal/Lake Tuva sponges. In "Biosilica in Evolution, Morphogenesis, and Nanobiotechnology" (W. E. G. Müller and M. A. Grachev, eds), Marine Molecular Biotechnology, **47,** pp. 111–144. Springer-Verlag, Berlin Heidelberg.

Wiens, M., Wang, X. H., Schröder, H. C., Kolb, U., Schloßmacher, U., Ushijima, H., and Müller, W. E. G. (2010). The role of biosilica in the osteoprotegerin/RANKL ratio in human osteoblastlike cells. *Biomaterials* **31,** 7716–7725.

Wiens, M., Schröder, H. C., Wang, X. H., Link, T., Steindorf, D., and Müller, W. E. G. (2011). Isolation of the silicatein-α interactor silintaphin-2 by a novel solid-phase pull-down assay. *Biochemistry* **50,** 1981–1990.

Wöhler, F. (1828). Ueber künstliche Bildung des Harnstoffs. *Annalen Physik Chemie* **12,** 253–256.

Yamamoto, M., Hilgemann, D. H., Feng, S., Bito, H., Ishihara, H., Shibasaki, Y., and Yin, H. L. (2001). Phosphatidylinositol 4,5-bisphosphate induces actin stress-fiber formation and inhibits membrane ruffling in CV1 cells. *The Journal of Cell Biology* **152,** 867–876.

CHAPTER SIX

Cultivation of Sponges, Sponge Cells and Symbionts: Achievements and Future Prospects

Klaske J. Schippers*[,1], Detmer Sipkema†[,1], Ronald Osinga‡[,1], Hauke Smidt†, Shirley A. Pomponi§, Dirk E. Martens* and René H. Wijffels*[,2]

Contents

* Bioprocess Engineering, Wageningen University, Wageningen, The Netherlands
† Laboratory of Microbiology, Wageningen University, Wageningen, The Netherlands
‡ Aquaculture and Fisheries, Wageningen University, Wageningen, The Netherlands
§ Harbor Branch Oceanographic Institute/Florida Atlantic University, Fort Pierce, FL, USA
[1] Authors contributed equally.
[2] Corresponding author: Email: rene.wijffels@wur.nl

Advances in Marine Biology, Volume 62
ISSN 0065-2881, DOI: 10.1016/B978-0-12-394283-8.00006-0

Abstract

Marine sponges are a rich source of bioactive compounds with pharmaceutical potential. Since biological production is one option to supply materials for early drug development, the main challenge is to establish generic techniques for small-scale production of marine organisms. We analysed the state of the art for cultivation of whole sponges, sponge cells and sponge symbionts. To date, cultivation of whole sponges has been most successful *in situ*; however, optimal conditions are species specific. The establishment of sponge cell lines has been limited by the inability to obtain an axenic inoculum as well as the lack of knowledge on nutritional requirements *in vitro*. Approaches to overcome these bottlenecks, including transformation of sponge cells and using media based on yolk, are elaborated. Although a number of bioactive metabolite-producing microorganisms have been isolated from sponges, and it has been suggested that the source of most sponge-derived bioactive compounds is microbial symbionts, cultivation of sponge-specific microorganisms has had limited success. The current genomics revolution provides novel approaches to cultivate these microorganisms.

Key Words: sponge; cultivation; cell culture; symbiosis; natural product; microorganisms; secondary metabolite; genomics

1. Introduction

The use of sponges dates back many centuries. As early as 700 BC, Homer and Aristotle described the use of sponges in medicine (Voultsiadou, 2007). Until the end of the nineteenth century, sponges had been widely used in surgery (Müller *et al.*, 2004b). For example, sponges soaked with extracts of opium were used to anaesthetize patients prior to surgery, and extracts of *Spongia tosta* were found to be effective in the treatment of scrofula (Müller *et al.*, 2004b). Besides their biomedical application, sponges also had commercial value as bath sponges, because of their absorptive properties. Until 1960, commercial bath sponge farming was a lucrative business spreading from the Mediterranean to the Atlantic and Pacific Ocean (Bernard, 1968; Duckworth, 2009). From 1960, due to the invention of the less expensive synthetic bath sponge, farming of the natural bath sponge decreased and now only occupies a small niche in the market for high-quality applications (Hogg *et al.*, 2010). Current biotechnological applications of sponges are more directed towards the production of marine natural products and biomaterials.

Screening for bioactive compounds in sponges began in the mid-twentieth century. Bergmann and Feeney isolated some of the first nucleosides from the Caribbean sponge *Cryptotethya crypta* in 1951 (Bergmann and

Feeney, 1951). These nucleosides appeared to have antiviral properties and synthesis of derivatives of these nucleosides resulted in the production of Ara-A (active against herpes) and Ara-C (effective in leukaemia treatment). As a result, Roche Research Institute of Marine Pharmacology started pioneering research in marine natural product drug discovery in the mid-1970s, and scientists began to explore the chemical diversity found in marine organisms and its potential for drug discovery (Pomponi, 2001). More than 20,000 novel, marine-derived chemicals have been discovered since 1965, of which hundreds have potential pharmaceutical applications (Blunt *et al.*, 2009). Marine sponges have continuously been the most prolific source of newly discovered bioactive compounds, with more than 7000 sponge-derived, novel molecules (Blunt *et al.*, 2009). Table 6.1 lists sponge-derived metabolites that have advanced to clinical trials.

Sponges are benthic and sessile as adults, and they have evolved a sophisticated chemical communication and defence system based on secondary metabolites (Müller *et al.*, 2004b; Hogg *et al.*, 2010). These compounds protect against fouling (Proksch, 1994), predation (Becerro *et al.*, 1997) and microbial infection (Sipkema *et al.*, 2005a).

For most sponge-derived bioactive compounds, it is not clear whether they are produced by the sponge or the symbiont (which can contribute up to 40% of the sponge volume). Metabolites known to be produced by the sponge are, for example, avarol, found in *Dysidea avara* (Uriz *et al.*, 1996) and stevensine, found in *Axinella corrugata* (Andrade *et al.*, 1999). On the other hand, it has been demonstrated that *Oscillatoria spongelia*, a cyanobacterial symbiont, produces antimicrobial polybrominated biphenyl ethers and might keep the sponge free of other bacteria (Unson *et al.*, 1994).

Besides pharmaceutical products many other applications of sponge-derived biomaterials have been identified. For example, the silicon skeleton of glass sponges may serve as a blueprint for the production of very efficient fibre optics (Sundar *et al.*, 2003) and biosilica-producing enzymes from sponges have been applied in nanotechnology (Schröder *et al.*, 2007). In fact, sponges are the only animal able to polymerize silica to generate massive skeletal elements in a single reaction at ambient temperature and pressure (biosintering) (Müller *et al.*, 2009).

Collagen is among the most promising sponge-derived biomaterials. Sponge collagen (e.g. from the marine sponge *Chondrosia reniformis*; Swatschek *et al.*, 2002) offers advantages in medical and cosmetic applications compared to mammalian connective tissue-extracted collagens because it is free of risks associated with Bovine Spongiform Encephalopathy (BSE) (Heinemann *et al.*, 2007).

Despite their enormous biotechnological potential, only a few marine natural products have been successfully developed into products. A major obstacle is the lack of sufficient supply of biological material for preclinical and clinical development. Although controlled chemical synthesis of the

Table 6.1 Sponge metabolites resulting from drug discovery, modified from Pomponi (2001), Newman and Cragg (2004), Molinski *et al.* (2009) and Sashidhara *et al.* (2009)

Metabolite	Source	Status	Target	Developed by
In clinical use				
E7389 (halichondrin B derivative)	*Lissodendoryx* sp. and *Halichondria okadai*	Phase III (cancer)	Tubulin	Eisai
HMR-4011A (aka IPL-576,092, contignasterol derivative)	*Neopetrosia contignata*	Phase II (anti-asthmatic)	Histamine	Aventis
Zalypsis (renieramycin derivative)	*Xestospongia* sp.	Phase I (cancer)	DNA	Pharmamar
E7974 (hemiasterlin derivative)	*Cymbastela* sp. and *Auletta* sp.	Phase I (cancer)	Tubulin	Eisai
Taltobulin (aka HTI-286, hemiasterlin derivative)	*Cymbastela* sp. and *Auletta* sp.	Phase I (cancer)	Tubulin	Wyeth-Andersen
KRN-7000 (agelasphin derivative)	*Agelas mauritianus*	Phase I (cancer)	NKT	Koezuka-Kirin
LBH 589 (psammaplin derivative)	*Psammaplinaplysilla* sp.	Phase I (cancer)	HDAC	Novartis
Debromohymenialdisine	*Stylotella aurantium*	Phase I (Alzheimer's)	Kinases	Genzyme Tissue Repair
Peloruside A	*Mycale hentscheli*	Preclinical (cancer)	Tubulin	Reata Pharmaceuticals
Discontinued from clinical trials				
Cryptophycin 52 (=arenastatin)	*Dysidea arenaria*	Phase II (cancer)	Tubulin	Lilly, Valeriote
Manoalide	*Luffariella variabilis*	Phase II (antipsoriatic)	Calcium channel	Allergan
Discodermalide	*Discodermia dissoluta*	Phase I (cancer)	Tubulin	Novartis, HBOI
LAF 389 (bengamide derivative)	*Jaspis* sp.	Phase I (cancer)	MetAP	Novartis, Crews
LAQ 824 (psammaplin derivative)	*Psammaplinaplysilla* sp.	Phase I (cancer)	HDAC	Novartis, Crews
Girolline (aka girodazole)	*Cymbastela cantharella*	Phase I (cancer)	Protein synthesis	Potier

bioactive molecules (or their analogues) is the preferred production method for pharmaceutical application, most bioactive natural compounds are chemically complex structures that cannot be synthesized easily. Investments in establishing a synthesis route for complex molecules are usually not done until the potency of the compound has been sufficiently proven (i.e. after completion of preclinical development studies or early clinical trials). As a result, material obtained from natural sources is often needed for preclinical development. Since many of the marine-derived compounds are found in nature in small quantities tons or thousands of tons of fresh material may be needed to supply sufficient amounts. It is obvious that such large quantities of biomass can never be harvested from nature without risking serious damage to the sponge population or even the extinction of the respective species. Culture of marine organisms that produce the bioactive compounds may overcome the supply issue (e.g. Dumdei *et al.*, 1998; Mendola, 2003; Mendola *et al.*, 2006), but at present, many marine organisms are considered unculturable. This paradox (early drug development being dependent on biological production methods that do not exist), which is often referred to as "The Supply Issue" (e.g. Osinga *et al.*, 1998; Munro *et al.*, 1999; Faulkner, 2000; Fusetani, 2000), emphasizes the need to establish appropriate culture techniques. Since biological production is particularly needed for early drug development, the main challenge for marine biotechnologists is to establish generic techniques for efficient, small- to medium-scale production of large numbers of marine organisms.

Depending on the product of interest, different cultivation approaches may be applied to overcome "The Supply Issue". When product concentrations are high inside the sponge or if the sponge itself is the product (e.g. bath sponges), cultivation of adult sponges may be the best option (Sipkema *et al.*, 2005b). However, many secondary metabolites are present only at low concentrations. In this case, the amount of sponge biomass required is simply too high, and to obtain an economically feasible production system, the product concentrations have to be increased. The use of *in vitro* cell cultures may then be the method of choice. *In vitro* cultivation allows for control of culture conditions and manipulation to increase product concentrations. If the compound of interest is produced by a microbial symbiont, the production system could consist of only the symbiont, if it is possible to culture the symbionts separately. If this is not possible or if the compound is partly synthesized by the sponge and partly by the symbiont, co-culture may be required. Thus, different molecules will require different production systems, and it is important to have a versatile set of cultivation tools.

In this overview, the state of the art of cultivation of whole sponges, sponge cells and sponge symbionts is analysed, and an outlook is presented for development or integration of different cultivation methods.

2. Sponge Aquaculture

In the past 15 years, more than 10 review papers have been published on the production of sponge materials for biotechnological purposes and for the trade in bath sponges (Osinga *et al.*, 1998, 1999; Belarbi *et al.*, 2003; Müller *et al.*, 2004a; Mendola *et al.*, 2006; Sipkema *et al.*, 2005b; De Caralt *et al.*, 2007b; Pronzato and Manconi, 2008; Duckworth, 2009; Koopmans *et al.*, 2009). Two of these reviews integrated over 100 years of sponge aquaculture activities (Pronzato and Manconi, 2008; Duckworth, 2009). It is not the aim of this section to repeat the previously reviewed information. We will limit the current review to recent findings and new developments related to *in situ* and *ex situ* aquaculture of sponges. In addition, design principles for sea-based sponge aquaculture will be outlined and we will look in more detail to metabolite production in aquaculture.

2.1. *In situ* culture

Sea-based culture of sponges (also termed mariculture or *in situ* culture) refers to growing sponges in their natural environment on man-made constructions such as longlines with scallop lanterns, vertical ropes, underwater platforms, frames and cages (Duckworth, 2009). To start up a sponge mariculture, donor sponges obtained from the wild are usually fragmented artificially by cutting them into smaller pieces termed explants. The explants are then secured to an artificial substrate, for example, ropes, nylon lines or plastic plates, or are positioned into nets or cages.

The idea to fragment sponges into explants originates from the nineteenth century (see historical perspective by Pronzato and Manconi, 2008). Schmidt (1863) was the first to publish a scientific report on mariculture using fragmented sponges. He described cultures of commercial bath sponges in the Northern Adriatic Sea. Since then, many case studies on sponge mariculture have been published (Duckworth, 2009). Despite this long historical record, there are still many problems associated with the concept. Mariculture is vulnerable to biological factors such as predation and fouling. Other potential biological and non-biological risks include diseases (Webster *et al.*, 2002), and damage to the cultures due to rough weather, vandalism and boats. Some real-life examples of these risks will be described.

Table 6.2 summarizes recently published data on survival and growth for maricultured sponge species. Growth rates reported range between −157% (i.e. a decrease in size) and 2437% per year, with a record in growth rate still being the 5000% increase per month reported by Battershill and Page (1996) for *Lissodendoryx* sp., which has been included in Table 6.2 for comparison. Most authors determined growth as the increase in size between the start

Table 6.2 Growth rates and survival percentages of sponges in mariculture

Species	Method	Location	Survival (%)	Growth rate reported	Estimated SGR (day^{-1})	Reference
Discodermia dissoluta	Net pockets	Santa Marta, Colombia, 18 m	93	20–39% in 6 months	0.001–0.00183	Zea *et al.* (2010)
Dysidea avara	Plastic pins	Eastern Mediterranean, under fish farm, 21 m	100	137% in 4 months	0.0072	Osinga *et al.* (2010)
Dysidea avara	Horizontal ropes	Eastern Mediterranean, under fish farm, 21 m	90–100	100% $year^{-1}$	0.0019	Osinga *et al.* (2010)
Dysidea avara	Horizontal ropes	Eastern Mediterranean, 18 m	90–100	0% $year^{-1}$	0	Osinga *et al.* (2010)
Dysidea avara	Horizontal ropes	Eastern Mediterranean, 28 m	90–100	0% $year^{-1}$	0	Osinga *et al.* (2010)
Dysidea avara	Horizontal ropes	Eastern Mediterranean, 38 m	90–100	20–50% $year^{-1}$	0.0005–0.0011	Osinga *et al.* (2010)
Dysidea avara	Glued to frames	Western Mediterranean	40	468.9% in 10 months	0.0057	de Caralt *et al.* (2010)
Dysidea avara	Ropes	Western Mediterranean	11	145.25% in 10 months	0.003	de Caralt *et al.* (2010)
Dysidea avara	Cages	Western Mediterranean	70	166.75% in 10 months	0.0033	de Caralt *et al.* (2010)
Haliclona oculata	Bound to tiles	Scheldt Estuary, Netherlands	n.d.	0.82% day^{-1}	0.0082	Koopmans and Wijffels (2008)
Diacarnus erythraenus	Attached with steel wires to PVC plates on frames	Red Sea, 10 m	72	64.5% $year^{-1}$	0.0014	Bergman *et al.* (2011)

(continued)

Table 6.2 *(continued)*

Species	Method	Location	Survival (%)	Growth rate reported	Estimated SGR (day^{-1})	Reference
Diacarnus erythraenus	Attached with steel wires to PVC plates on frames	Red Sea, 20 m	70	79.3% $year^{-1}$	0.0016	Bergman *et al.* (2011)
Lissodendoryx sp.	Scallop lanterns	Wellington Harbour, New Zealand	n.d.	5000% $month^{-1}$	0.131	Battershill and Page (1996)
Mycale cecilia	Bound to ceramic tiles	Mexican Pacific, 3–6 m	95	207% in 60 days	0.0187	Carballo *et al.* (2010)
Mycale hentcheli	Horizontal ropes, F0 generation explants	Pelorus Sound, New Zealand, near mussel farm, 5–10 m	n.d.	1355–2437% $year^{-1}$	0.0073–0.0089	Page *et al.* (2011)
Mycale hentcheli	Horizontal ropes, F1 generation explants	Pelorus Sound, New Zealand, near mussel farm, 5–10 m	n.d.	359% $year^{-1}$	0.0042	Page *et al.* (2011)
Mycale hentcheli	Horizontal ropes, F2 generation explants	Pelorus Sound, New Zealand, near mussel farm, 5–10 m	n.d.	−157% $year^{-1}$	Negative	Page *et al.* (2011)

The abbreviation n.d. means "not determined". Estimated specific growth rates (SGR) per day were calculated assuming first-order kinetics (Sipkema *et al.*, 2006).

and the end of the culture trial. This makes it difficult to compare the growth rates. We have estimated the specific growth rate (SGR) per day for all experiments listed in Table 6.2, under the assumption that all sponges grew exponentially following first-order kinetics. The highest SGR reported was 0.0187 day^{-1} for *Mycale cecilia* (Carballo *et al.*, 2010). For comparison, the record of 5000% per month reported for *Lissodendoryx* sp. (Battershill and Page, 1996) equals an SGR of 0.131 day^{-1}, that is, one order of magnitude higher.

Recently, much effort has been dedicated to develop culture methods for *D. avara* (Schmidt, 1863) in order to produce the bioactive compound avarol. In a joint research project, *D. avara* was cultured both in the Western Mediterranean (De Caralt *et al.*, 2010) and in the Eastern Mediterranean (Osinga *et al.*, 2010) using comparable methods. Stainless steel frames (height: 50 cm) were positioned on the seafloor and explants of *D. avara* were secured to the frames on horizontal nylon lines (both studies), by gluing (De Caralt *et al.*, 2010) and by putting the explants on plastic pins (Osinga *et al.*, 2010). De Caralt *et al.* (2010) also grew the explants in stainless steel cages. Survival rates were much higher in the Eastern Mediterranean (90–100%) than in the Western Mediterranean (11–70%, depending on the method used). Surprisingly, survival in the Eastern Mediterranean was also nearly 100% for explants that had been prepared during summer, despite the general consensus in the literature that cold water is better for explant survival after cutting, due to a slower response of infectious microorganisms (Duckworth, 2009). In both studies (de Caralt *et al.*, 2010; Osinga *et al.*, 2010), growth rates depended on the methodology applied and varied from 40% to 800% increase per year. De Caralt *et al.* (2010) found specific benefits for all three methods that were applied: cage culture displayed the lowest mortality, explants glued to frames showed the highest growth rates and culturing on ropes promoted bioactivity. Osinga *et al.* (2010) found that growth of *D. avara* explants was highest when plastic pins were applied (Fig. 6.1). This method also proved to be far less labour intensive than establishing rope cultures, and hence, Osinga *et al.* (2010) concluded that plastic pins provided an economically feasible production method for avarol. Duckworth *et al.* (2007) had applied a comparable technique termed "spike method" to the bath sponge species *Rhopaloeides odorabile* and *Coscinodermia* sp. and concluded that these spikes could not be recommended to culture these species (growth rates being much lower than for other techniques such as culture in nets and on ropes), confirming once again that optimization of sponge aquaculture is a species-specific process (Duckworth, 2009). The study by Osinga *et al.* (2010) also demonstrated the vulnerability of sponge mariculture to external factors: the culture frame was found upside down on the seafloor when it was revisited after 1 year.

One of the most comprehensive studies on sponge aquaculture so far was published recently by Page and co-workers (Page *et al.*, 2005, 2011). These

Figure 6.1 Explants of *Dysidea avara* on plastic pins, which were horizontally mounted into a metal frame, approximately 50 cm above the seafloor. (A) Explants at the start of the culture trial. (B) Explants after 4 months in culture. Explants had fused into sponge sausages (cf. Page *et al.*, 2011). (For colour version of this figure, the reader is referred to the Web version of this chapter.)

authors reviewed 7 years of aquaculture trials on *Mycale hentcheli*, targeting the production of the bioactive compound peloruside A for preclinical research. Their studies included many aspects related to culture success, including culture methods, location selection, broodstock selection, seasonality and metabolite production. Page *et al.* (2011) were the first to thoroughly document the effects of repeated cloning on culture success. After doing three repeated trials with explants obtained from wild parent sponges (termed F0), two subsequent cultures were made by preparing explants from sponges harvested from the previous culture trials (termed F1 and F2). Although variability between years was considerable for F0 explants, growth rates obtained were at the top end of the range shown in Table 6.2. Annual growth decreased dramatically for F1 explants and was even negative for F2 explants. The F2 series was confronted with an outbreak of a predatory nudibranch (*Hoplodoris nodulosa*) that consumed a considerable proportion of the explanted biomass. It remains to be studied whether the outbreak of the predator was a coincidence or the result of decreased fitness of the sponges due to repeated cloning, but this case study clearly shows that mariculture remains vulnerable to biological pests.

Fouling is another biological threat. According to Page *et al.* (2011), rope culture is the least vulnerable to fouling. This is presumably due to the fact that when explants are positioned close to each other, they will rapidly cover the entire rope surface (forming structures that the authors termed sponge sausages), thus preventing the settlement of fouling organisms onto the ropes. Similar sponge sausages also formed on the plastic pins used by Osinga *et al.* (2010) for the culture of *D. avara* (Fig. 6.1B). Explants obtained from the same genetic individual tend to fuse when approaching each other, whereas different genotypes will reject each other (Kruse *et al.*, 1999). Hence, in order to get proper sausage formation, materials from the same genetic individual should be used.

Both Osinga *et al.* (2010) and Page *et al.* (2011) positioned their cultures in the vicinity of other aquacultures (fish and mussels, respectively), which was observed to be beneficial for sponge growth. The benefit was most likely due to an increased availability of organic food for the sponges. Osinga *et al.* (2010) reported a 5- to 10-fold lower visibility at the fish farm site, which is indicative for a higher concentration of particulate organic matter.

In a study on *Diacarnus erythraenus*, Bergman *et al.* (2011) studied changes in communities of sponge-associated microorganisms between wild and maricultured sponges, another previously neglected aspect of sponge culture. No striking differences were detected. Comparable results were obtained by Webster *et al.* (2011) for *R. odorabile*. Apparently, maricultured sponges are capable of maintaining a natural bacterial composition within their tissue, which is particularly relevant when the associated microorganisms are involved in the production of the targeted secondary metabolites (see also Section 5).

2.2. Design principles for sponge mariculture

Several studies have demonstrated the feasibility of mariculture for the production of sponges on a commercial scale, from which it can be concluded that mariculture is a suitable method for supply. Another conclusion is that successful culture procedures are to a large extent species specific (Duckworth, 2009) and site specific, the latter being nicely illustrated by the conflicting results obtained for *D. avara* at two different parts of the Mediterranean (De Caralt *et al.*, 2010; Osinga *et al.*, 2010). For starting up a sponge mariculture on a new species, it is therefore recommended to first conduct a 1-year trial to evaluate productivity at different locations using different methods. A further point of consideration is seasonality: explant growth may vary considerably throughout the year. Hence, initial aquaculture trials should preferably comprise at least one complete annual cycle. Examples of recent studies describing growth throughout the year include works on *Haliclona oculata* (Koopmans and Wijffels, 2008) *Corticium candelabrum* (De Caralt *et al.*, 2008), *Discodermia dissoluta* (Zea *et al.*, 2010), *D. erythraenus* (Bergman *et al.*, 2011) and *Mycale hentscheli* (Page *et al.*, 2011). Effects of seasonality appear to be limited to temperate and subtropical zones. The growth of temperate and subtropical species (*H. oculata*, *C. candelabrum* and *M. hentcheli*) was positively correlated with temperature, whereas growth of the two tropical species (*D. dissoluta* and *D. erythraenus*) did not show significant seasonal effects.

Surprisingly, not many studies have discussed design principles for sponge aquaculture. A simple design model is represented by

$$Y = N_{\mathrm{exp}} \times P_{\mathrm{exp}}, \tag{6.1}$$

in which Y is the yield (the desired harvest, expressed, for example, in kg wet weight), N_{exp} is the number of explants to be seeded and P_{exp} is the average harvest per explant during the production period (kg/explant). From this, it follows that the initial amount of explants to be seeded can be calculated as

$$N_{\mathrm{exp}} = Y/P_{\mathrm{exp}} \tag{6.2}$$

In order to determine P_{exp} for the targeted species, growth rate and survival percentage must be determined from an initial culture trial. P_{exp} can then be calculated as

$$P_{\mathrm{exp}} = (1 - M)E_t, \tag{6.3}$$

in which E_t is the size of the explant (kg) at end of a culture period with duration t, and M is the mortality factor, which is calculated as (100 − percentage survival)/100.

In case of repeated production (i.e. when part of the harvest is used as broodstock for the next production cycle), Eq. (6.2) can be extended to

$$P'_{\mathrm{exp}} = (1 - M)E_t - (Y2/Y \times E_0), \tag{6.4}$$

where $Y2$ represents the desired harvest for the subsequent year and P'_{exp} is the harvest per explant after subtraction of the part of the harvest that is needed as broodstock to seed the next culture cycle. For example, the targeted sponge grows from an initial explant size of 0.1 kg to a final explant size of 0.8 kg in 1 year, during which time 25% of the explants die. For a repeated annual production of 100 kg, P'_{exp} equals (1–0.25)0.8 − (100/100)0.1 = 0.5 kg. From Eq. (6.2), it follows that N_{exp} = 100/0.5 = 200. If these 200 explants are seeded, 150 will survive, representing a total biomass after 1 year of

$150 \times 0.8 = 120$ kg. Since the desired yield is 100 kg, the remaining 20 kg of harvested biomass can be used to seed the next culture cycle. The 20 kg of biomass is enough for another 200 initial explants of 0.1 kg so that the cycle can be repeated. It should be noted here that productivity in subsequent years may decrease due to effects of repetitive cloning (Page *et al.*, 2011).

The calculation principles outlined above can be applied irrespective of the kinetic principles underlying growth but are in that case limited to the duration of the culture period that is chosen for the initial trial (expressed as t in the equations). Predicted production (yield) should preferably be based upon a good understanding of the growth kinetics of the targeted species (Sipkema *et al.*, 2006). In practice, this means that repeated measurements of size should be done during regular intervals. When growth kinetics are known, it becomes possible to translate previously found growth rates to different culture periods. A commonly made mistake in this respect is that a linear growth model is applied to calculate a dimensionless daily growth rate (% day^{-1}) based upon a longer culture interval (e.g. 365% increase per year $= 365/365 = 1$% per day). This incorrect approach suggests an SGR that is much higher than the specific growth would be when based upon an exponential growth model (first-order kinetics; see Sipkema *et al.*, 2006). Linear growth should be expressed in size units per day, for example, cm day^{-1}, mg day^{-1}, etc.

Koopmans and Wijffels (2008) give a convenient overview of growth rates of sponges in aquaculture and corresponding kinetics. This type of information is crucial for a proper design of commercial sponge maricultures.

2.3. *Ex situ* culture

Keeping sponges in *ex situ* systems such as aquaria remains a challenging task. Although success stories remain scarce (see Table 6.3 for an overview of growth rates of sponges in aquaria), some progress has been realized in better understanding the cultivation requirements of sponges. Osinga and Kotterman (2007) discovered that continuous addition of ferric iron (Fe^{3+}, supplied as ferric citrate) to aquarium water promoted oscule formation and pumping activity in the Mediterranean sponges *C. reniformis* and *Acanthella acuta*. This finding is in agreement with the work of Krasko *et al.* (2002), who showed that ferric iron upregulates genes involved in channel formation in primmorphs of *Suberites domuncula*. Xue and Zhang (2009) further elaborated on this aspect and found that variations in Fe^{3+} concentrations significantly affected culture success of juveniles of *Hymeniacidon perlevis*.

Mendola (2008) studied the role of water flow on the pumping behaviour of *D. avara*. This species exhibits the formation of fragile exhalant structures, which are flattened by drag forces caused by flow rates higher than 7.5 cm s^{-1}. In nature, *D. avara* was found to grow only at sheltered habitats where the average (oscillating) flow did not exceed 5.9 cm s^{-1} (Mendola, 2008). Mendola (2008) concluded that flow rates in tanks

Table 6.3 Growth and survival of sponges in *ex situ* systems

Species	Food	Survival (%)	Growth rate reported	Estimated SGR (day^{-1})	Reference
Axinella corrugata	Four bacterial strains, *Isochrysis galbana*, *Thallasiosira weissflogii* and *S. cerevisiae*	100	22% in 8 weeks	0.0036	Duckworth *et al.* (2003)
Chondrosia reniformis	Algae powder from *Phaeodactylum tricornotum* and fish powder	100	0.0067 day^{-1}	0.0067	Sipkema *et al.* (2006)
Crambe crambe	RPMI 1640 commercial medium enriched with amino acids and 20% extract of *Octopus vulgaris*	n.d.	400% in 100 days	0.0161	Garcia Camacho *et al.* (2006b)
Dysidea avara	*Phaeodactylum tricornotum*	100	0.1–0.028 day^{-1}	0.01–0.028	Sipkema *et al.* (2006)
Halichondria melanadocia	Four bacterial strains: *M. halophilus*, *V. alginolyticus*, *E. coli* and *B. subtilis*	50–100	0.08% day^{-1}	0.0008	Duckworth and Pomponi (2005)
Halichondria melanadocia	Two bacterial strains, *Isochrysis galbana* and *Saccharomyces cerevisiae*	50–100	0.60% day^{-1}	0.006	Duckworth and Pomponi (2005)
Hymeniacidon perleve (juveniles)	*Isochrysis galbana* and photosynthetic bacteria	65	2900% in 43 days	0.078	Xue and Zhang (2009)
Mycale cecilia	*Isochrysis* sp., *Tetraselmis* sp., powdered yeast	65	65% in 60 days	0.0084	Carballo *et al.* (2010)
Pseudosuberites andrewsi	*Phaeodactylum tricornotum* or an extract of *P. tricornotum*	100	2% day^{-1}	0.02	Osinga *et al.* (2003)

Estimated specific growth rates (SGR) per day were calculated assuming first-order kinetics (Sipkema *et al.*, 2006).

designed for the culture of *D. avara* should not exceed 8–10 cm s^{-1}. Further evidence for the role of flow was provided by Carballo *et al.* (2010), who successfully cultured explants of *M. cecilia* in airlift aquaria under continuous flow (the flow rate was not specified by the authors). However, growth of the sponges in the aquarium system was threefold lower than growth in mariculture: 65% and 207% in 60 days for aquarium culture and mariculture, respectively (Carballo *et al.*, 2010).

Xue and Zhang (2009) studied effects of temperature on the growth of juvenile specimen of *H. perlevis*. Increasing the temperature from 18 to 23 °C throughout the incubation resulted in a 29-fold increase in size within 43 days, whereas a decreasing temperature caused the juvenile sponges to cease. Juveniles maintained under a constant temperature showed a moderate growth (6.5-fold increase). Natural fluctuations in temperature are crucial for the initiation of sexual reproduction in temperate sponges (Ereskovsky, 2000) and may also influence juvenile development. It remains to be studied to what extent adult sponges are influenced by changes in temperature.

A persistent question with respect to aquaculture success relates to the dietary requirements of sponges. A short synopsis of developments in our understanding of sponge nutrition and productivity may aid further attempts to design feeding regimes for sponges in aquaria. Sponges were traditionally regarded as particle feeders, a view that is still common among sponge scientists. Evidence is growing that particle feeding alone is often not sufficient to sustain the metabolic requirements of sponges (Osinga *et al.*, 2002). Dissolved organic carbon (DOC) appears to be a major source of food for some sponges (Yahel *et al.*, 2003; de Goeij and van Duyl, 2007), comprising up to 90% of the sponge diet. Other studies show that particle feeding alone can be sufficient to sustain sponge growth, both in nature (Hadas *et al.*, 2009, Koopmans *et al.*, 2010) and in *ex situ* culture (Duckworth *et al.*, 2003; Osinga *et al.*, 2003; Duckworth and Pomponi, 2005; Xue and Zhang, 2009; Carballo *et al.*, 2010), although Duckworth and Pomponi (2005) had to apply five times the natural concentration of planktonic food to obtain positive growth rates. Koopmans *et al.* (2010) presented a carbon balance for the temperate sponge *H. oculata*, based upon measurements of growth, respiration and particle feeding. *H. oculata* used only 34% of the ingested (particulate) carbon for growth and respiration. It was concluded that this sponge grows inefficiently. A recent paper by De Goeij *et al.* (2009) sheds a different light on productivity and efficiency of resource utilization in sponges. In the Caribbean sponge *Halisarca caerulea*, these authors found extremely high rates of cell division in choanocytes, and these are the cells responsible for the pumping activity of sponges. However, a large proportion of the cellular material was released into the exhalant channels, a process termed shedding. As such, *H. caerulea* exhibited a high cellular turnover without a corresponding increase in biomass.

Observations on rapid somatic growth, such as reported for *Lissodendoryx* sp. (Battershill and Page, 1996), indicate that sometimes, rapid cell proliferation does lead to a rapid increase in biomass.

Further evidence for the potential role of DOC as food for sponges in culture was provided by Garcia Camacho *et al.* (2006b), who successfully cultured small fragments of *Crambe crambe* on an commercial medium for cell culture enriched with amino acids and a marine extract obtained from *Octopus vulgaris* (see Section 4.3 for a more detailed analysis of this study). Growth rates declined in time (a phenomenon that had also been observed by Duckworth *et al.*, 2003) but remained positive for a 100-day period. The only other successful attempts to grow sponges *ex situ* on DOC were studies by Osinga *et al.* (2003) and Sipkema *et al.* (2006), who managed to grow *Pseudosuberites andrewsi* and *C. reniformis* on an extract of the marine diatom *Phaeodactylum tricornotum*.

Since the presence of associated microorganisms in sponges can be of crucial importance for their primary and secondary metabolism, it is of importance to know whether the microbial community in sponges shifts after transfer of the animals into aquarium systems. Analogous to the *in situ* studies by Bergman *et al.* (2011) and Webster *et al.* (2011) described in Section 2.1, Isaacs *et al.* (2009) and Mohamed *et al.* (2008a,b) compared the bacterial communities of wild specimen of *Clathria prolifera*, *Mycale laxissima* and *Ircinia strobilina* to *ex situ* cultured specimen of each species and found profound differences in community structure. Webster *et al.* (2011) also found changes in microbial composition when explants of *R. odorabile* were transferred from the sea to aquarium systems. Although some species do not exhibit such changes upon transfer to aquaria (Gerce *et al.*, 2009), the aquarium effect appears to be a commonly occurring phenomenon among marine animals, since several aquarium corals also show dramatic changes in bacterial composition when compared to their wild conspecifics (Kooperman *et al.*, 2007). Culturing sponges and other invertebrates outside their natural environment may disturb symbiotic relationships and may impede natural product formation, since this may be related to the secondary metabolism of the sponge–symbiont association.

Although commercial *ex situ* sponge culture may not be realized in the near future, we provide a short perspective on future design studies. The design principles outlined in Section 2.2 can be applied to *ex situ* culture systems as well. However, in *ex situ* culture systems, seasonality effects can be reduced or even entirely excluded. Therefore, *ex situ* systems may be operated in a semi-continuous mode (regular seeding and harvesting of a small part of the culture) rather than in a batch mode. This has an advantage in terms of volumetric productivity and makes maintenance routines easier because feeding and maintenance regimes do not have to be adapted to changes in standing stock in the system (see analogy for coral culture by Osinga *et al.*, 2011).

2.4. Metabolite production in aquaculture

It is crucial for commercial production of sponge metabolites through aquaculture that the cultured sponges retain their ability to produce the metabolites of interest. Some of the studies described above provide data on the production of secondary metabolites by the cultured sponges. Osinga *et al.* (2010) and Bergman *et al.* (2011) found comparable levels of avarol and norsesterterpene peroxide acids in wild and maricultured sponges (*D. avara* and *D. erythraenus*, respectively). Carballo *et al.* (2010) did a qualitative assay on the presence of mycalazal-type of metabolites in crude extract from wild, maricultured and *ex situ* cultured sponges. The metabolites were present under all circumstances. Klöppel *et al.* (2008) reported a slight increase in metabolite production in *Aplysina aerophoba* when cultured *ex situ*, but the sponges did not grow under the aquarium settings applied. A very promising result was obtained by Zea *et al.* (2010), who found a fourfold increase in the discodermolide concentration in specimen of *D. dissoluta* that had been maricultured for 6 months. Table 6.4 summarizes recent and earlier data on metabolite production in sponge aquacultures. The results in Table 6.4 indicate that metabolite production is maintained in cultured sponges, both *in situ* and *ex situ*. Variability was high, reported values ranging from a fourfold increase to a complete absence of bioactivity.

Koopmans *et al.* (2009) have recently reviewed factors determining metabolite production in sponges. In many species, concentrations of the targeted metabolites are variable in time and space, and production of the metabolites can be induced by stimuli such as simulated predation (wounding) and changes in temperature. In contrast, manoalide production in *Luffariella variabilis* is not at all affected by season, depth and location, and attempts to induce increased manoalide production were not successful (Ettinger-Epstein *et al.*, 2008). Similarly, *Acanthella caversona* cultured in aquaria showed no change in metabolite production upon exposure to light, tissue damage, squeezing of tissue and combinations thereof (Mendola, 2003). In contrast, additional feeding did promote bioactivity in aquarium cultures of *A. corrugata* (Duckworth *et al.*, 2003). Farming techniques have also been reported to influence metabolite production (Duckworth and Battershill, 2003a,b). Hence, optimizing metabolite production in aquaculture is another species-specific process.

3. *In Vitro* Sponge Culture

3.1. Monodispersed cells

Cell or tissue culture (both subsequently referred to as *in vitro* culture) is a preferred method because the system can be defined and controlled. Sponges have great potential for cell culture because of the presence of

Table 6.4 Records of metabolite formation in sponge aquaculture

Sponge species	Metabolite formation	Reference
Acanthella cavernosa	Similar levels of kalihinols in mariculture and wild type, no effects of stress treatments on metabolite production	Mendola (2003)
Aplysina aerophoba	Different alkaloid metabolites were present in similar or slightly higher concentration when compared to wild-type sponges; light slightly inhibited metabolite formation	Klöppel *et al.* (2008)
Axinella corrugata	Stevensine levels in aquarium culture were 110–157% of the levels found in wild type	Duckworth *et al.* (2003)
Callyspongia biru	Similar levels of amphitoxin in mariculture and wild type; high variability among individuals	De Voogd (2007)
Diacarnus erythraenus	Similar levels of norsesterterpene peroxide acids in mariculture and wild type	Bergman *et al.* (2011)
Discodermia dissoluta	Fourfold increase in discodermolide in maricultured sponges compared to levels in wild type	Zea *et al.* (2010)
Dysidea avara	Similar levels of avarol in mariculture and wild type	Osinga *et al.* (2010)
Latrunculia sp. / *wellingtonensis*	Maricultured sponges exhibited bioactivity, levels were similar or higher than in wild type; toxicity of the sponge tissue varied among the different farming methods applied	Duckworth and Battershill (2003a,b)
Lyssodendorix sp.	Halichondrins in mariculture were 25–50% of the levels found in wild type	Dumdei *et al.* (1998)
Mycale cecilia	Mycalazals present in wild type, mariculture and aquarium culture (qualitative analysis)	Carballo *et al.* (2010)
Mycale hentscheli	Mycalamide A and pateamine were present in similar quantities in donor sponges and cultured sponges; the presence of Peluroside A was strongly dependent on location	Page *et al.* (2005)
Negombata magnifica	Concentration of latrunculin B in maricultured sponges was up three times higher than previously reported in the literature	Hadas *et al.* (2005)
Polymastia croceus	Maricultured sponges exhibited bioactivity, levels were similar or higher than in wild type; toxicity of the sponge tissue varied among the different farming methods applied	Duckworth and Battershill (2003a,b)
Verongia aerophoba	Production and release of aeroplysinin by *ex situ* cultured sponge fragments, mainly under influence of light and aeration	Kreuter *et al.* (1992)

totipotent stem cells and because cells can be easily dissociated from its tissue, due to their loosely cellular organization. Despite efforts by several research groups (see Table 6.5), a continuous[1] sponge cell line has not yet been developed, and the number of primary[1] sponge cell cultures developed is very limited (Rinkevich, 2005). The only report of a continuous sponge cell culture was Klautau *et al.* (1993, 1994); however, this cell line was subsequently identified as a protozoan (Custodio *et al.*, 1995).

3.1.1. Sponge cell dissociation

The first step in obtaining a primary sponge cell culture involves dissociating the cells into a monodisperse suspension. Several dissociation techniques, mechanical, chemical, enzymatic and spontaneous, have been studied (reviewed by Pomponi, 2006). The most successful dissociation method is a combination of mechanical and chemical dissociation (Pomponi and Willoughby, 1994; Rinkevich *et al.*, 1998). If a monodisperse cell suspension is needed, the method described by Pomponi and Willoughby (1994) is generally used (see Fig. 6.2), whereas the method described by Custodio *et al.* (1998) is applied when re-aggregation of sponge cells (i.e. to form primmorphs) is needed (see Fig. 6.5).

The dissociation method, described in Fig. 6.2, will result in a mixture of cell types, for example, archaeocytes, choanocytes, pinacocytes, collencytes. Archaeocytes and choanocytes are considered to be the pluripotent stem cells in sponges. These stem cell-like cells have the capacity to proliferate and to differentiate into other cell types (Funayama, 2010; Funayama *et al.*, 2010). Most cell types in a sponge are expected to be terminally differentiated and not able to divide. Consequently, selecting and enriching for archaeocytes and choanocytes may result in a proliferating sponge cell culture (reviewed by Pomponi, 2006). Sun *et al.* (2007) developed a novel four-step protocol for the purification of archaeocytes from the marine sponge *H. perlevis*. By using a process involving differential centrifugation, selective aggregation in low Ca^{2+}/Mg^{2+} seawater, differential adherence in artificial seawater and Ficoll-Urografin density gradient centrifugation, they obtained an archaeocyte-enriched fraction greater than 80%, and proliferation was demonstrated by BrdU incorporation.

3.1.2. Sponge cell culture medium

Development of a culture medium is necessary to support the growth of sponge cells. Generally, an axenic inoculum is used to initiate a primary cell culture from animal tissue (Freshney, 2005). It is not possible to isolate such an axenic part from sponges, since both eukaryotic and prokaryotic

[1] Primary cell cultures contain cells which are cultured directly from the tissue and have a limited lifespan. Continuous cell cultures are cells with an unlimited lifespan, are often transformed or are derived from tumours.

Table 6.5 Publications about primary sponge cell cultures (1993–2010)

Sponge species	Reference	Proliferation observed?	Comments
Clathrina aurea	Klautau *et al.* (1993, 1994)	Yes, but...	Later identified as protozoa (Custodio *et al.*, 1995)
Hymeniacidon heliophila	Pomponi and Willoughby (1994)	Yes	PHA-stimulated cells doubled after 72 h
Ephydatia muelleri	Imsiecke *et al.* (1995)	Yes	Cells were taken from hatched gemmules
Latrunculia (=*Negombata*) *magnifica*	Ilan *et al.* (1996) and Rinkevich *et al.* (1998)	n.d.	Eventually, cultures were dominated by thraustochytrids (Ilan *et al.*, 1996)
Teichaxinella morchella=*Axinella corrugata*	Pomponi *et al.* (1997, 1998) and Pomponi and Willoughby (2000)	Yes	PHA-stimulated cells doubled after 36 h
Acanthella acuta	Nickel *et al.* (2000)	No	
Dysidea avara	Nickel *et al.* (2000) and De Rosa *et al.* (2003)	No (Nickel *et al.*, 2001) n.d. (De Rosa *et al.*, 2003)	
Ircinia muscarum=*Sarcotragus foetidus*	De Rosa *et al.* (2001, 2003)	Yes, but...	No data of sponge cell growth over time are shown
Suberites domuncula	De Rosa *et al.* (2003) and Zhang *et al.* (2004)	n.d.	Cell viability was measured using an MTT assay (Zhang *et al.*, 2004)
Xestospongia muta	Richelle-Maurer *et al.* (2003)	n.d.	
Hymeniacidon perleve (=perlevis)	Zhao *et al.* (2005) and Sun *et al.* (2007)	n.d. (Zhao *et al.*, 2005) Yes (Sun *et al.*, 2007)	Purified archaeocytes doubled after approximately 3 days (Sun *et al.*, 2007)
Haliclona oculata	Schippers *et al.* (2011)	No	

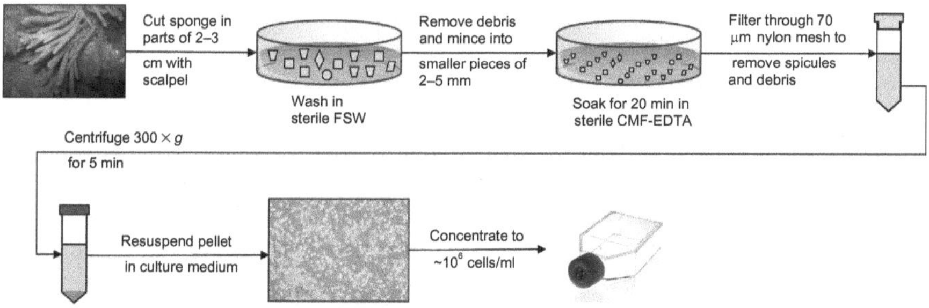

Figure 6.2 Dissociation method described by Pomponi and Willoughby (1994). Photograph of *Haliclona oculata* taken by M. Koopmans. (For colour version of this figure, the reader is referred to the Web version of this chapter.)

symbionts are intermingled with the sponge in the tissue. This has made it very difficult to develop nutrient media for sponge cells, since adding nutrients for sponge cell growth will also stimulate growth of contaminants. Therefore, measures have been taken to limit the growth of contaminants, such as bacteria, yeasts, protozoa and thraustochytrids. To prevent growth of these contaminants, antibiotics and antimycotics can be added to the medium, but most of these only temporarily inhibit growth of bacteria and fungi. This may be due to the fact that some bacteria in sponges may have antibiotic resistance. In addition, high concentrations of antibiotics and antimycotics can also have a negative effect on the viability of the sponge cells.

Despite these difficulties, researchers have tried to develop culture media for sponge cells. The most basic culture medium consists of filtered seawater (Richelle-Maurer *et al.*, 2003; Schippers *et al.*, 2011), which provides the correct salinity and pH, and contains some dissolved nutrients and trace elements which are generally present in the sea. Sponge cells could be maintained in filtered seawater for several weeks without contaminants overgrowing the culture. However, since no growth of the sponges cells was observed (Schippers *et al.*, 2011), it is concluded that the dissolved nutrients in seawater are not sufficient to support long-term growth of sponge cells.

Other sponge cell culture media (reviewed by Pomponi, 2006) have mostly been derived from existing animal cell culture media, which contain amino acids, vitamins and glucose. Since these media are rich in nutrients, the use of antibiotics and antimycotics is also essential. Despite the use of antibiotics and antimycotics, contamination can still occur (Ilan *et al.*, 1996), and the addition of cell culture media alone does not support sponge growth. Growth factors are essential. For example, the addition of PHA (phytohemagglutinin), a mitogenic lectin, stimulated cell division (Pomponi *et al.*, 1997; Sun *et al.* 2007) and the addition of cholesterol resulted in increased cell density compared to cell cultures without cholesterol (De Rosa *et al.*, 2001, 2003). Evidently, cells *in vitro* are unable to biosynthesize sterols, and the addition of sterols is needed to satisfy membrane requirements. Ferric iron was also found to significantly improve cell viability (Zhang *et al.*, 2004; Zhao *et al.*, 2005). Table 6.6 provides a more detailed overview.

3.1.3. Sponge cell enumeration and viability

Several methods have been used to study sponge cell growth and viability. Counting cells with a haemocytometer has been widely used to measure cell growth (Klautau *et al.*, 1994; Ilan *et al.*, 1996; Pomponi *et al.*, 1997). This method is laborious and time consuming and makes high-throughput screening difficult. Automatic cell counting systems based on image analysis are available but have not been applied to sponge cell culture, most likely because sponge cells have very different sizes and shapes, and contaminants and symbionts need to be discriminated from the sponge cells.

To discriminate between viable and dead sponge cells, trypan blue exclusion from live cells has been used (De Sutter and Van de Vyver, 1977; Pomponi and Willoughby, 1994; Ilan *et al.*, 1996; De Rosa *et al.*, 2003; Richelle-Maurer *et al.*, 2003). However, Sipkema *et al.* (2004) found that trypan blue diffusion into dead sponge cells can be disturbed, which makes trypan blue an unreliable marker to measure sponge cell viability. A more reliable method is the combined use of fluorescein diacetate (FDA) and propidium iodide (PI), which, respectively, stain viable and dead eukaryotic cells (Sipkema *et al.*, 2004). The dead and viable cells can be visualized by fluorescence microscopy (see Fig. 6.3) or can be quantified using flow cytometry, as described by Sipkema *et al.* (2004).

Another method that has been used to measure viability of sponge cells is the MTT assay (Zhang *et al.*, 2004; Zhao *et al.*, 2005). An advantage of this method is the possibility of using 96-well plates, which makes it possible to screen many samples simultaneously. However, since the cells need to be lysed for the MTT assay, it is no longer possible to discriminate between sponge cells and contaminants. To determine whether sponge cells are dying through apoptosis, caspase activity, an indicator of early apoptosis, can be measured. Schippers *et al.* (2011) measured caspase activities of sponge cells before and after dissociation and found a threefold increase of caspase activity of the sponge cells after dissociation, indicating that the dissociation process evidently induced apoptosis.

To get more insight into the proliferative state of the sponge cells, it is also possible to measure the different phases of the cell cycle (Schippers *et al.*, 2011). The cell cycle distribution of sponge cells was measured via flow cytometry after staining the DNA with PI. Figure 6.4 shows an example of a single-parameter histogram of cells from *D. avara*. It is possible to distinguish between the apopotic, G1/G0 and G2/M fraction. The S-phase fraction is hardly visible, indicating that the cells were not actively dividing. Schippers

Table 6.6 Sponge cell culture media (1994–2011)

Media	Antibiotics	Supplements	Contamination	Sponge species
SW	–	–	No	*Xestospongia muta*[a] *Haliclona oculata*[b]
M199 in CMF	Rifampicin	5% FBS, 1.5% PHA*	No	*Hymeniacidon heliophila*[c]
M199 in CMF	Rifampicin, fungizone	5% FBS, 1.5% PHA,* glutamine, sodium selenite	No	*Axinella corrugata*[d,e]
Iscove's MDM	–	Epidermal growth factor, bovine pituitary extract,* arachidonic acid, insulin, prostaglandin E2, 1.5% PHA*	n.d.	*Axinella corrugata*[d]
Leibowitz L15/ Landureau in SW	Gentamycin, penicillin, streptomycin, amphotericin, nystatin	10% FBS[f,g] (fibroblast growth factor, epidermal growth factor, insulin,* concavalin A, wheat germ agglutinin,* *Ulex europaeus* agglutinin,* sponge extract)[g]	Yes[f] n.d.[g]	*Negombata magnifica*[f,g]
DMEM in ASW	Nystatin, rifampicin, gentamycin, Penicillin, Ciprofloxacin	–	No	*Acanthella acuta*[h] *Dysidea avara*[h]
DMEM (half dose) in SW	Ampicillin, gentamycin, kanamycin, tylosin, tetracycline (removed after 2 weeks)	Cholesterol*[,i] (oleic acid, linoleic acid, retinol, cyclodextrin)[j]	No	*Ircinia muscarum*[i] *Dysidea avara*[j] *Suberites domuncula*[j]
SW	Penicillin, streptomycin	Glutamine, pyruvate,* iron citrate,* silicon, RPMI 1640, Marine Broth 2216	n.d.	*Suberites domuncula*[k]

(*continued*)

Table 6.6 *(continued)*

Media	Antibiotics	Supplements	Contamination	Sponge species
Modified ASW (Zhao's medium[l])	–	Inorganic salts (e.g. ferric iron,* NaCl,*) amino acids, sugars, vitamin C,* DMEM, RPMI 1640, 1.5% PHA	n.d.	*Hymeniacidon perlevis*[l]
Zhao's medium[l]	Gentamycin	2.5% PHA, insulin, transferrin	n.d.	*Hymeniacidon perlevis*[m]

n.d., not determined; SW, sterile seawater; CMF, calcium- and magnesium-free seawater; ASW, artificial seawater; DMEM, Dulbecco's minimum essential medium; FBS, foetal bovine serum.

* Improved viability of sponge cells.

[a] Richelle-Maurer *et al.* (2003).

[b] Schippers *et al.* (2011).

[c] Pomponi and Willough by (1994).

[d] Pomponi *et al.* (1997).

[e] Willoughby and Pomponi (2000).

[f] Ilan *et al.* (1996).

[g] Rinkevich *et al.* (1998).

[h] Nickel *et al.* (2000).

[i] De Rosa *et al.* (2001).

[j] De Rosa *et al.* (2003).

[k] Zhang *et al.* (2004).

[l] Zhao *et al.* (2005).

[m] Sun *et al.* (2007).

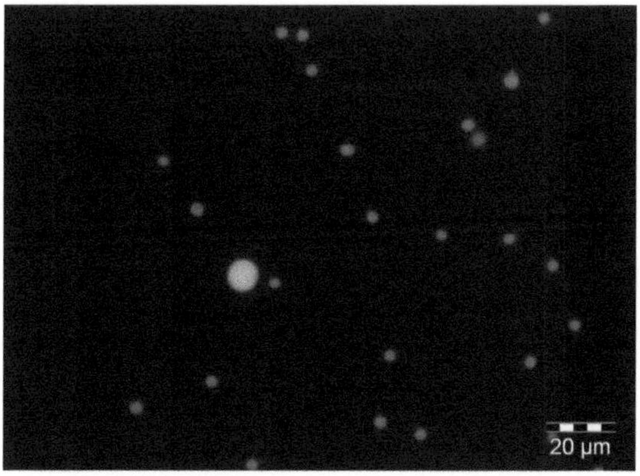

Figure 6.3 Viability of cryopreserved *Dysidea avara* cells. Red indicates dead cells and green indicates viable cells. (For interpretation of the references to colour in this figure legend, the reader is referred to the Web version of this chapter.)

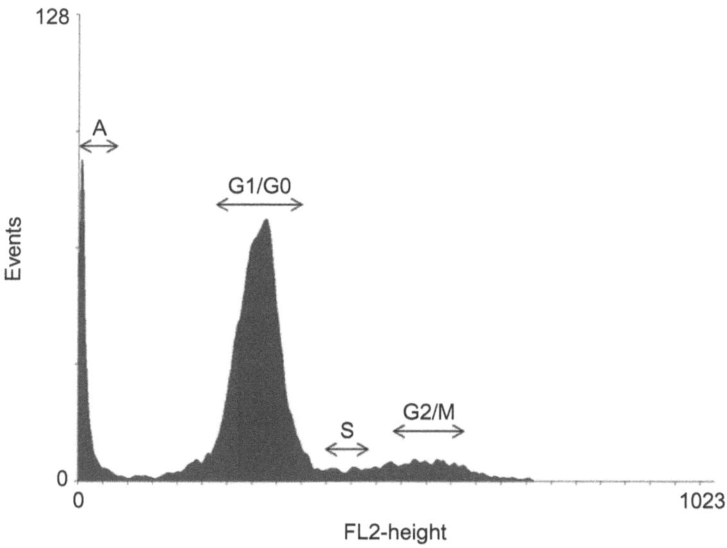

Figure 6.4 Flow cytometric cell cycle analysis, single-parameter histogram of FL2-height (red fluorescence, DNA content) of *Dysidea avara* cells. A, apoptotic cells; G1, cells in gap1; G0, quiescent cells; G2, cells in gap2; M, mitotic cells; S-phase, cells synthesizing DNA (Schippers *et al.*, 2011). (For interpretation of the references to colour in this figure legend, the reader is referred to the Web version of this chapter.)

et al. (2011) used this method in combination with measurement of caspase activity to quickly determine the proliferative status of sponge cell culture inoculum or starting material.

Cell proliferation can also be detected by means of bromodeoxyuridine (BrdU) incorporation. BrdU is an analogue of thymidine that is incorporated in the DNA during DNA synthesis and can be detected with a fluorescent antibody. The antibody can be used on thin sections of the whole sponge or on fixed cells, which has the advantage that one can still discriminate between cell types and in the case of thin sections also between the location in the sponge where proliferation occurs. It can also be used with the above mentioned PI staining of suspended cells and flow cytometry, with the advantage that the S-phase is better separated from the G1 and G2/M phases. BrdU incorporation is stained with a green fluorescent-labelled antibody. Incorporation of BrdU has not been used for dissociated sponge cells but has been used for measuring cell proliferation in primmorphs, gemmules, tissue fragments and adult sponges (Custodio *et al.*, 1998; Nickel and Brummer, 2003; De Goeij *et al.*, 2009; Funayama *et al.*, 2010).

3.1.4. Sponge cell-type verification

Contamination of sponge cell cultures remains a problem, which is mainly caused by the fact that it is impossible to obtain an axenic sponge inoculum. Therefore, if growth of sponge cells is observed, it is important to verify if this growth is caused by dividing sponge cells or dividing contaminants, hence the need for cell-type verification. Klautau *et al.* (1994) performed isoenzyme analysis to confirm that they were working with sponge cells; however, since it appeared later that they were cultivating protozoa, isoenzyme analysis seems unreliable for sponge cell identification. A more reliable technique is to identify the sponge cell culture based on unique DNA sequences. Lopez *et al.* (2002) were the first to use molecular techniques to identify sponge cell cultures. They applied amplified fragment length polymorphism (AFLP) to compare fingerprints of the internal transcribed tracer (ITS) region of adult sponges and cell cultures of *A. corrugata*. They could successfully detect sponge cells; however, they could not detect the presence of other eukaryotic contaminants. Sipkema *et al.* (2003a) developed a method based on 18S rDNA, which made it possible to identify sponge cells and eukaryotic contaminants. Another method that could be applied to sponge cell-type verification is fluorescent *in situ* hybridization (FISH). FISH uses fluorescent probes that can bind to specific DNA sequences such as 18S rDNA. By using fluorescent microscopy or flow cytometry, it would be possible to detect if cells are from sponge origin or not. Until now, however, there are no published reports of a FISH-detection method for sponge cells.

3.2. Primmorphs

Koziol *et al.* (1998) found that monodisperse sponge cells lack the capacity to proliferate. A possible explanation could be the loss of cell adhesion factors, such as cell–cell and cell–matrix contact, when cells are dissociated using calcium- and magnesium-free seawater (Koziol *et al.*, 1998). They found that the activity of telomerase, a bio-marker for cell proliferation, in tissue of *S. domuncula* and *Geodia cydonium* was very high, while the telomerase activity of dissociated sponge cells was almost zero. When the dissociated sponge cells re-aggregated, the telomerase activity was recovered, confirming the hypothesis that cell adhesion factors are important for cell proliferation (Custodio *et al.*, 1998). These multicellular aggregates from dissociated sponge cells are termed primmorphs. Primmorphs are dense sphere-shaped aggregates, 1–2 mm in diameter. They differ from aggregates that occur after dissociation of sponge tissue in that they have an external layer of pinacocytes and a central zone with primarily spherulous cells (Custodio *et al.*, 1998; Sipkema *et al.*, 2003c). The formation of primmorphs is based on the discovery of Wilson (1907) that sponge cells have re-aggregative properties and can form a newly functional sponge.

3.2.1. Primmorph formation

Custodio *et al.* (1998) were the first to describe the application of this three-dimensional (3D)-sponge cell culture system for the production of bioactive compounds. They developed a method for creating primmorphs from sponges, and this method has been widely applied by other researchers for generating primmorphs from various sponge species. Figure 6.5 shows the procedure for generating primmorphs. The main difference between this method and that of Pomponi and Willoughby (1994) is the high cell concentration during incubation and the presence of calcium and magnesium resulting in re-aggregation of cells.

Primmorph formation depends on several factors. First, not every sponge species has the ability to generate primmorphs. Valisano *et al.* (2006b) screened 21 Mediterranean sponge species, of which 13 species were able to produce primmorphs. Nickel *et al.* (2001) screened nine sponge species, of which three produced primmorphs. Sipkema *et al.* (2003c) screened seven sponge species and all seven were able to generate primmorphs. Table 6.7 lists all species which have been demonstrated to generate primmorphs. It is not known why certain sponge species are able to produce primmorphs and others are not. Holmes and Blanch (2008) investigated if there is a taxonomic trend in the success of primmorph formation. Although their sample set was small (33 sponges of 9 different orders), they found that sponges from the Poecilosclerida, Suberitidae (Hadromerida), Agelasida and the family Axinellidae I (Clade C) have more success in the formation of primmorphs, while sponges from the Astrophorida, Verongida, Chondrosida, Dendroceratida and the family Axinellidae II show limited or no potential for primmorph formation. These findings, however, are not unambiguous, since contradictory results have been published. For example, Valisano *et al.* (2006b) could not obtain primmorphs from *Axinella polypoides*, while Sipkema *et al.* (2003c) could. Also,

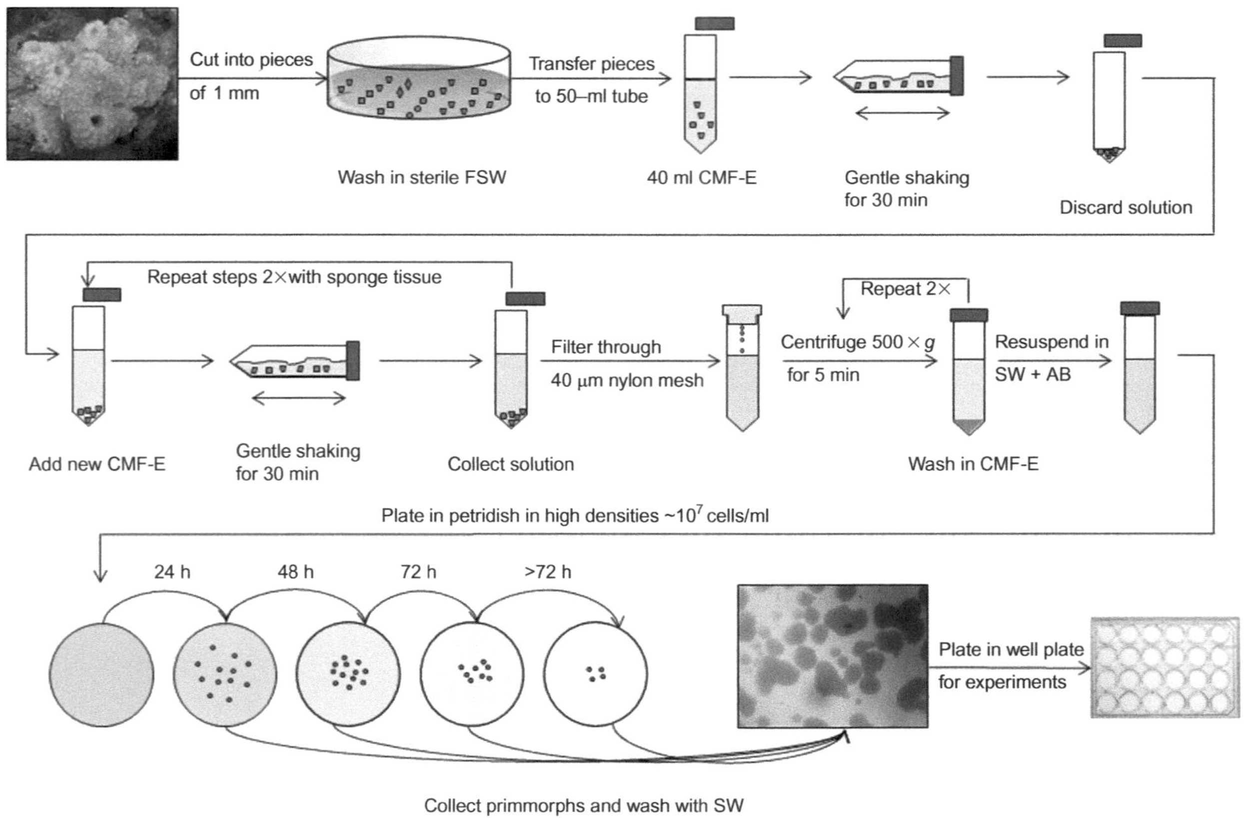

Figure 6.5 Method to generate primmorphs, described by Custodio *et al.* (1998). Abbreviations: FSW, filtered seawater; CMF-E, calcium- and magnesium-free seawater with EDTA; SW, seawater; AB, antibiotics. (For colour version of this figure, the reader is referred to the Web version of this chapter.)

Table 6.7 Publications about primmorph cultures (1998–2010)

Sponge species	Reference	Proliferation observed?	Lifespan
Suberites domuncula	Custodio *et al.* (1998), Muller *et al.* (1999), Nickel *et al.* (2001), and Sipkema *et al.* (2003c)	Yes, with BrdU incorporation (Custodio *et al.*, 1998; Muller *et al.*, 1999) n.d. (Nickel *et al.*, 2001; Sipkema *et al.*, 2003c)	>5 months (Custodio *et al.*, 1998; Muller *et al.*, 1999) >3 months (Nickel *et al.*, 2001) >4 months (Sipkema *et al.*, 2003c)
Dysidea avara	Muller *et al.* (2000) and Valisano *et al.* (2006b)	Yes, with [3H]dThd incorporation (Muller *et al.*, 2000) n.d. (Valisano *et al.*, 2006b)	0.8 month (Muller *et al.*, 2000) >1 month (Valisano *et al.*, 2006b)
Petrosia ficiformis	Nickel *et al.* (2001), Pozzolini *et al.* (2004) and Valisano *et al.* (2006a,b)	n.d.	>2 months (Nickel *et al.*, 2001; Pozzolini *et al.*, 2004) >1 month (Valisano *et al.*, 2006a,b)
Axinella verrucosa	Nickel *et al.* (2001))	n.d.	>1 month
Stylissa massa	Sipkema *et al.* (2003c)	n.d.	>2 months
Pseudosuberites aff. *andrewsi*	Sipkema *et al.* (2003c)	n.d.	0.3 months
Geodia cydonium	Sipkema *et al.* (2003c)	n.d.	>2 months
Axinella polypoides	Sipkema *et al.* (2003c)	n.d.	>5 months
Halichondria panicea	Sipkema *et al.* (2003c)	n.d.	>5 months
Stylotella agminata	Zhang *et al.* (2003a)	Yes, with cell density	>7 days
Hymediacidon perlevis	Zhang *et al.* (2003b)	Yes, with BrdU incorporation and MTT assay	>1 month

(continued)

Table 6.7 *(continued)*

Sponge species	Reference	Proliferation observed?	Lifespan
Spirastrella cunctatrix	Valisano *et al.* (2006b)	n.d.	>1 month
Cliona celata	Valisano *et al.* (2006b)	n.d.	0.2 month
Hemimycale columella	Valisano *et al.* (2006b)	n.d.	0.5 month
Phorbas fictitius	Valisano *et al.* (2006b)	n.d.	0.5 month
Batzella inops	Valisano *et al.* (2006b)	n.d.	>1 month
Axinella damicornis	Valisano *et al.* (2006b)	n.d.	>1 month
Axinyssa aurantiaca	Valisano *et al.* (2006b)	n.d.	0.3 month
Acanthella acuta	Valisano *et al.* (2006b)	n.d.	>1 month
Agelas oroides	Valisano *et al.* (2006b)	n.d.	>1 month
Haliclona fulva	Valisano *et al.* (2006b)	n.d.	>1 month
Pleraplysilla spinifera	Valisano *et al.* (2006b)	n.d.	>1 month
Hymeniacidon heliophila	Vilanova *et al.* (2010)	n.d.	>0.3 month
Lubomirskia baikalensis	Chernogor *et al.* (2011)	Yes, by measuring optical density	>10 months

Nickel *et al.* (2001) could not obtain primmorphs from *Axinella damicornis* and *A. acuta*, while Valisano *et al.* (2006b) were able to obtain primmorphs from these sponges.

Another factor related to the success of primmorph formation is seasonal variability. Valisano *et al.* (2006a) found that for *Petrosia ficiformis*, a Mediterranean sponge, the amount of primmorph biomass formed (size of primmorph × number of primmorphs) is high from April to July, and negligible in the other months. These results indicate that there is not only interspecific variability in the production of primmorphs but also intraspecific variability with respect to the seasonal cycle.

The size of the aggregates is influenced by the inoculum cell density. Both Zhang *et al.* (2003a) and Sipkema *et al.* (2003c) found that higher initial cell concentrations resulted in an increased size of the primmorphs. However, Chernogor *et al.* (2011) found exactly the opposite, that is, a higher cell concentration resulted in smaller primmorphs. A fundamental difference between both experiments is fact that Zhang *et al.* (2003a) and Sipkema *et al.* (2003c) used marine sponges and Chernogor *et al.* (2011) used a freshwater species. These results could indicate that the formation of primmorphs between freshwater and marine sponges is different. However, no decisive conclusions can be drawn, since no other reports of primmorph formation in freshwater sponges have been published.

3.2.2. Effect of silica and iron

Another factor influencing primmorph formation is the concentration of silica and iron. By increasing the dissolved silica concentrations from 1 to 60 μM, the formation of silicatein and myotrophin in *S. domuncula* increased (Krasko *et al.*, 2000), resulting in enhanced spiculogenesis and the production of collagen, respectively (see Fig. 6.6). Subsequently, when silica concentrations increased from 1 to 60 μM, the primmorphs tripled in size from 2 to 6 mm (Le Pennec *et al.*, 2003). Attachment and development of spicules in primmorphs were also induced by the presence of high silica concentrations (>70 μM) (Sipkema *et al.*, 2003b). For primmorph formation in *P. ficiformis,* the optimal silica concentration was found to be 120 μM, and these high silica concentrations had a positive effect on primmorph size and spiculogenesis (Valisano *et al.*, 2007). Increased concentrations of iron showed a positive influence on proliferation levels of cells in primmorphs from *S. domuncula* (Krasko *et al.*, 2002). When the concentration of iron was increased to 30 μM in combination with 60 μM silica, the size of the primmorphs almost doubled to 10 mm in diameter. However, for primmorphs from *P. ficiformis,* an increase in iron concentration did not influence the size of the primmorphs, indicating that the influence of iron on primmorph formation is species dependent (Valisano *et al.*, 2007).

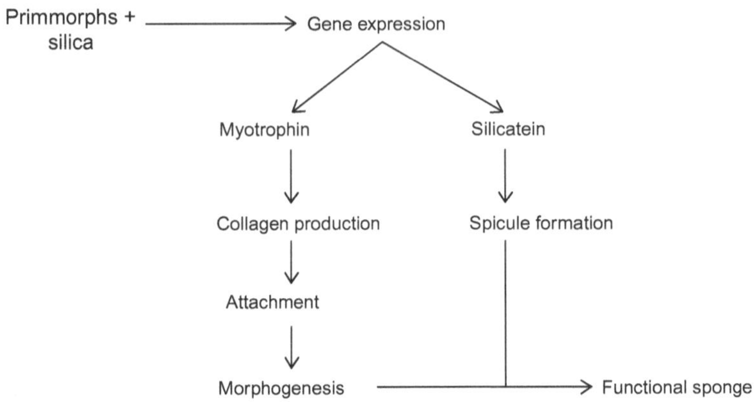

Figure 6.6 Hypothesized influence of silica on morphogenesis and spiculogenesis of primmorphs. Modified scheme from Krasko *et al.* (2000) and Sipkema *et al.* (2003b).

3.2.3. Effect of antibiotics

To prevent contamination in primmorphs, antibiotics were added to the culture medium in most studies. Penicillin and streptomycin have been commonly used in primmorph cultures (Custodio *et al.*, 1998; Muller *et al.*, 1999; Muller *et al.*, 2000; Zhang *et al.*, 2003a,b). Sipkema *et al.* (2003c) tested the efficacy and effects of several antibiotics on the formation of primmorphs of seven species of marine sponges. The addition of penicillin, streptomycin and gentamycin did not have an effect on the formation of primmorphs, but the addition of amphotericin and KTTG (antibiotic cocktail by De Rosa *et al.*, 2001 containing kanamycin, tylosin, tetracycline and gentamycin) inhibited primmorph formation and is therefore expected to have a negative effect on the viability of the sponge cells. It is remarkable that when no antibiotics were added to the seawater, all primmorph cultures were free of contamination, while almost half of the cultures were contaminated when antibiotics were added. Sipkema *et al.* (2003c) hypothesize that addition of antibiotics in the seawater kills the endogenous bacterial population of the sponge, which could protect the sponge against microorganisms that are pathogenic to the sponge.

3.2.4. Cell proliferation

Cell proliferation has been observed in primmorphs by means of BrdU incorporation (Custodio *et al.*, 1998; Zhang *et al.*, 2003b); however, an increase in total biomass over time has not been observed. Possibly there is an equilibrium in the primmorphs between proliferating and apoptotic cells (Koziol *et al.*, 1998). A possible explanation for the lack of increase in biomass is the deficiency of nutrients in the media. Primmorphs were generally cultivated in nutrient-poor medium, such as natural seawater or artificial seawater without the addition of any nutrients (Custodio *et al.*, 1998; Nickel *et al.*, 2001; Sipkema *et al.*, 2003c; Zhang *et al.*, 2003a,b; Valisano *et al.*, 2006b). Only Muller *et al.* (2000) and Krasko *et al.* (2002) added 0.1% (v/v) marine broth 2216 or 0.2% RPMI 1640, respectively, to their media, but they do not show data for nutrient uptake by the primmorphs. To be able to obtain continuous biomass growth, it is important to supply appropriate nutrients and growth factors.

3.3. Tissue fragments

Tissue fragments have been used for cultivating sponges *in situ* and *ex situ* as described in Sections 2.1 and 2.3 but can also be used in an *in vitro* approach (De Caralt *et al.*, 2003; Nickel and Brummer, 2003; Garcia Camacho *et al.*, 2006b; Gunda and Janapala, 2009). The main differences are the sizes of the fragments, which in *in vitro* culture range from 25 to 2500 mm^3—considerably smaller than the tissue fragments used for *in situ* and *ex situ* cultivation, and the use of a closed system, which allows control of the cultivation parameters. An important advantage of using tissue fragments instead of dissociated cells and primmorphs for *in vitro* culture is the avoidance of cellular stress caused by dissociation and the fact that the cell–cell and cell–matrix contacts are kept intact (Nickel and Brummer, 2003). Although this is a promising method, only five reports of *in vitro* sponge fragment cultivation have been published (see Table 6.8).

Nickel and Brummer (2003) cultured fragments of *C. reniformis* in six-well plates with filter sterilized seawater, supplemented with silicon and gentamycin to prevent bacterial contamination and fed the cultures with lyophilized *Escherichia coli*. During the first weeks, reorganization of tissue took place, and proliferation was detected by BrdU incorporation. However, the amount of proliferating cells was below 1% of the total cell

Table 6.8 Publications about tissue fragment *in vitro* culture (2003–2009)

Sponge species	Reference	Proliferation observed?	Lifespan
Chondrosia reniformis	Nickel and Brummer (2003)	Yes, by using BrdU labelling	>19 months
Corticium candelabrum	De Caralt *et al.* (2003)	Yes (*ex situ*), no (*in vitro*), by measuring the area by digital imaging	>3 months
Crambe crambe	Garcia Camacho *et al.* (2006a,b)	Yes, by measuring the volume by digital imaging	>4 months
Haliclona pigmentifera	Gunda and Janapala (2009)	n.d.	>1 month

number, indicating a very low proliferation rate. After a couple of months, proliferation could no longer be observed, but the sponge fragments still showed active silica metabolism and fragments could be kept viable for more than 19 months.

De Caralt *et al.* (2003) cultured fragments of *C. candelabrum* in six-well plates with seawater and monitored explant survival under several experimental conditions. They found that fragments with a high ectosome/choanosome ratio collected from sponges during winter/spring and cultured at a low temperature (14 °C) with antibiotics (penicillin/streptomycin) survived longest. Growth rates were highest when explants were cultured in aquaria without addition of antibiotics. De Caralt *et al.* (2003) hypothesize that antibiotics are important during the critical first steps when fragments are sensitive for infections but for long-term culture antibiotics have a negative effect on growth. Also, growth rates were higher when explants were fed with algae than with bacteria, which is possibly due to the higher carbon content of algae.

Garcia Camacho *et al.* (2006b) cultured fragments of *C. crambe* in glass vials with RPMI, enriched with salts, amino acids and *Octopus* extract. Before starting cultivation, they dipped the sponge fragments in 5% sodium hypochlorite and in media with gentamycin, nystatin and penicillin to prevent contamination. The sponge fragments increased 400% in volume in 120 days when 20% (v/v) of *Octopus* extract was added to the medium. If *Octopus* extract was omitted from the medium, no growth took place. This indicates that the *Octopus* extract apparently contains certain nutrients and/or growth factors which stimulate growth of the sponge tissue fragments. Also they showed that the sponge fragments were able to take up DOC (dissolved organic matter), since DOC levels decreased rapidly during the first 6 h after adding nutrients.

Gunda and Janapala (2009) cultured fragments of *Haliclona pigmentifera* in flasks with filtered seawater without supplementing nutrients because they only wanted to observe the effect of dissolved oxygen (DO) on the viability of the sponge fragments. Under hypoxia conditions (1.5–2.0 ppm DO), the sponge fragments survived >1 month and the fragments exhibited adhesive growth. Under anoxia conditions (<0.3 ppm DO), the sponge fragments survived only a couple of days. These results indicate that not only nutrients but also DO levels are essential for *in vitro* culture of sponge fragments.

4. Cultivation of Symbionts

Recent technological developments in genomics have changed the way we study genes, species and communities and offer new possibilities for exploiting sponge-derived natural products and sponge–microbial symbioses. The revolution in DNA sequencing with the establishment of pyro- and illumina sequencing has changed the way complex microbial communities, or even whole biotopes can be studied and exploited. Therefore, one might wonder why we would need microbial cultivation after all, especially since the majority of microorganisms seem not capable of growth under artificial conditions. However, there are important reasons why cultivation of target microorganisms is of added value to studying their genes in a (meta)genomic context:

- Isolation of the target microorganisms enables studies of regulation and functionality of (secondary) metabolic pathways in their natural genetic environment. In addition, genomes derived from pure cultures present a simpler data set when compared to complex environmental metagenomes.
- Isolation of the target organism allows preliminary chemical characterization, bioactivity analysis and small-scale production of secondary metabolites with potential pharmaceutical applications.
- Prokaryotic strain collections are (yet) the cornerstone of cellular biodiversity research and conservation.

Marine sponges are host to a very diverse range of microbial symbionts[2] and a source of discovery for many new candidate species, genera and even phyla of microorganisms (Fieseler *et al.*, 2004; Sipkema and Blanch, 2010). Although some of these novel microorganisms have been obtained in culture (e.g. Sipkema *et al.*, 2011), most of them are no exception to the rule that the large majority of the microorganisms from environmental samples do not thrive *in vitro*. The relationship of sponges with their microbial symbionts escapes from the classical paradigm of symbiosis that one host and one symbiont perform a complicated dance (Haygood, 2006). Rather, the sponge–microbe conglomerate can be seen as an ecological niche in which complex interactive networks between the host and a variety of partners exist.

A digest of studies on microbial diversity in sponges has yielded the following rules of thumb:

1. Generally, bacteria make up for the bulk of the microorganisms in sponges, and they are often complemented by archaea, fungi, unicellular algae or protozoa (Taylor *et al.*, 2007). In this review, we restrict ourselves to bacteria, archaea and fungi as only a few papers have been published on the cultivation of sponge-associated algae or protozoa.
2. Sponges harbour mostly sponge-specific bacteria and archaea, that is, species belonging to a phylogenetic clade that has no representatives from other environments, such as seawater (Taylor *et al.*, 2007).

[2] Symbiosis is used in the broad sense according de Bary's definition as "a close association between organisms of different species" De Bary, A. (1879).

3. Two classes of sponges exist with respect to microbial density: high-microbial-abundance sponges, harbouring up to 10^8–10^{10} bacteria per gram sponge, which may account for up to 40% of their body mass and low-microbial-abundance sponges, which typically harbour 10^5–10^6 bacteria per gram sponge which is within the range of concentrations in natural seawater (Hentschel *et al.*, 2006).
4. The majority of sponge symbionts are located in the mesohyl, but they are also found on the outer surface of sponges and in a number of cases intracellularly (Hentschel *et al.*, 2006).
5. Sponge symbionts are particularly rich in microorganisms that produce secondary metabolites with potential pharmaceutical applications (e.g. Piel, 2004).

The latter issue has often been the source of inspiration for studies directed towards the isolation of sponge symbionts, and the cumulative number of microbial species that have been isolated from marine sponges in different studies is impressive. In addition, the number of bioactive metabolites that have been detected in these isolates is large and their chemical structures are versatile. However, the number of microbial species that are associated with sponges and has not been obtained in any of these cultivation studies is far more impressive. In addition, none of the sponge-derived lead compounds in clinical trials that are of known or suspected microbial origin, such as halichondrin B and discodermolide, have been obtained via cultivation of their putative symbiotic bacteria.

In the next three sections, we will review microorganisms (bacteria, archaea and fungi) that have been isolated from sponges against a background of microorganisms that have been identified in sponges using molecular techniques. It is important to mention that the microbial symbionts from different kingdoms in sponges have currently a different status with respect to the specificity in their association to sponges. For sponge-associated bacteria, it has become clear that most are sponge specific (Hentschel *et al.*, 2002), which is, to a lesser extent, also true for sponge-associated archaea (Holmes and Blanch, 2007). For sponge-associated fungi, no such selective interactions have yet been identified.

4.1. Culturable bacteria from sponges

New papers about bacteria that have been isolated from a marine sponge and are the producer of a previously unknown bioactive compound are being published at a high frequency (e.g. Muscholl-Silberhorn *et al.*, 2008; Schneemann *et al.*, 2010b). However, of these isolated strains, neither the ecological significance of the symbiosis nor their specificity to the sponge as their preferred habitat is generally known. In this review, we illustrate the state of the art in cultivation of sponge-derived bacteria by analysing bacterial isolates from four well-studied sponges for which substantial cultivation-dependent *and* cultivation-independent data are available: the high-microbial-abundance sponges *A. aerophoba* (Hentschel *et al.*, 2001; Pabel *et al.*, 2003; Pimentel-Elardo *et al.*, 2003, 2010; Ahn *et al.*, 2009; Abdelmohsen *et al.*, 2010) and *R. odorabile* (Webster and Hill, 2001; Webster *et al.*, 2001, 2002) and the low-microbial-abundance sponges *Halichondria panicea* (Perovic *et al.*, 1998; Wichels *et al.*, 2006; Lee *et al.*, 2007, 2009; Schneemann *et al.*, 2010a) and *Haliclona* (*Gellius*) sp. (Sipkema *et al.*, 2009, 2011). We analysed which part of the bacterial communities could be isolated against the background of cultivation-independent assessments of the bacterial communities from the same species.

At the higher taxonomic level (phylum, but Proteobacteria is analysed at the class level), differences are obvious for all model sponge species when cultivation-dependent and cultivation-independent bacterial profiles are compared (Fig. 6.7). Firstly, the diversity of phyla/classes obtained via cultivation-independent methods (13.3 ± 5.0 phyla/classes) is higher than the diversity obtained via cultivation-dependent methods (6.5 ± 2.4) for all four sponge species. Secondly, for none of the four species are there obvious similarities in the dominant phyla between cultivation-independent and cultivation-dependent bacterial profiles. The latter should be no surprise as the "great plate count anomaly" coined by Staley and Konopka (1985) has only been increasing. Despite new developments in cultivation methods of microorganisms, they cannot keep up with the discovery of new phyla by cultivation-independent methods and currently only 30% of the bacterial phyla have cultivated representatives (Achtman and Wagner, 2008).

A more detailed analysis of sponge-associated bacteria (at the species level) via cultivation-independent and cultivation-dependent methods significantly increases the gap between cultivated and uncultivated species (Fig. 6.8). While for *Haliclona (Gellius)* sp. there is limited overlap between bacterial 16S rRNA gene sequences obtained by cultivation-dependent and cultivation-independent methods; for all other model sponges there is not a single isolate-derived 16S rRNA gene sequence that matches a sequence obtained via cultivation-independent methods (based on a 97% identity cut-off). These results are not limited to the four model sponges but have been confirmed by other studies where sponge-derived cultivation-dependent and cultivation-independent bacterial profiles were compared (Isaacs *et al.*, 2009; Lee *et al.*, 2009; Sun *et al.*, 2010). Although these discrepancies alone are not unusual (see also for comparisons of phylotypes obtained via cultivation-dependent and cultivation-independent methods in gut (Rajilic-Stojanovic *et al.*, 2007) or soil (Zhang *et al.*, 2007)), they underline the alarming inadequacy of our current cultivation methods and the large gap between the discovery of bioactive compounds in sponges and their production in bacterial cultures.

The fact that sponge-derived isolates do not reflect the microbial profiles obtained via cultivation-independent studies implies that although these

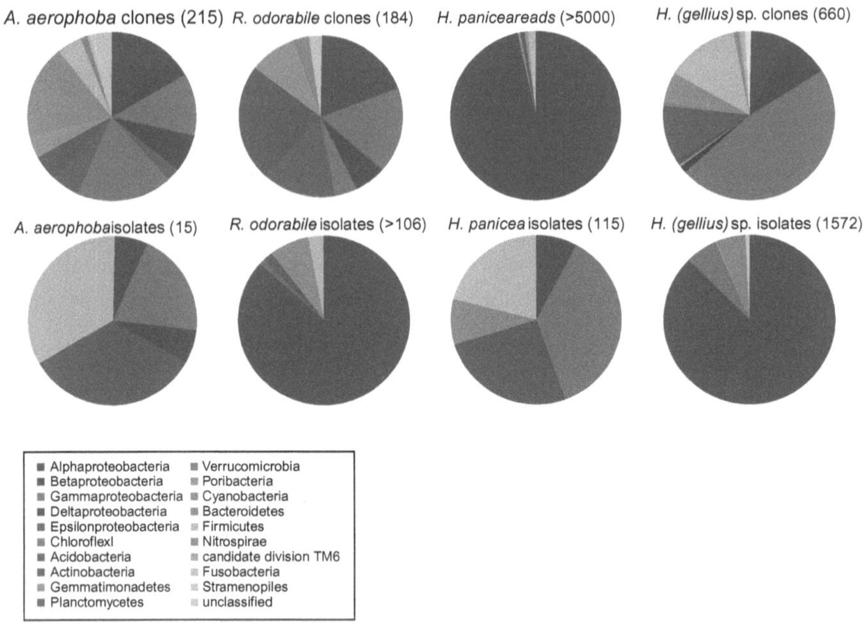

Figure 6.7 Cultivation-independent (upper halve) and cultivation-dependent (lower halve) bacterial profiles from four model sponge species: *Aplysina aerophoba, Rhopaloeides odorabile, Halichondria panicea* and *Haliclona* (*gellius*) sp. per phylum (the phylum Proteobacteria is divided into classes). 16S rRNA gene sequences present in the NCBI GenBank were downloaded at 18 February 2011 using the sponge species name AND "16S" as keywords for the different species. The data were manually corrected for the presence of non-target sponge-derived sequences. In addition, 16S rRNA gene fragments from DGGE bands and diseased *R. odorabile* sponges were excluded from the analysis. As only few 16S rRNA gene sequences from *H. panicea* clones were present in the in the NCBI GenBank, we used the 20 most abundant phylotypes (based on 97% identity) with a length of approximately 350 bp from a 454-pyrosequencing study (Morillo-Perez *et al.*, in progress). The numbers of clones and isolates (or 454-pyrosequencing reads) are in parenthesis for all categories. The number of *R. odorabile* isolates was with a very conservative estimate set at 106, as the exact number of clones could not be derived from the data from Webster and Hill (2001). (For colour version of this figure, the reader is referred to the Web version of this chapter.)

recovered isolates must be present in the sponge tissue, they are most likely there in low numbers only. We have identified five cultivation clusters from sponge-derived bacteria (i.e. clusters of sponge isolates from at least three of the four model sponges that have no related sequences obtained via cultivation-independent methods) in α-Proteobacteria, γ-Proteobacteria, Firmicutes and Actinobacteria (Fig. 6.8). Representatives of these clusters may be present in many sponges, as, for example, NW001 (in CC Alpha), the most

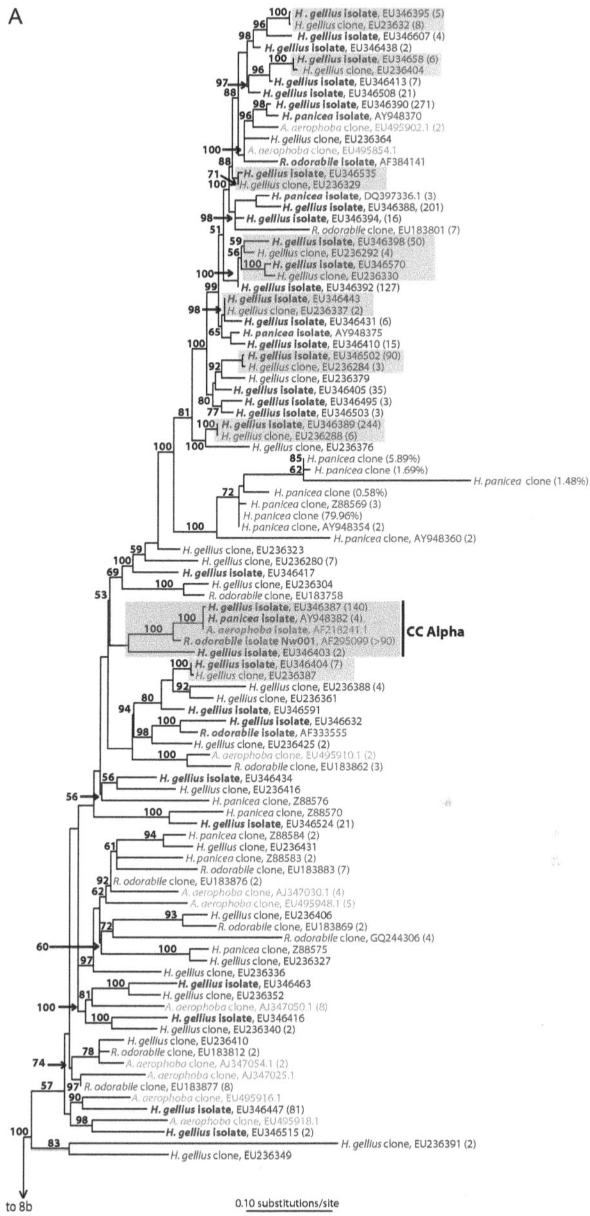

Figure 6.8 Continued

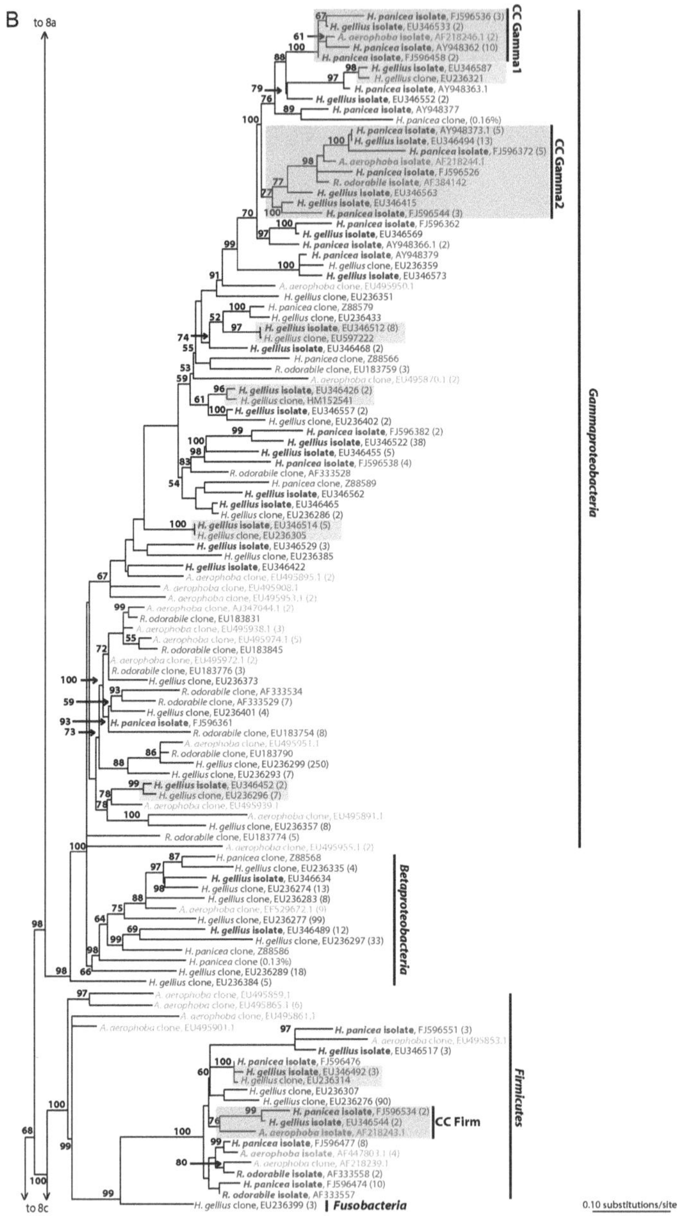

Figure 6.8 Continued

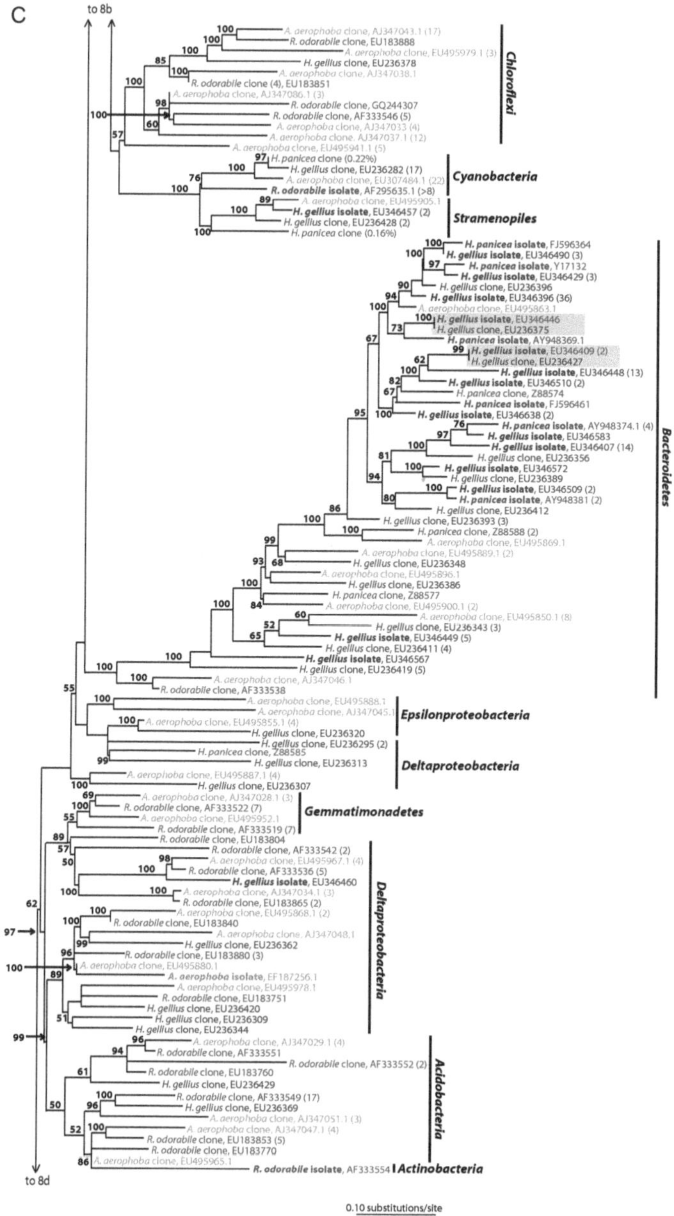

Figure 6.8 Continued

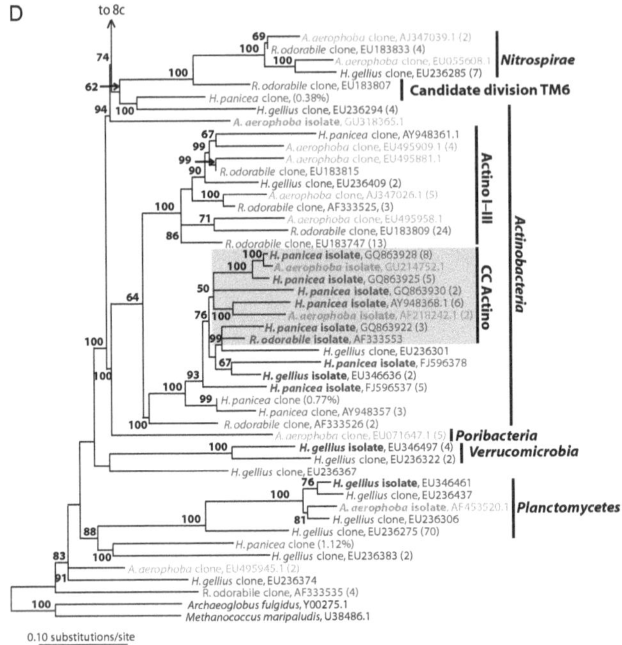

Figure 6.8 Bayesian phylogram of 16S rRNA gene sequences of all clones and isolates from *A. aerophoba*, R. *odorabile*, *H. panicea* and *H.* (*gellius*) sp. Isolates are depicted in bold. For the selection of the data, see Fig. 6.7. If sequences within "isolates" or "clones" from the same sponge species were closer related to each other than to other sequences, they were grouped with the sequence with the longest read length. For these sequences, the number of sequences represented by the clone or isolate is included in parentheses. For the 20 most abundant phylotypes from *H. panicea*, a 454-pyrosequencing study of the 16S rRNA gene the % of the reads within the phylotypes is included in parenthesis. Beige boxes indicate clones and isolates from the same sponge species that are 97% or more than 97% identical. Pink boxes indicate cultivation clusters: a cluster of sequences obtained from isolates from three of the four model sponges without sequences obtained via cultivation-independent techniques. "Actino I–III" are sponge-specific *Actinobacteria* as defined by Hentschel *et al.* (2002). The numbers above or below the branches correspond to posterior probability (PP) values of the Baysian analysis. Nodes with PP values of < 50 are not indicated. *Archaeoglobes fulgidus* and *Methanococcus maripaludis* were used as outgroup. Data treatment: all 16S rRNA gene sequences were imported from the Silva database (Pruesse *et al.*, 2007) into the ARB software package (Ludwig *et al.*, 2004) and the alignments were refined manually. Ambiguous regions of the alignment were systematically removed using the program Gblocks v.0.91b (Castresana, 2000). The default programme parameters were used, except allowing a minimum block length of 5 and gaps in 50% of positions. Phylogenetic trees were created by Bayesian analysis, using MrBayes v3.0b4 (Huelsenbeck and Ronquist, 2001) at the freely available Bioportal server (www.bioportal.uio.no). All parameters were treated as unknown variables with uniform prior-probability densities at the beginning of each run and their values were estimated from the data during the analysis (Ronquist *et al.*, 2005). All Bayesian analyses were initiated with random starting trees and were run for 1×10^7 generations. (For colour version of this figure, the reader is referred to the Web version of this chapter.)

common bacterial isolate from *R. odorabile* (Webster and Hill, 2001), relatives of which have also been isolated from many other sponge species worldwide (Enticknap *et al.*, 2006). The only related sponge-derived cultivation-independent sequence that has been obtained (EF092248.1) was present in a very large clone library of *A. corrugata* of 192 clones per sample (Holmes, 2006), which increases the chance of picking up rare bacteria. A similar situation exists for the actinobacterial cultivation cluster (Fig. 6.8). Representatives have been isolated from a large number of sponges, but they do not coincide with the actinobacterial sponge-specific clades ActinoI-III identified by Hentschel *et al.* (2002). Thus, although *Actinobacteria* isolated from marine sponges are a very prolific source of new bioactive compounds, these bacteria or their compounds are rarely detected in the sponge itself (Schneemann *et al.*, 2010b; Sun *et al.*, 2010). It is conceivable that many of these isolated bacteria are only present in the canal system when the sponge is collected. Although most researchers rinse the sponge specimens several times with sterile artificial seawater before initiating cultivation experiments, it is very difficult to remove bacteria that are present inside the canal system of the sponge because the sponge most likely ceases pumping upon collection. The presence of more than a million bacteria per millilitre of seawater (Watson *et al.*, 1977) still yields many potential "sponge-derived isolates" selected from a pool of opportunistic passers-by.

When cultivation methods targeting associated bacteria from the four model sponges are assessed, for a substantial number of studies only rich media (e.g. marine agar 2216) were used (Perovic *et al.*, 1998; Webster *et al.*, 2002; Pabel *et al.*, 2003; Lee *et al.*, 2007). In addition, temperatures significantly above ambient temperatures were used (Lee *et al.*, 2007; Ahn *et al.*, 2009; Abdelmohsen *et al.*, 2010), and in some cases, incubation times were 1 week or less (Perovic *et al.*, 1998; Pabel *et al.*, 2003; Lee *et al.*, 2007; Pimentel-Elardo *et al.*, 2010). These conditions favour isolation of opportunistic bacteria, which may then easily outcompete sponge-specific bacteria.

The only cultivation study of the four model sponge species that resulted in the isolation of bacteria that were confirmed by cultivation-independent data from the same species (*H.* (*Gellius*) sp.) differed from the other studies in the way that in addition to traditional agar plate cultures also liquid cultures according to the dilution-to-extinction principle and cultivation on floating polycarbonate filters were applied (Sipkema *et al.*, 2011). The use of these "alternative" cultivation methods yielded six isolates that corresponded to cultivation-independently obtained 16S rRNA gene sequences from the same sponge species that were not recovered from agar plates.

The use of environmental extracts (e.g. sponge extracts) has been frequently advocated as a successful strategy to augment growth of bacteria specific to the particular environment. Sponge extracts have indeed led to the isolation of a different set of microorganisms compared to cultivation conditions without the use of sponge extracts (e.g. Olson *et al.*, 2000; Selvin *et al.*, 2009). Although the latter was also true for the use of sponge extracts to isolate bacteria from the four model sponge species (Webster *et al.*, 2001; Pabel *et al.*, 2003; Wichels *et al.*, 2006; Abdelmohsen *et al.*, 2010; Sipkema *et al.*, 2011), these "extra isolates" were neither sponge-specific nor detected by cultivation-independent techniques in any of the sponge species. This finding may imply that the "dead interactions" supplied with the sponge extract are not sufficient to mimic the natural environment of sponge-specific bacteria, but that live interactions are required.

Thus, sponges serve as a very rich reservoir for the isolation of novel bacteria and novel bioactive compounds, but we seem yet incapable in isolating the microorganisms that are important to the sponge.

4.2. Culturable archaea from sponges

No archaea from sponges have been obtained in culture. In contrast, archaea belonging to the recently proposed third archaeal phylum Thaumarchaeota (Brochier-Armanet *et al.*, 2008) have been detected in considerable sized populations in many marine sponges (Holmes and Blanch, 2007). A number of cultivation trials have been done for the sponge-associated Thaumarchaeota, but little information has been published due to the lack of success (e.g. Sipkema *et al.*, 2011). Sponges are not a unique case in that respect as there are only three cases of the successful isolations of thaumarchaeotes (Konneke *et al.*, 2005; Simon *et al.*, 2005; Tourna *et al.*, 2011). They were all a result of long-term experiments with enrichment strategies that lasted for up to 2 years.

The presence of thaumarchaeotes in sponges was first established in *Axinella mexicana* where *Cenarchaeum symbiosum* comprises 65% of the microbial biomass (Preston *et al.*, 1996). A sponge-specific clade of *Thaumarchaeota* that is predominantly represented by symbionts from the genus *Axinella* has been proposed (Holmes and Blanch, 2007). Genome analysis of *C. symbiosum* has indicated that it can function as a strict autotroph, or as a mixotroph utilizing both carbon dioxide and organic material as carbon sources, while energy is generated by ammonia oxidation (Hallam *et al.*, 2006b). These are important leads for the set-up of cultivation experiments of sponge-associated archaea. At this point, it is important to note that the Thaumarchaeota that were obtained in culture were initially (Konneke *et al.*, 2005; Tourna *et al.*, 2011) or only (Simon *et al.*, 2005) obtained in enrichment cultures. Therefore, successful isolation of sponge-derived Thaumarchaeota may require co-cultivation strategies in addition to a lot of patience and perseverance.

4.3. Culturable fungi from sponges

Fungi are at the opposite end of the spectrum with respect to culturability when compared to sponge-associated archaea: hundreds of fungi associated with sponges have been obtained from cultivation studies, while nearly none exist from cultivation-independent studies (Holler *et al.*, 2000; Morrison-Gardiner, 2002; Pivkin *et al.*, 2006; Wang *et al.*, 2008; Baker *et al.*, 2009; Wiese *et al.*, 2011). In addition, sponge-derived fungi are a particularly rich source of bioactive compounds compared to other fungi (Jensen and Fenical, 2002).

It is yet unclear how consistent these sponge–fungus associations are due to limited experimentation. There is some evidence that the species composition of fungal isolates is related to the sponge species (Holler *et al.*, 2000; Pivkin *et al.*, 2006); however, results are not consistent. The sponge *Myxilla incrustans* has been the target of two independent studies and was characterized as very rich (Holler *et al.*, 2000) or poor (Pivkin *et al.*, 2006) source for the isolation of fungi. This difference may be related to the different geographical origins of the samples (Helgoland and Sakhalin, respectively), but on the other hand, sponge-associated bacterial populations are generally stable across large geographical distances (Hentschel *et al.*, 2006). A second issue is that nearly none of the fungi isolated from sponges are characterized as obligate marine fungi (i.e. fungi that grow and sporulate exclusively in marine or brackish water; Kohlmeyer, 1974), which leaves it unclear whether the isolated strains are consistently associated fungi or passers-by that were trapped in the sponge's filtration system. Fungal spores have a diameter in the order of micrometres and sponges are highly efficient at filtering particles of that size category from the environment (Osinga *et al.*, 2001). Thus, there may be a prolific reservoir of fungi present inside sponges that lead to the isolation of many fungi that are highly related to fungi from terrestrial sources. Terrestrial fungi are generally well able to cope with media containing salt concentrations equal to concentrations in the marine environment, which increases the chance of their isolation from marine sources. Thirdly, when reviewing studies that identify sponge-associated fungi, both by cultivation-dependent and by cultivation-independent methods, cultivated representatives do not reflect data obtained with molecular methods and the paucity of molecular data from sponge-associated fungi leaves it currently unclear whether fungi are present at ecologically relevant concentrations in sponges. Gao *et al.* (2008) assessed the sponges *Suberites zeteki* and *Mycale armata* by molecular biology-based tools and established the presence of two phyla (Basidiomycota and Ascomycota) and 11 orders of fungi in these sponges, most of them belonging to Malasseziales. The major issue with molecular studies of sponge-associated fungi is that PCR primers used to amplify the 18S rRNA gene or the ITS region generally also amplify the sponge host's 18S rRNA gene or ITS region resulting in only a small fraction of fungal target products. Interestingly, no representatives of Malasseziales were obtained in a cultivation study of fungi derived from the same sponge species (Wang *et al.*, 2008), which is an indication that sponge-associated fungal isolates do not represent the most abundant fungi present in the sponge. Baker *et al.* (2009) compared fungal isolates from the sponge *Haliclona simulans* with fungal sequences obtained via cultivation-independent methods and found little overlap between fungal species obtained with the two different methods. However, because of the before-mentioned issues with PCR primer specificity, they used primers that predominantly amplify Pezizomycotina and Agaricomycotina, and therefore, they obtained a highly biased molecular profile of *H. simulans*-associated fungi. The scarcity in molecular data of sponge-associated fungi is accompanied by anecdotal microscopic confirmations of the presence of fungi in sponges. The first record is a close association between a coral reef-inhabiting *Koralionastes* species and an unidentified encrusting sponge that grows on top of it (Kohlmeyer and Volkmann-Kohlmeyer, 1990). Another observation of a maternally transmitted endosymbiotic yeast in the marine sponge *Chondrilla* sp. implies a long-standing relationship between the sponge and the yeast (Maldonado *et al.*, 2005).

In summary, there is currently a large discrepancy between fungal species isolated from sponges and fungal species detected in sponges by cultivation-independent methods, not in the last place due to very a limited data set on cultivation-independent sponge-associated fungi. The sheer number of different sponge species requires a more robust data set before conclusions on the existence of sponge-specific fungi can be drawn. The rapidly ongoing discovery of novel bioactive compounds from sponge-derived fungi will be a strong driving force that will yield more conclusive information on this issue within the next decade.

5. Future Directions

5.1. Sponge aquaculture

Recent progress in sea-based aquaculture confirms the value of this method for the production of bath sponges and natural compounds. There is renewed interest in multiple use of sponge mariculture as compartments of integrated multitrophic aquaculture systems (see Fig. 6.9), a trend that has also been recognized by Duckworth (2009), Osinga *et al.* (2010) and Page *et al.* (2011). Targeted sponges grew faster in the vicinity of sea farms growing sea bass (Osinga *et al.*, 2010) and mussels (Page *et al.*, 2011) when compared to pristine sites, where the visibility (a qualitative measure for the availability of organic food) was 5- to 10-fold higher (Osinga *et al.*, 2010). The effect of the sponges on the environment was not evaluated in these studies.

Figure 6.9 Mariculture of *Dysidea avara* at 21 m depth underneath a floating-cage sea bass farm in Southwest Turkey. The visibility at the culture site is less than 3 m. (For colour version of this figure, the reader is referred to the Web version of this chapter.)

A general overview on zooremediation—the potential use of sponges and other marine animals in water treatment—is provided by Gifford *et al.* (2007). More specifically, Fu *et al.* (2006) and Zhang *et al.* (2010) tested the ability of the sponge *H. perlevis* to remove pathogenic bacteria from seawater. Although the removal potential of the sponges was considerable, removal efficiencies for the different types of pathogens tested were highly variable. Hence, the application of such selective filter feeders in wastewater treatment may lead to competitive advantages for those pathogens that are not effectively being grazed upon (Maldonado *et al.*, 2010).

Despite the good progress that has been made towards controlled sponge culture in aquarium systems (e.g. Garcia Camacho *et al.*, 2006b; Xue and Zhang, 2009; Carballo *et al.*, 2010), *ex situ* culture has currently no competitive advantage over sea-based culture, which is still more productive and cost effective. A further complicating aspect with regard to aquarium culture is the apparent inability of the cultured sponges to maintain a natural composition of associated microorganisms. It remains to be studied to what extent this change in microbiology is reflected in primary and secondary metabolism of the sponge host. In addition, further understanding of the dietary requirements (in particular DOC vs. particles) and the cellular biology of sponges (De Goeij *et al.*, 2009; Koopmans *et al.*, 2010) is likely to turn *ex situ* sponge culture into a feasible technique within the forthcoming years.

5.2. *In vitro* sponge culture

5.2.1. Stem cells

Stem cells possess unlimited proliferative capacities *in vivo* and have the ability to differentiate into other cell types, which make them a good starting material for cell culture (Rinkevich, 2011). In adult sponges, certain cell types are present with stem cell properties, such as archaeocytes, which are considered to be pluripotent and to support both sexual and asexual reproduction in sponges. Recent molecular studies also suggest that choanocytes have pluripotent stem cell-like potential (Funayama *et al.*, 2005, 2010; Funayama, 2010). Funayama *et al.* (2010) found expression of Piwi (genes essential for stem cell self-renewal) homologues in archaeocytes as well as in choanocytes in a freshwater sponge, *Ehypdatia fluviatilis*, indicating that both cell types have stem cell properties.

Since adult sponges maintain large pools of adult stem cells, it is possible to isolate fractions of stem cells (De Sutter and Buscema, 1977; Pomponi and Willoughby, 1994; Sun *et al.*, 2007). Rinkevich (2011) also suggest targeting resting somatic sponge cells and manipulating them to dedifferentiate and become induced pluripotent stem (iPS) cells. These dedifferentiation techniques have already been performed on mammalian cells (Takahashi and Yamanaka, 2006). Besides selecting adult stem cells from sponges, it is also possible to select stem cells from embryos, larvae or gemmules (Imsiecke *et al.*, 1995; Rinkevich *et al.*, 1998; De Caralt *et al.*, 2007b).

5.2.2. Immortalization

Most cell types undergo a limited number of divisions and then enter senescence. The establishment of cell cultures with unlimited cell divisions, that is, a transformed or immortalized cell line, has been accomplished in mammalian cells by inserting immortalizing genes by means of transfection or transduction. Immortalizing genes that have been extensively used for mammalian cells are listed in Table 6.9. To achieve immortalization using immortalizing genes the following prerequisites need to be met: viable primary cells, a proper promoter that is recognized by the RNA polymerase of the sponge, a proper transfection or transduction method, functional immortalizing genes and finally a method to obtain stable insertion of the genes.

Since it is difficult to obtain viable and proliferative sponge cells in primary cell cultures established from dissociated cells, an alternative approach would be to select other sponge material than dissociated sponge cells for immortalization, for example, gemmules or larvae. Gemmules and

Table 6.9 Genes used for immortalization, modified from Freshney (2005)

Gene abbreviation	In full
EBV	Epstein–Bar virus
SV40LT	Simian virus 40 large T
hTERT	Human telomerase reverse transcriptase
HPV	Human papilloma virus
Ad5 E1a	Adenovirus E1a

larvae both feed on yolk reserves, which provide enough nutrients for the first stages of development into a juvenile sponge. Gemmules have an additional advantage because the outside can be sterilized using 1% H_2O_2, which makes contamination less of a problem. It has already been demonstrated that gemmules and larvae can be developed into juvenile sponges *in vitro* (De Caralt *et al.*, 2007a; Funayama *et al.*, 2010).

Successful transfection of freshwater juvenile sponges has already been proven. Pfannkuchen and Brummer (2009) demonstrated the expression of a red fluorescent protein (DsRed2) under the control of a CMV (cytomegalovirus) promoter in *Spongilla lacustris*, using particle bombardment to insert the plasmid. Pomponi *et al.* (2007) demonstrated expression of telomerase reverse transcriptase (*tert*) under the control of an MPSV (myeloproliferative sarcoma virus) promoter in primary cell cultures of *A. corrugata*, using lipofection. As transfection has proven to be possible and a promoter that is functional in sponges has been found, the next step towards immortalization is inserting and expressing immortalizing genes.

One candidate gene for sponge cell immortalization is *tert*. Introduction of a human telomerase reverse transcriptase (*htert*) gene into normal human cells showed an extension of the lifespan of the culture, and a portion of the cells became immortal (Bodnar *et al.*, 1998). Telomerase expression has already been demonstrated in adult sponges and primmorphs, as well as primary cultures established from dissociated cells (Pomponi *et al.* 2007), although expression was not stable and the cells did not become immortalized. It suggests, however, that insertion and expression of a telomerase gene could lead to an immortalized sponge cell line.

5.2.3. Growth media

Nutritional requirements of sponge cells are still poorly understood. Since the basic building blocks for biomass (protein, lipids, DNA, RNA) are the same for sponge cells and mammalian cells, mammalian cell culture media may be expected to supply the basic nutrients for these polymers also in sponges. Some sponge-specific compounds, for example, fatty acids, may need to be supplemented. Furthermore, sponge cells may be able to synthesize certain precursors, and some components of mammalian media may not be necessary. Growth factor supplements may be more problematic because it is not likely that mammalian and sponge cells have similar requirements. The use of invertebrate (*Octopus*) serum (Garcia Camacho *et al.*, 2006b) has promoted growth in *C. crambe* tissue fragments. Analysis of the contents of the yolk in gemmules and larvae may also provide insights into improving sponge cell culture medium. Inorganic components, for example, iron and silicate, have promoted cell proliferation Krasko *et al.* (2002); the effects of other relevant inorganic components should also be analysed.

Genomics, transcriptomics and proteomics are tools that should be used to define culture conditions and to develop rational approaches to immortalization. The genome of one sponge, *Amphimedon queenslandica*, has been sequenced, and the draft genome sequence reveals that it is remarkably similar to other animal genomes in relation to content, structure and organization (Srivastava *et al.*, 2010). Subsequent genome annotation and comparative genomics will allow for the identification of an extended set of genes and thus reveal the metabolic capabilities and signalling pathways of the sponge. This information can be used to define culture media and provide insights into the regulation of genes related to cell cycling, and perhaps even production of bioactive compounds.

5.2.4. 3D culture systems

Normal, untransformed cells from multicellular animals cannot grow in suspension (except for hematopoietic cells) and need to attach to a surface in order to survive and proliferate (Freshney, 2005). Koziol *et al.* (1998) and Custodio *et al.* (1998) proved that this is also true for sponge cells in suspension. Isolated sponge cells lacked telomerase activity, while re-aggregated sponge cells (primmorphs) regained telomerase activity. These primmorphs could be kept in culture for several months and were able to hatch and start spreading. Other 3D culture systems could be the cultivation of tissue fragments, gemmules and larvae (Garcia Camacho *et al.*, 2006b; De Caralt *et al.*, 2007a; Funayama *et al.*, 2010). 3D culture systems could be used as starting material for cell culture, since they can be maintained *in vitro* and proliferation has been demonstrated. Ideally, transformation of sponge cells in 3D cultures could lead directly to sponge cell growth in suspension.

5.3. Symbiont culture

The metagenomics revolution will provide an enormously rich resource of genomics, transcriptomics and proteomics information of presently unculturable microorganisms. Practical knowledge on nutrient cycles, amino acid, cofactor and vitamin synthesis potential of individual members of a community is of key importance for the design of tailor-made cultivation conditions for microorganisms that are yet unculturable. For example, based on carbon balances, it has been shown that many sponge species must obtain the major part of their nutrients from DOC rather than from "food bacteria" present in the seawater (Reiswig, 1974; de Goeij *et al.*, 2008). It is expected that a large fraction of DOC in seawater is quite recalcitrant to degradation, but its composition is unknown. Identification of putative enzymes that are involved in the degradation of recalcitrant carbon sources by transcriptome analysis of the sponge community will provide new ideas for growth medium design.

A few recent studies have now resulted in the publication of partial genomes of an unculturable sponge-specific poribacterium (Siegl *et al.*, 2011), an unculturable sponge-specific δ-proteobacterium (Liu *et al.*, 2011) and a sponge-specific archaeon (Hallam *et al.*, 2006a). These studies revealed important information about the metabolism of these sponge symbionts: the poribacterium has the genomic repertoire for autotrophic CO_2-fixation through the Wood–Ljungdahl pathway and is capable of sterol synthesis, which is almost exclusively restricted to eukaryotes. These sterols most likely play a role in the formation of DNA-containing organelles that are found in *Poribacteria* (Siegl *et al.*, 2011). Genome analysis of the sponge-specific archaeon has indicated that it can function as a strict autotroph utilizing carbon dioxide as carbon source and ammonia as energy source. The autotrophic nature of these two sponge-specific microorganisms makes them good candidates for isolation in enrichment cultures because highly selective conditions can be applied. The sponge-specific δ-proteobacterium was found to possess a glutathione porter that is crucial for growth on glutathione as the sole sulphur source. In addition, its genome encodes a tetracycline resistance protein (Liu *et al.*, 2011). These characteristics are very useful to create specific cultivation conditions for this δ-proteobacterium. Moreover, this δ-proteobacterium seemed to be specifically associated with (most probably) a cyanobacterium within the sponge tissue and may live off its photosynthetically fixed carbon (Liu *et al.*, 2011). For that reason, co-cultivation with the cyanobacterium may be the best way to obtain this δ-proteobacterium in culture. Co-culture with other microorganisms has proven a fruitful strategy for the isolation of "difficult" bacteria and archaea. For example, an organohalide-respiring a *Dehalobacter* sp. could only be grown in the presence of a *Sedimentibacter* sp. most likely due to the inability of *Dehalobacter* sp. to synthesize its own vitamins (Van Doesburg *et al.*, 2005). Other bacteria could be grown in pure cultures, but only in diffusion growth chambers, where they were exposed to excreted metabolites from other bacteria (Kaeberlein *et al.*, 2002), and one of the three thaumarchaeotes that could be cultivated was only obtained in a mixed culture (Simon *et al.*, 2005).

The set-up of highly selective cultivation conditions, based on new input by next-generation sequencing technology, is a most promising method to isolate sponge-specific microbes and the microbes that are the putative producers of halichondrin B, discodermolide and many other potent bioactive compounds that have been discovered in marine sponges, but have never been obtained by cultivation of the actual producer.

In addition to genomic information about bacterial and archaeal symbionts from marine sponges, next-generation sequencing will also, perhaps inadvertently, lead to a large number of fungus-derived sequences from sponges that will prove new insights in whether fungi are specifically associated with sponges, and if so, which type of fungi they are.

5.4. Co-cultivation of sponge cells and symbionts

Despite significant efforts, axenic primary sponge cell cultures have not been established and very few sponge-associated microbes that are sponge-specific have been obtained in pure cultures. For most interactions between sponges and associated microbes, it is currently not clear whether they are commensal or mutualistic. Moreover, it is unclear whether they are obligate or facultative relationships. The mere existence of low-microbial-abundance sponges indicates that many sponge species are able to sustain life on their own without the help of microbial symbionts suggesting that the symbiotic relationships may not be obligate, but of a more facultative nature. On the other hand, many symbiotic relationships between sponges and microbes seem to be persistent in time and space (e.g. Friedrich *et al.*, 2001; Lee *et al.*, 2009; Sipkema *et al.*, 2009) and have become very intimate with the presence of bacteria, archaea and yeasts inside the egg cells of sponges (e.g. Maldonado *et al.*, 2005; Sharp *et al.*, 2007; Schmitt *et al.*, 2008). Attempts have been made to study the effect of eradicating sponge-associated bacteria in high-microbial-abundance sponges, but it has been shown that even with an extensive antibiotic treatment under starvation conditions, a large part of the bacteria cannot be cleared (Friedrich *et al.*, 2001; Detmer Sipkema and Sonia de Caralt, unpublished). Therefore, the possibility to cultivate sponges without their associated bacteria could not be tested. For the cultivation of sponge-associated bacteria, sponge extracts have been used as a medium supplement to enhance growth of the bacteria (see Section 4.1); however, these experiments have only been used to somehow supply the bacteria with an artificial sponge host and not the other way around. In addition, only "dead interactions" can be simulated by the use of sponge extracts, while for symbiosis between insects and bacteria, it has been found that many metabolic exchanges exist (Zientz *et al.*, 2004).

Interactions for both primary and secondary metabolism have been hypothesized for sponges and symbionts. Bacteria have been suggested to be involved in the uptake of DOC, which makes up for a large part of the sponge–microbe conglomerate diet (Yahel *et al.*, 2003; de Goeij *et al.*, 2008). Uptake of nitrogen by bacteria and translocation to the sponge host has recently been confirmed (Bayer *et al.*, 2008), and translocation of the photosynthate from endosymbiotic algae to the sponge has also been documented (Wilkinson and Garrone, 1980). Evidence of the exchange of secondary metabolites is scarce, but it has been reported that a sponge accumulates polybrominated biphenyl ethers that are produced by its cyanobacterial endosymbiont (Unson *et al.*, 1994). Due to the mostly uncharacterized interactions between host and microbial symbionts, *in vitro* co-cultivations could provide an intermediate step between the natural environment and axenic cultures. There are many examples of cells that can only be cultivated or indeed be better cultivated in the presence of other cells. For the cultivation of many mammalian cells (e.g. human embryonic stem cells), a layer of feeder cells (e.g. fibroblasts) is used to release nutrients and growth factors into the medium to stimulate growth of the target cells (Unger *et al.*, 2009). For *in vitro* cultivation of lichens, it has been observed that lichen spores germinate but cannot develop into mycelium without contact with a compatible photobiont (Stocker-Worgotter, 1995).

This does not necessarily have to result in co-cultivation as a final solution, but as an initial step that can be used for further development. Also for *in vitro* cultivation of lichens, it transpired that after initial co-cultivation, the fungus could be cultivated without the photobiont when medium design was aided by information from co-cultivation (Stocker-Worgotter, 1995). For those reasons, co-cultivation of sponge cells and associated microbes is an as yet surprisingly underexplored alternative.

ACKNOWLEDGEMENTS

We wish to acknowledge José Morillo-Perez for sharing his preliminary data of 16S rRNA gene pyrosequencing of *H. panicea*.

REFERENCES

Abdelmohsen, U. R., Pimentel-Elardo, S. M., Hanora, A., Radwan, M., Abou-El-Ela, S. H., Ahmed, S., and Hentschel, U. (2010). Isolation, phylogenetic analysis and anti-infective activity screening of marine sponge-associated actinomycetes. *Marine Drugs* **8,** 399–412.

Achtman, M., and Wagner, M. (2008). Microbial diversity and the genetic nature of microbial species. *Nature Reviews Microbiology* **6,** 431–440.

Ahn, Y.-B., Kerkhof, L. J., and Häggblom, M. M. (2009). *Desulfoluna spongiiphila* sp. nov., a dehalogenating bacterium in the *Desulfobacteraceae* from the marine sponge *Aplysina aerophoba. International Journal of Systematic and Evolutionary Microbiology* **59,** 2133–2139.

Andrade, P., Willoughby, R., Pomponi, S. A., and Kerr, R. G. (1999). Biosynthetic studies of the alkaloid, stevensine, in a cell culture of the marine sponge *Teichaxinella morchella. Tetrahedron Letters* **40,** 4775–4778.

Baker, P., Kennedy, J., Dobson, A., and Marchesi, J. (2009). Phylogenetic diversity and antimicrobial activities of fungi associated with *Haliclona simulans* isolated from Irish coastal waters. *Marine Biotechnology* **11,** 540–547.

Battershill, C. N., and Page, M. J. (1996). Sponge aquaculture for drug production. *Aquaculture Update* **16,** 5–6.

Bayer, K., Schmitt, S., and Hentschel, U. (2008). Physiology, phylogeny and *in situ* evidence for bacterial and archaeal nitrifiers in the marine sponge *Aplysina aerophoba*. *Environmental Microbiology* **10,** 2942–2955.

Becerro, M. A., Turon, X., and Uriz, M. J. (1997). Multiple functions for secondary metabolites in encrusting marine invertebrates. *Journal of Chemical Ecology* **23,** 1527–1547.

Belarbi, E. H., Contreras Gomez, A., Chisti, Y., Garcia Camacho, F., and Molina Grima, E. (2003). Producing drugs from marine sponges. *Biotechnology Advances* **21,** 585–598.

Bergman, O., Haber, M., Mayzel, B., Anderson, M., Shpigel, M., Hill, R., and Ilan, M. (2011). Marine-based cultivation of Diacarnus sponges and the bacterial community composition of wild and maricultured sponges and their larvae. *Marine Biotechnology* 1–14.

Bergmann, W., and Feeney, R. J. (1951). Contributions to the study of marine products. XXXII. The nucleosides of sponges. *I. Journal of Organic Chemistry* **16,** 981–987.

Bernard, H. R. (1968). Kalymnos: Economic and Cultural Change on a Greek Sponge Fishing Island. University of Illinois Press, USA.

Blunt, J. W., Copp, B. R., Hu, W. P., Munro, M. H. G., Northcote, P. T., and Prinsep, M. R. (2009). Marine natural products. *Natural Product Reports* **26,** 170–244.

Bodnar, A. G., Ouellette, M., Frolkis, M., Holt, S. E., Chiu, C. P., Morin, G. B., Harley, C. B., Shay, J. W., Lichtsteiner, S., and Wright, W. E. (1998). Extension of life-span by introduction of telomerase into normal human cells. *Science* **279,** 349–352.

Brochier-Armanet, C., Boussau, B., Gribaldo, S., and Forterre, P. (2008). Mesophilic crenarchaeota: Proposal for a third archaeal phylum, the Thaumarchaeota. *Nature Reviews Microbiology* **6,** 245–252.

Carballo, J. L., Yanez, B., Zubia, E., Ortega, M. J., and Vega, C. (2010). Culture of explants from the sponge *Mycale cecilia* to obtain bioactive mycalazal-type metabolites. *Marine Biotechnology* **12,** 516–525.

Castresana, J. (2000). Selection of conserved blocks from multiple alignments for their use in phylogenetic analysis. *Molecular Biology and Evolution* **17,** 540–552.

Chernogor, L., Denikina, N., Belikov, S., and Ereskovsky, A. (2011). Long-term cultivation of primmorphs from freshwater Baikal sponges; *Lubomirskia baikalensis*. *Marine Biotechnology* 1–11.

Custodio, M. R., Imsiecke, G., Borojevic, R., Rinkevich, B., Rogerson, A., and Muller, W. E. (1995). Evolution of cell adhesion systems: Evidence for Arg-Gly-Asp-mediated adhesion in the protozoan *Neoparamoeba aestuarina*. *The Journal of Eukaryotic Microbiology* **42,** 721–724.

Custodio, M. R., Prokic, I., Steffen, R., Koziol, C., Borojevic, R., Brummer, F., Nickel, M., and Muller, W. E. G. (1998). Primmorphs generated from dissociated cells of the sponge *Suberites domuncula*: A model system for studies of cell proliferation and cell death. *Mechanisms of Ageing and Development* **105,** 45–59.

De Bary, A. (1879). Die Erscheinung der Symbiose. Verlag Trubner, Straβburg.

De Caralt, S., Agell, G., and Uriz, M. J. (2003). Long-term culture of sponge explants: Conditions enhancing survival and growth, and assessment of bioactivity. *Biomolecular Engineering* **20,** 339–347.

De Caralt, S., Otjens, H., Uriz, M., and Wijffels, R. (2007a). Cultivation of sponge larvae: Settlement, survival, and growth of juveniles. *Marine Biotechnology* **9,** 592–605.

De Caralt, S., Sanchez-Fontenla, J., Uriz, M. J., and Wijffels, R. H. (2010). *In situ* aquaculture methods for *Dysidea avara* (Demospongiae, Porifera) in the Northwestern Mediterranean. *Marine Drugs* **8,** 1731–1742.

De Caralt, S., Uriz, M. J., and Wijffels, R. H. (2007b). Cell culture from sponges: Pluripotency and immortality. *Trends in Biotechnology* **25,** 467–471.

De Caralt, S., Uriz, M. J., and Wijffels, R. H. (2008). Grazing, differential size-class dynamics and survival of the Mediterranean sponge *Corticium candelabrum*. *Marine Ecology Progress Series* **360,** 97–106.

De Goeij, J. M., De Kluijver, A., Van Duyl, F. C., Vacelet, J., Wijffels, R. H., De Goeij, A., Cleutjens, J. P. M., and Schutte, B. (2009). Cell kinetics of the marine sponge *Halisarca caerulea* reveal rapid cell turnover and shedding. *Journal of Experimental Biology* **212,** 3892–3900.

De Goeij, J. M., Van Den Berg, H., Van Oostveen, M. M., Epping, E. H. G., and Van Duyl, F. C. (2008). Major bulk dissolved organic carbon (DOC) removal by encrusting coral reef cavity sponges. *Marine Ecology Progress Series* **357,** 139–151.

De Goeij, J. M., and Van Duyl, F. C. (2007). Coral cavities are sinks of dissolved organic carbon (DOC). *Limnology and Oceanography* **52,** 2608–2617.

De Rosa, S., De Caro, S., Iodice, C., Tommonaro, G., Stefanov, K., and Popov, S. (2003). Development in primary cell culture of demosponges. *Journal of Biotechnology* **100,** 119–125.

De Rosa, S., De Caro, S., Tommonaro, G., Slantchev, K., Stefanov, K., and Popov, S. (2001). Development in a primary cell culture of the marine sponge *Ircinia Muscarum* and analysis of the polar compounds. *Marine Biotechnology* **3,** 281–286.

De Sutter, D., and Buscema, M. (1977). Isolation of a highly pure archeocyte fraction from the fresh-water sponge *Ephidatia fluviatilis*. *Wilhelm Roux's Archives* **183,** 149–153.

De Sutter, D., and Van De Vyver, G. (1977). Aggregative properties of different cell types of the fresh-water sponge *Ephydatia fluviatilis* isolated on ficoll gradients. *Wilhelm Roux's Archives* **181,** 151–161.

De Voogd, N. J. (2007). The mariculture potential of the Indonesian reef-dwelling sponge *Callyspongia (Euplacella) biru*: Growth, survival and bioactive compounds. *Aquaculture* **262,** 54–64.

Duckworth, A. (2009). Farming sponges to supply bioactive metabolites and bath sponges: A review. *Marine Biotechnology* **11,** 669–679.

Duckworth, A. R., and Pomponi, S. A. (2005). Relative importance of bacteria, microalgae and yeast for growth of the sponge *Halichondria melanadocia* (De Laubenfels, 1936): A laboratory study. *Journal of Experimental Marine Biology and Ecology* **323,** 151–159.

Duckworth, A. R., Samples, G. A., Wright, A. E., and Pomponi, S. A. (2003). *In vitro* culture of the tropical sponge *Axinella corrugata* (Demospongia): Effect of food cell concentration on growth, clearance rate and biosynthesis of stevensine. *Marine Biotechnology* **5,** 519–527.

Duckworth, A. R., Wolff, C., and Evans-Illidge, E. (2007). Developing methods for commercially farming bath sponges in tropical Australia. In "Porifera Research—Biodiversity, Innovation and Sustainability" (M. R. Custódio, G. Lôbo-Hajdu, E. Hajdu and G. Muricy, eds), pp. 297–302. Museu Nacional, Rio de Janeiro.

Duckworth, A. R., and Battershill, C. N. (2003a). Developing farming structures for production of biologically active sponge metabolites. *Aquaculture* **217,** 139–156.

Duckworth, A. R., and Battershill, C. N. (2003b). Sponge aquaculture for the production of biologically active metabolites: The influence of farming protocols and environment. *Aquaculture* **221,** 311–329.

Dumdei, E. J., Blunt, J. W., Munro, M. H. G., Battershill, C. N., and Page, M. J. (1998). The whys and whats of sponge chemistry: Why chemists extract sponges and what problems does this cause? In "Sponge Sciences; Multidisciplinary Perspectives" (Y. Watanabe and N. Fusetani, eds). Springer-Verlag, Tokyo.

Enticknap, J. J., Kelly, M., Peraud, O., and Hill, R. T. (2006). Characterization of a culturable alphaproteobacterial symbiont common to many marine sponges and evidence for vertical transmission via sponge larvae. *Applied and Environmental Microbiology* **72,** 3724–3732.

Ereskovsky, A. V. (2000). Reproduction cycles and strategies of the cold-water sponges *Halisarca dujardini* (Demospongiae, Halisacida), *Myxilla incrustans* and *Iophin piceus* (Demospongiae, Poecilosclerida) from the White Sea. *Biological Bulletin* **198,** 77–87.

Ettinger-Epstein, P., Tapiolas, D. M., Motti, C. A., Wright, A. D., Battershill, C. N., and De Nys, R. (2008). Production of manoalide and its analogues by the sponge *Luffariella variabilis* is hardwired. *Marine Biotechnology* **10,** 64–74.

Faulkner, D. J. (2000). Marine pharmacology. *Antonie van Leeuwenhoek* **77,** 135–145.

Fieseler, L., Horn, M., Wagner, M., and Hentschel, U. (2004). Discovery of the novel candidate phylum "*Poribacteria*" in marine sponges. *Applied and Environmental Microbiology* **70,** 3724–3732.

Freshney, R. I. (2005). Culture of Animal Cells: A Manual of Basic Technique. Wiley-Liss, Hoboken, NJ.

Friedrich, A. B., Fischer, I., Proksch, P., Hacker, J., and Hentschel, U. (2001). Temporal variation of the microbial community associated with the Mediterranean sponge *Aplysina aerophoba. FEMS Microbiology Ecology* **38,** 105–115.

Fu, W. T., Sun, L. M., Zhang, X. C., and Zhang, W. (2006). Potential of the marine sponge *Hymeniacidon perleve* as a bioremediator of pathogenic bacteria in integrated aquaculture ecosystems. *Biotechnology and Bioengineering* **93,** 1112–1122.

Funayama, N. (2010). The stem cell system in demosponges: Insights into the origin of somatic stem cells. *Development, Growth and Differentiation* **52,** 1–14.

Funayama, N., Nakatsukasa, M., Hayashi, T., and Agata, K. (2005). Isolation of the choanocyte in the fresh water sponge, *Ephydatia fluviatilis* and its lineage marker, Ef annexin. *Development, Growth and Differentiation* **47,** 243–253.

Funayama, N., Nakatsukasa, M., Mohri, K., Masuda, Y., and Agata, K. (2010). Piwi expression in archeocytes and choanocytes in demosponges: Insights into the stem cell system in demosponges. *Evolution and Development* **12,** 275–287.

Fusetani, N. (2000). Drugs from the Sea. Karger, Basel, Switzerland.

Gao, Z., Li, B. L., Zheng, C. C., and Wang, G. Y. (2008). Molecular detection of fungal communities in the Hawaiian marine sponges *Suberites zeteki* and *Mycale armata. Applied and Environmental Microbiology* **74,** 6091–6101.

Garcia Camacho, F., Chileh, T., Ceron Garcia, M. C., Sanchez Miron, A., Belarbi, E. H., Chisti, Y., and Molina Grima, E. (2006a). A bioreaction-diffusion model for growth of marine sponge explants in bioreactors. *Applied Microbiology and Biotechnology* **73,** 525–532.

Garcia Camacho, F., Chileh, T., Ceron García, M. C., Sanchez Mirón, A., Belarbi, E. H., Contreras Gómez, A., and Molina Grima, E. (2006b). Sustained growth of explants from Mediterranean sponge *Crambe crambe* cultured *in vitro* with enriched RPMI 1640. *Biotechnology Progress* **22,** 781–790.

Gerce, B., Schwartz, T., Voigt, M., Ruhle, S., Kirchen, S., Putz, A., Proksch, P., Obst, U., Syldatk, C., and Hausmann, R. (2009). Morphological, bacterial, and secondary metabolite changes of *Aplysina aerophoba* upon long-term maintenance under artificial conditions. *Microbial Ecology* **58,** 865–878.

Gifford, S., Dunstan, R. H., O'connor, W., Koller, C. E., and Macfarlane, G. R. (2007). Aquatic zooremediation: Deploying animals to remediate contaminated aquatic environments. *Trends in Biotechnology* **25,** 60–65.

Gunda, V., and Janapala, V. (2009). Effects of dissolved oxygen levels on survival and growth *in vitro* of *Haliclona pigmentifera* (Demospongiae). *Cell and Tissue Research* **337,** 527–535.

Hadas, E., Shpigel, M., and Ilan, M. (2005). Sea ranching of the marine sponge (Demospongiae, Latrunculiidae) as a first step for latrunculin B mass production. *Aquaculture* **244L,** 159–169.

Hadas, E., Shpigel, M., and Ilan, M. (2009). Particulate organic matter as a food source for a coral reef sponge. *Journal of Experimental Biology* **212,** 3643–3650.

Hallam, S. J., Konstantinidis, K. T., Putnam, N., Schleper, C., Watanabe, Y., Sugahara, J., Preston, C., De La Torre, J., Richardson, P. M., and Delong, E. F. (2006a). Genomic analysis of the uncultivated marine crenarchaeote *Cenarchaeum symbiosum*. *Proceedings of the National academy of Sciences of the United States of America* **103,** 18296–18301.

Hallam, S. J., Mincer, T. J., Schleper, C., Preston, C. M., Roberts, K., Richardson, P. M., and Delong, E. F. (2006b). Pathways of carbon assimilation and ammonia oxidation suggested by environmental genomic analyses of marine *Crenarchaeota*. *PLoS Biology* **4,** 520–536.

Haygood, M. G. (2006). Identification of a vertical transmitted microbial assemblage in the tropical sponge *Corticium candelabrum*. In "7th International Sponge Symposium. Buzios, Brazil." .

Heinemann, S., Ehrlich, H., Douglas, T., Heinemann, C., Worch, H., Schatton, W., and Hanke, T. (2007). Ultrastructural studies on the collagen of the marine sponge *Chondrosia reniformis* Nardo. *Biomacromolecules* **8,** 3452–3457.

Hentschel, U., Hopke, J., Horn, M., Friedrich, A. B., Wagner, M., Hacker, J., and Moore, B. S. (2002). Molecular evidence for a uniform microbial community in sponges from different oceans. *Applied and Environmental Microbiology* **68,** 4431–4440.

Hentschel, U., Schmid, M., Wagner, M., Fieseler, L., Gernert, C., and Hacker, J. (2001). Isolation and phylogenetic analysis of bacteria with antimicrobial activities from the Mediterranean sponges *Aplysina aerophoba* and *Aplysina cavernicola*. *FEMS Microbiology Ecology* **35,** 305–312.

Hentschel, U., Usher, K. M., and Taylor, M. W. (2006). Marine sponges as microbial fermenters. *FEMS Microbiology Ecology* **55,** 167–177.

Hogg, M. M., Tendal, O. S., Conway, K. W., Pomponi, S. A., Van Soest, R. W. M., Gutt, J., Krautter, M., and Roberts, J. M. (2010). Deep-sea sponge grounds: Reservoirs of biodiversity. (UNEP-WCMC, ed.)., Biodiversity Series No. 32, Cambridge, UK.

Holler, U., Wright, A. D., Matthee, G. F., Konig, G. M., Draeger, S., Aust, H. J., and Schulz, B. (2000). Fungi from marine sponges: diversity, biological activity and secondary metabolites. *Mycological Research* **104,** 1354–1365.

Holmes, B., and Blanch, H. (2007). Genus-specific associations of marine sponges with group I crenarchaeotes. *Marine Biology* **150,** 759–772.

Holmes, B., and Blanch, H. (2008). Possible taxonomic trends in the success of primary aggregate formation in marine sponge cell cultures. *Marine Biotechnology* **10,** 99–109.

Holmes, B. M. (2006). Marine Sponges: Systematics, Symbiosis and Cell Culture. UC Berkeley, Berkeley, USA.

Huelsenbeck, J. P., and Ronquist, F. (2001). MRBAYES: Bayesian inference of phylogenetic trees. *Bioinformatics* **17,** 754–755.

Ilan, M., Contini, H., Carmeli, S., and Rinkevich, B. (1996). Progress towards cell cultures from a marine sponge that produces bioactive compounds. *Journal of Marine Biotechnology* **4,** 145–149.

Imsiecke, G., Steffen, R., Custodio, M. R., Borojevic, R., and Muller, I. M. (1995). Formation of spicules by sclerocytes from the freshwater sponge *Ephydatia muelleri* in short-term cultures *in vitro*. *In Vitro Cellular and Developmental Biology—Animal* **31,** 528–535.

Isaacs, L. T., Kan, J. J., Nguyen, L., Videau, P., Anderson, M. A., Wright, T. L., and Hill, R. T. (2009). Comparison of the bacterial communities of wild and captive sponge *Clathria prolifera* from the Chesapeake Bay. *Marine Biotechnology* **11,** 758–770.

Jensen, P. R., and Fenical, W. (2002). Secondary metabolites from marine fungi. *Fungi in Marine Environments*. Fungal Diversity Press Ctr Res Fungal Diversity, Hong Kong.

Kaeberlein, T., Lewis, K., and Epstein, S. S. (2002). Isolating "uncultivable" microorganisms in pure culture in a simulated natural environment. *Science* **296,** 1127–1129.

Klautau, M., Custodio, M. R., and Borojevic, R. (1993). Cell cultures of sponges *Clathrina* and *Polymastia*. *In Vitro Cellular and Developmental Biology* **29A,** 97–99.
Klautau, M., Custodio, M. R., and Borojevic, R. (1994). *In vitro* culture of primary cell lines from marine sponges. In "Sponges in Time and Space" (R. Van Soest, T. M. G. Van Kempen and J. C. Braekman, eds), pp. 401–406. Rotterdam, Balkema.
Klöppel, A., Pfannkuchen, M., Putz, A., Proksch, P., and Brümmer, F. (2008). *Ex situ* cultivation of *Aplysina aerophoba* close to *in situ* conditions: Ecological, biochemical and histological aspects. *Marine Ecology* **29,** 259–272.
Kohlmeyer, J. (1974). On the definition and taxonomy of higher marine fungi. *Veröffentlichungen des Instituts für Meeresforschung in Bremerhave* (Suppl. 5), 263–286.
Kohlmeyer, J., and Volkmann-Kohlmeyer, B. (1990). New species of Koralionastes (Ascomycotina) from the Caribbean and Australia. *Canadian Journal of Botany/Revue Canadien de Botaniqe* **68,** 1554–1559.
Konneke, M., Bernhard, A. E., De La Torre, J. R., Walker, C. B., Waterbury, J. B., and Stahl, D. A. (2005). Isolation of an autotrophic ammonia-oxidizing marine archaeon. *Nature* **437,** 543–546.
Kooperman, N., Ben-Dov, E., Kramarsky-Winter, E., Barak, Z., and Kushmaro, A. (2007). Coral mucus-associated bacterial communities from natural and aquarium environments. *FEMS Microbiology Letters* **276,** 106–113.
Koopmans, M., Martens, D., and Wijffels, R. H. (2009). Towards commercial production of sponge medicines. *Marine Drugs* **7,** 787–802.
Koopmans, M., Martens, D., and Wijffels, R. H. (2010). Growth efficiency and carbon balance for the sponge *Haliclona oculata*. *Marine Biotechnology* **12,** 340–349.
Koopmans, M., and Wijffels, R. H. (2008). Seasonal growth rate of the sponge *Haliclona oculata* (demospongiae: Haplosclerida). *Marine Biotechnology* **10,** 502–510.
Koziol, C., Borojevic, R., Steffen, R., and Müller, W. E. G. (1998). Sponges (Porifera) model systems to study the shift from immortal to senescent somatic cells: The telomerase activity in somatic cells. *Mechanisms of Ageing and Development* **100,** 107–120.
Krasko, A., Lorenz, B., Batel, R., Schroder, H. C., Muller, I. M., and Muller, W. E. G. (2000). Expression of silicatein and collagen genes in the marine sponge *Suberites domuncula* is controlled by silicate and myotrophin. *European Journal of Biochemistry* **267,** 1–11.
Krasko, A., Schroder, H. C., Batel, R., Grebenjuk, V. A., Steffen, R., Muller, I. M., and Muller, W. E. G. (2002). Iron induces proliferation and morphogenesis in primmorphs from the marine sponge *Suberites domuncula*. *DNA and Cell Biology* **21,** 67–80.
Kreuter, M. H., Robitzki, A., Chang, S., Steffen, R., Michaelis, M., Kljajic, Z., Bachmann, M., Schröder, H. C., and Müller, W. E. G. (1992). Production of the cytostatic agent aeroplysinin by the sponge *Verongia aerophoba* in *in vitro* culture. *Comparative Biochemistry and Physiology Part C: Comparative Pharmacology* **101,** 183–187.
Kruse, M., Steffen, R., Batel, R., Muller, I. M., and Muller, W. E. (1999). Differential expression of allograft inflammatory factor 1 and of glutathione peroxidase during auto- and allograft response in marine sponges. *Journal of Cell Science* **112,** 4305–4313.
Le Pennec, G., Perovic, S., Ammar, M. S. A., Grebenjuk, V. A., Steffen, R., Brummer, F., and Muller, I. M. (2003). Cultivation of primmorphs from the marine sponge *Suberites domuncula*: Morphogenetic potential of silicon and iron. *Journal of Biotechnology* **100,** 93–108.
Lee, O. O., Tsoi, M. M. Y., Li, X., Wong, P.-K., and Qian, P.-Y. (2007). *Thalassococcus halodurans* gen. nov., sp. nov., a novel halotolerant member of the *Roseobacter* clade isolated from the marine sponge *Halichondria panicea* at Friday Harbor, USA. *International Journal of Systematic and Evolutionary Microbiology* **57,** 1919–1924.
Lee, O. O., Wong, Y. H., and Qian, P.-Y. (2009). Inter- and intraspecific variations of bacterial communities associated with marine sponges from San Juan Island, Washington. *Applied and Environmental Microbiology* **75,** 3513–3521.

Liu, M. Y., Kjelleberg, S., and Thomas, T. (2011). Functional genomic analysis of an uncultured [delta]-proteobacterium in the sponge *Cymbastela concentrica*. *The ISME Journal* **5,** 427–435.

Lopez, J. V., Peterson, C. L., Willoughby, R., Wright, A. E., Enright, E., Zoladz, S., Reed, J. K., and Pomponi, S. A. (2002). Characterization of genetic markers for *in vitro* cell line identification of the marine sponge *Axinella corrugata*. *The Journal of Heredity* **93,** 27–36.

Ludwig, W., Strunk, O., Westram, R., Richter, L., Meier, H., Yadhukumar, B. A., Lai, T., Steppi, S., Jobb, G., Forster, W., Brettske, I., Gerber, S. *et al.* (2004). ARB: A software environment for sequence data. *Nucleic Acids Research* **32,** 1363–1371.

Maldonado, M., Cortadellas, N., Trillas, M. I., and Rutzler, K. (2005). Endosymbiotic yeast maternally transmitted in a marine sponge. *Biological Bulletin* **209,** 94–106.

Maldonado, M., Zhang, X. C., Cao, X. P., Xue, L. Y., Cao, H., and Zhang, W. (2010). Selective feeding by sponges on pathogenic microbes: a reassessment of potential for abatement of microbial pollution. *Marine Ecology Progress Series* **403,** 75–89.

Mendola, D. (2003). Aquaculture of three phyla of marine invertebrates to yield bioactive metabolites: Process developments and economics. *Biomolecular Engineering* **20,** 441–458.

Mendola, D. (2008). The importance of water flow for culture of *Dysidea avara* sponges. Bioprocess Engineering Wageningen University, Wageningen.

Mendola, D., Naranjo Lozano, S. A., Duckworth, A. R., and Osinga, R. (2006). The promise of aquaculture for delivering sustainable supplies of new drugs from the sea: Examples from in-sea, and tank-based invertebrate culture projects from around the world. In "Frontiers in Marine Biotechnology" (P. Proksch and W. E. G. Muller, eds), pp. 21–72. Horizon Science Press, Norfolk, UK.

Mohamed, N. M., Enticknap, J. J., Lohr, J. E., Mcintosh, S. M., and Hill, R. T. (2008a). Changes in bacterial communities of the marine sponge *Mycale laxissima* on transfer into aquaculture. *Applied and Environmental Microbiology* **74,** 1209–1222.

Mohamed, N. M., Rao, V., Hamann, M. T., Kelly, M., and Hill, R. T. (2008b). Monitoring bacterial diversity of the marine sponge *Ircinia strobilina* upon transfer into aquaculture. *Applied and Environmental Microbiology* **74,** 4133–4143.

Molinski, T. F., Dalisay, D. S., Lievens, S. L., and Saludes, J. P. (2009). Drug development from marine natural products. *Nature Reviews* **8,** 69–85.

Morrison-Gardiner, S. (2002). Dominant fungi from Australian coral reefs. *Fungal Diversity* **9,** 105–121.

Muller, W. E. G., Bohm, M., Batel, R., De Rosa, S., Tommonaro, G., Muller, I. M., and Schroder, H. C. (2000). Application of cell culture for the production of bioactive compounds from sponges: Synthesis of avarol by primmorphs from *Dysidea avara*. *Journal of Natural Products* **63,** 1077–1081.

Müller, W. E. G., Grebenjuk, V. A., Le Pennec, G., Schröder, H.-C., Brümmer, F., Hentschel, U., Müller, I. M., and Breter, H.-J. (2004a). Sustainable production of bioactive compounds by sponges - cell culture and gene cluster approach: A review. *Marine Biotechnology* **6,** 105–117.

Müller, W. E. G., Schröder, H. C., Wiens, M., Perović-Ottstadt, S., Batel, R., and Müller, I. M. (2004b). Traditional and modern biomedical prospecting: Part II—The benefits, approaches for a sustainable exploitation of biodiversity (secondary metabolites and biomaterials from sponges). *Evidence-Based Complementary and Alternative Medicine* **1,** 133–144.

Müller, W. E. G., Wang, X., Cui, F.-Z., Jochum, K., Tremel, W., Bill, J., Schröder, H., Natalio, F., Schloßmacher, U., and Wiens, M. (2009). Sponge spicules as blueprints for the biofabrication of inorganic–organic composites and biomaterials. *Applied Microbiology and Biotechnology* **83,** 397–413.

Muller, W. E. G., Wiens, M., Batel, R., Steffen, R., Borojevic, R., and Custodio, M. R. (1999). Establishment of a primary cell culture from a sponge: Primmorphs from *Suberites domuncula*. *Marine Ecology Progress Series* **178,** 205–219.
Munro, M. H. G., Blunt, J. W., Dumdei, E. J., Hickford, S. J. H., Lill, R. E., Li, S., Battershill, C. N., and Duckworth, A. R. (1999). The discovery and development of marine compounds with pharmaceutical potential. *Journal of Biotechnology* **70,** 15–25.
Muscholl-Silberhorn, A., Thiel, V., and Imhoff, J. F. (2008). Abundance and bioactivity of cultured sponge-associated bacteria from the Mediterranean Sea. *Microbial Ecology* **55,** 94–106.
Newman, D. J., and Cragg, G. M. (2004). Marine natural products and related compounds in clinical and advanced preclinical trials. *Journal of Natural Products* **67,** 1216–1238.
Nickel, M., and Brummer, F. (2003). *In vitro* sponge fragment culture of *Chondrosia reniformis* (Nardo, 1847). *Journal of Biotechnology* **100,** 147–159.
Nickel, M., Leininger, S., Proll, G., and Brummer, F. (2001). Comparative studies on two potential methods for the biotechnological production of sponge biomass. *Journal of Biotechnology* **92,** 169–178.
Nickel, M., Proll, G., and Brummer, F. (2000). Natural products of marine sponges—From ecology to biomass. In "Proceedings of the 4th International Congress on Biochemical Engineering" (H. Bruner, ed.), pp. 194–198. Fraunhofer IRB Verlag, Stuttgart.
Olson, J. B., Lord, C. C., and Mccarthy, P. J. (2000). Improved recoverability of microbial colonies from marine sponge samples. *Microbial Ecology* **40,** 139–147.
Osinga, R., Belarbi, E. H., Molina Grima, E., Tramper, J., and Wijffels, R. H. (2003). Progress towards a controlled culture of the marine sponge *Pseadosuberites andrewsi* in a bioreactor. *Journal of Biotechnology* **100,** 141–146.
Osinga, R., Belarbi, E. H., and Wijffels, R. H. (2002). Feeding sponges in bioreactors: What do they eat? *Bollettino dei Musei e degli Istituti Biologici dell'Universita di Genova* **66–67,** 147.
Osinga, R., Kleijn, R., Groenendijk, E., Niesink, P., Tramper, J., and Wijffels, R. H. (2001). Development of *in vivo* sponge cultures: Particle feeding by the tropical sponge *Pseudosuberites aff. andrewsi*. *Marine Biotechnology* **3,** 544–554.
Osinga, R., and Kotterman, M. (2007). Ferric iron promotes the formation of oscules: Observations on sponges in aquaria. In "*Porifera Research—Biodiversity, Innovation and Sustainability*" (M. R. Custodio, G. Lôbo-Hajdu, E. Hajdu and G. Muricy, eds), pp. 497–502. Museo Nacional, Rio de Janeiro.
Osinga, R., Schutter, M., Griffioen, B., Wijffels, R., Verreth, J., Shafir, S., Henard, S., Taruffi, M., Gili, C., and Lavorano, S. (2011). The biology and economics of coral growth. *Marine Biotechnology* 1–14.
Osinga, R., Sidri, M., Cerig, E., Gokalp, S. Z., and Gokalp, M. (2010). Sponge aquaculture trials in the East-Mediterranean sea: New approaches to earlier ideas. *The Open Marine Biology Journal* **4,** 74–81.
Osinga, R., Tramper, J., and Wijffels, R. H. (1998). Cultivation of marine sponges for metabolite production: Applications for biotechnology? *Trends in Biotechnology* **16,** 130–134.
Osinga, R., Tramper, J., and Wijffels, R. H. (1999). Cultivation of marine sponges. *Marine Biotechnology* **1,** 509–532.
Pabel, C. T., Vater, J., Wilde, C., Franke, P., Hofemeister, J., Adler, B., Bringmann, G., Hacker, J., and Hentschel, U. (2003). Antimicrobial activities and matrix-assisted laser desorption/ionization mass spectrometry of *Bacillus* isolates from the marine sponge *Aplysina aerophoba*. *Marine Biotechnology* **5,** 424–434.
Page, M. J., Handley, S. J., Northcote, P. T., Cairney, D., and Willan, R. C. (2011). Successes and pitfalls of the aquaculture of the sponge *Mycale hentscheli*. *Aquaculture* **312,** 52–61.
Page, M. J., Northcote, P. T., Webb, V. L., Mackey, S., and Handley, S. J. (2005). Aquaculture trials for the production of biologically active metabolites in the New

Zealand sponge *Mycale hentscheli* (Demospongiae: Poecilosclerida). *Aquaculture* **250,** 256–269.

Perovic, S., Wichels, A., Schutt, C., Gerdts, G., Pahler, S., Steffen, R., and Muller, W. E. G. (1998). Neuroactive compounds produced by bacteria from the marine sponge *Halichondria panicea*: Activation of the neuronal NMDA receptor. *Environmental Toxicology and Pharmacology* **6,** 125–133.

Pfannkuchen, M., and Brummer, F. (2009). Heterologous expression of DsRed2 in young sponges (Porifera). *International Journal of Developmental Biology* **53,** 1113–1117.

Piel, J. (2004). Metabolites from symbiotic bacteria. *Natural Product Reports* **21,** 519–538.

Pimentel-Elardo, S., Wehrl, M., Friedrich, A. B., Jensen, P. R., and Hentschel, U. (2003). Isolation of planctomycetes from *Aplysina* sponges. *Aquatic Microbial Ecology* **33,** 239–245.

Pimentel-Elardo, S. M., Kozytska, S., Bugni, T. S., Ireland, C. M., Moll, H., and Hentschel, U. (2010). Anti-parasitic compounds from *Streptomyces* sp. strains isolated from Mediterranean sponges. *Marine Drugs* **8,** 373–380.

Pivkin, M., Aleshko, S., Krasokhin, V., and Khudyakova, Y. (2006). Fungal assemblages associated with sponges of the southern coast of Sakhalin Island. *Russian Journal of Marine Biology* **32,** 207–213.

Pomponi, S. A. (2001). The oceans and human health: The discovery and development of marine-derived drugs. *Oceanography* **14,** 28–42.

Pomponi, S. A. (2006). Biology of the Porifera: Cell culture. *Canadian Journal of Zoology* **84,** 167–174.

Pomponi, S. A., and Willoughby, R. (1994). Sponge cell culture for the production of bioactive metabolites. In "Sponges in Time and Space" (R. Van Soest, T. M. G. Van Kempen and J. C. Braekman, eds). Balkema, Rotterdam.

Pomponi, S. A., and Willoughby, R. (2000). Development of sponge cell cultures for biomedical applications. In "Aquatic Invertebrate Cell Culture" (B. Austin and C. Mothersill, eds). Springer, London.

Pomponi, S. A., Willoughby, R., Kaighn, M. E., and Wright, A. E. (1997). Development of techniques for *in vitro* production of bioactive natural products from marine sponges. (K. Maramorosch and J. Mitsuhashi, eds), pp. 231–237. Science Publishers, USA.

Pomponi, S. A., Willoughby, R., and Kelly-Borges, M. (1998). Sponge cell culture. In "Molecular Approaches to the Study of the Ocean" (K. E. Cooksey, ed.), pp. 423–429. Chapman and Hall, London.

Pomponi, S. A., Thompson, J., Bottesch, J. H., and Sennett, S. H. (2007). Methods for extending the life span of marine invertebrate cells. US patent number: 20070254367.

Pozzolini, M., Sturla, L., Cerrano, C., Bavestrello, G., Camardella, L., Parodi, A. M., Raheli, F., Benatti, U., Muller, W. E. G., and Giovine, M. (2004). Molecular cloning of silicatein gene from marine sponge *Petrosia ficiformis* (Porifera, Demospongiae) and development of primmorphs as a model for biosilicification studies. *Marine Biotechnology* **6,** 594–603.

Preston, C. M., Wu, K. Y., Molinski, T. F., and Delong, E. F. (1996). A psychrophilic crenarchaeon inhabits a marine sponge: *Cenarchaeum symbiosum* gen. nov., sp. nov. *Proceedings of the National Academy of Sciences of the United States of America* **93,** 6241–6246.

Proksch, P. (1994). Defensive roles for secondary metabolites from marine sponges and sponge-feeding nudibranchs. *Toxicon* **32,** 639–655.

Pronzato, R., and Manconi, R. (2008). Mediterranean commercial sponges: Over 5000 years of natural history and cultural heritage. *Marine Ecology* **29,** 146–166.

Pruesse, E., Quast, C., Knittel, K., Fuchs, B. M., Ludwig, W., Peplies, J., and Glöckner, F. O. (2007). SILVA: A comprehensive online resource for quality checked and aligned ribosomal RNA sequence data compatible with ARB. *Nucleic Acids Reseach* **35,** 7188–7196.

Rajilic-Stojanovic, M., Smidt, H., and De Vos, W. M. (2007). Diversity of the human gastrointestinal tract microbiota revisited. *Environmental Microbiology* **9,** 2125–2136.

Reiswig, H. W. (1974). Water transport, respiration and energetics of three tropical marine sponges. *Journal of Experimental Marine Biology and Ecology* **14,** 231–249.

Richelle-Maurer, E., Gomez, R., Braekman, J. C., Van De Vyver, G., Soest, R. V. M., and Devijver, C. (2003). Primary cultures from the marine sponge *Xestospongia muta* (Petrosiidae, Haposclerida). *Journal of Biotechnology* **100,** 169–176.

Rinkevich, B. (2005). Marine invertebrate cell cultures: New millennium trends. *Marine Biotechnology* **7,** 429–439.

Rinkevich, B. (2011). Cell cultures from marine invertebrates: New insights for capturing endless stemness. *Marine Biotechnology* **13,** 345–354.

Rinkevich, B., Blisko, R., and Ilan, M. (1998). Further steps in the initiation of cell cultures from embryos and adult sponge colonies. *In Vitro Cellular and Developmental Biology—Animal* **34,** 753–756.

Ronquist, F., Huelsenbeck, J. P., and Van Der Mark, O. (2005). MrBayes 3.1 Manual. School of Computational Science. Florida State University, USA.

Sashidhara, K. V., White, K. N., and Crews, P. (2009). A selective account of effective paradigms and significant outcomes in the discovery of inspirational marine natural products. *Journal of Natural Products* **72,** 588–603.

Schippers, K. J., Martens, D. E., Pomponi, S. A., and Wijffels, R. H. (2011). Cell cycle analysis of primary sponge cell cultures. *In Vitro Cellular and Developmental Biology—Animal* **47,** 302–311.

Schmidt, O. (1863). Auszug aus dem Berichte des Professor Dr. O. Schmidt uber die im Auftrage der Triester Handels- und Gewerbekammer in Dalmatien angestellten Versuche uber die kunstliche Schwammzucht. *Zeitschrift fur Akklimatisation* **1,** 290–295.

Schmitt, S., Angermeier, H., Schiller, R., Lindquist, N., and Hentschel, U. (2008). Molecular microbial diversity survey of sponge teproductive stages and mechanistic insights into vertical transmission of microbial symbionts. *Applied and Environmental Microbiology* **74,** 7694–7708.

Schneemann, I., Nagel, K., Kajahn, I., Labes, A., Wiese, J., and Imhoff, J. F. (2010a). Comprehensive investigation of marine Actinobacteria associated with the sponge *Halichondria panicea*. *Applied and Environmental Microbiology* **76,** 3702–3714.

Schneemann, I., Ohlendorf, B., Zinecker, H., Nagel, K., Wiese, J., and Imhoff, J. F. (2010b). Nocapyrones A–D, γ-pyrones from a Nocardiopsis strain isolated from the marine sponge *Halichondria panicea*. *Journal of Natural Products* **73,** 1444–1447.

Schröder, H., Brandt, D., Schloßmacher, U., Wang, X., Tahir, M., Tremel, W., Belikov, S., and Müller, W. (2007). Enzymatic production of biosilica glass using enzymes from sponges: Basic aspects and application in nanobiotechnology (material sciences and medicine). *Naturwissenschaften* **94,** 339–359.

Selvin, J., Gandhimathi, R., Kiran, G. S., Priya, S. S., Ravji, T. R., and Hema, T. A. (2009). Culturable heterotrophic bacteria from the marine sponge *Dendrilla nigra*: Isolation and phylogenetic diversity of actinobacteria. *Helgoland Marine Research* **63,** 239–247.

Sharp, K. H., Eam, B., Faulkner, D. J., and Haygood, M. G. (2007). Vertical transmission of diverse microbes in the tropical sponge *Corticium* sp. *Applied and Environmental Microbiology* **73,** 622–629.

Siegl, A., Kamke, J., Hochmuth, T., Piel, J., Richter, M., Liang, C. G., Dandekar, T., and Hentschel, U. (2011). Single-cell genomics reveals the lifestyle of Poribacteria, a candidate phylum symbiotically associated with marine sponges. *The ISME Journal* **5,** 61–70.

Simon, H. M., Jahn, C. E., Bergerud, L. T., Sliwinski, M. K., Weimer, P. J., Willis, D. K., and Goodman, R. M. (2005). Cultivation of mesophilic soil crenarchaeotes in enrichment cultures from plant roots. *Applied and Environmental Microbiology* **71,** 4751–4760.

Sipkema, D., and Blanch, H. (2010). Spatial distribution of bacteria associated with the marine sponge *Tethya californiana*. *Marine Biology* **157,** 627–638.
Sipkema, D., Franssen, M. C. R., Osinga, R., Tramper, J., and Wijffels, R. H. (2005a). Marine sponges as pharmacy. *Marine Biotechnology* **7,** 142–162.
Sipkema, D., Heilig, H. G. H. J., Akkermans, A. D. L., Osinga, R., Tramper, J., and Wijffels, R. H. (2003a). Sponge cell culture? A molecular identification method for sponge cells. *Marine Biotechnology* **5,** 443–449.
Sipkema, D., Holmes, B., Nichols, S., and Blanch, H. (2009). Biological characterisation of *Haliclona (?gellius)* sp.: Sponge and associated microorganisms. *Microbial Ecology* **58,** 903–920.
Sipkema, D., Osinga, R., Schatton, W., Mendola, D., Tramper, J., and Wijffels, R. H. (2005b). Large-scale production of pharmaceuticals by marine sponges: Sea, cell, or synthesis? *Biotechnology and Bioengineering* **90,** 201–222.
Sipkema, D., Schippers, K., Maalcke, W. J., Yang, Y., Salim, S., and Blanch, H. W. (2011). Multiple approaches to enhance the cultivability of bacteria associated with the marine sponge *Haliclona (gellius)* sp. *Applied and Environmental Microbiology* **77,** 2130–2140.
Sipkema, D., Snijders, A. P. L., Schroen, C. G. P. H., Osinga, R., and Wijffels, R. H. (2004). The life and death of sponge cells. *Biotechnology and Bioengineering* **85,** 239–247.
Sipkema, D., Tramper, J., Osinga, R., and Wijffels, R.H. (2003). The influence of silicate on *Suberites domuncula* primmorphs. In: Colliec-Jouault, S., Bergé, J.P., Guézennec, J., Fleurence, J., Le Gal, Y., Roy, P. (eds.). *Ifremer actes de colloques* 36, Marine biotechnology: an overview of leading fields, 56-64.
Sipkema, D., Van Wielink, R., Van Lammeren, A. A. M., Tramper, J., Osinga, R., and Wijffels, R. H. (2003c). Primmorphs from seven marine sponges: Formation and structure. *Journal of Biotechnology* **100,** 127–139.
Sipkema, D., Yosef, N. A. M., Adamczewski, M., Osinga, R., Mendola, D., Tramper, J., and Wijffels, R. H. (2006). Hypothesized kinetic models for describing the growth of globular and encrusting Demosponges. *Marine Biotechnology* **8,** 40–51.
Srivastava, M., Simakov, O., Chapman, J., Fahey, B., Gauthier, M. E. A., Mitros, T., Richards, G. S., Conaco, C., Dacre, M., Hellsten, U., Larroux, C., Putnam, N. H. *et al.* (2010). The *Amphimedon queenslandica* genome and the evolution of animal complexity. *Nature* **466,** 720–726.
Staley, J. T., and Konopka, A. (1985). Measurement of *in situ* activities of nonphotosynthetic microorganisms in aquatic and terrestrial habitats. *Annual Review of Microbiology* **39,** 321–346.
Stocker-Worgotter, E. (1995). Experimental cultivation of lichens and lichen symbionts. *Canadian Journal of Botany/Revue Canadien de Botaniqe* **73,** S579–S589.
Sun, L. M., Song, Y. F., Qu, Y., Yu, X. J., and Zhang, W. (2007). Purification and *in vitro* cultivation of archaeocytes (stem cells) of the marine sponge *Hymeniacidon perleve* (Demospongiae). *Cell and Tissue Research* **328,** 223–237.
Sun, W., Dai, S., Jiang, S., Wang, G., Liu, G., Wu, H., and Li, X. (2010). Culture-dependent and culture-independent diversity of Actinobacteria associated with the marine sponge *Hymeniacidon perleve* from the South China Sea. *Antonie van Leeuwenhoek* **98,** 65–75.
Sundar, V. C., Yablon, A. D., Grazul, J. L., Ilan, M., and Aizenberg, J. (2003). Fibre-optical features of a glass sponge. *Nature* **424,** 899–900.
Swatschek, D., Schatton, W., Kellerman, J., Muller, I. M., and Kreuter, J. (2002). Marine sponge collagen: Isolation, characterization and effects on the skin parameters surface-pH, moisture and sebum. *European Journal of Pharmaceutics and Biopharmaceutics* **53,** 107–113.
Takahashi, K., and Yamanaka, S. (2006). Induction of pluripotent stem cells from mouse embryonic and adult fibroblast cultures by defined factors. *Cell* **126,** 663–676.

Taylor, M. W., Radax, R., Steger, D., and Wagner, M. (2007). Sponge-associated microorganisms: Evolution, ecology and biotechnological potential. *Microbiology and Molecular Biology Reviews* **71,** 295–347.

Tourna, M., Stieglmeier, M., Spang, A., Könneke, M., Schintlmeister, A., Urich, T., Engel, M., Schloter, M., Wagner, M., Richter, A., and Schleper, C. (2011). *Nitrososphaera viennensis*, an ammonia oxidizing archaeon from soil. *Proceedings of the National Academy of Sciences of the United States of America* **108,** 8420–8425.

Unger, C., Gao, S., Cohen, M., Jaconi, M., Bergstrom, R., Holm, F., Galan, A., Sanchez, E., Irion, O., Dubuisson, J. B., Giry-Laterriere, M., Salmon, P., Simon, C., Hovatta, O., and Feki, A. (2009). Immortalized human skin fibroblast feeder cells support growth and maintenance of both human embryonic and induced pluripotent stem cells. *Human Reproduction* **24,** 2567–2581.

Unson, M. D., Holland, N. D., and Faulkner, D. J. (1994). A brominated secondary metabolite synthesized by the cyanobacterial symbiont of a marine sponge and accumulation of the crystalline metabolite in the sponge tissue. *Marine Biology* **119,** 1–11.

Uriz, M. J., Turon, X., Galera, J., and Tur, J. M. (1996). New light on the cell location of avarol within the sponge *Dysidea avara* (Dendroceratida). *Cell and Tissue Research* **285,** 519–527.

Valisano, L., Bavestrello, G., Giovine, M., Arillo, A., and Cerrano, C. (2006a). Seasonal production of primmorphs from the marine sponge *Petrosia ficiformis* (Poiret, 1789) and new culturing approaches. *Journal of Experimental Marine Biology and Ecology* **337,** 171–177.

Valisano, L., Bavestrello, G., Giovine, M., Arillo, A., and Cerrano, C. (2007). Effect of iron and dissolved silica on primmorphs of *Petrosia ficiformis* (Poiret, 1789). *Chemistry and Ecology* **23,** 233–241.

Valisano, L., Bavestrello, G., Giovine, M., and Cerrano, C. (2006b). Primmorphs formation dynamics: A screening among Mediterranean sponges. *Marine Biology* **149,** 1037–1046.

Van Doesburg, W., Van Eekert, M. H. A., Middeldorp, P. J. M., Balk, M., Schraa, G., and Stams, A. J. M. (2005). Reductive dechlorination of [beta]-hexachlorocyclohexane ([beta]-HCH) by a *Dehalobacter* species in coculture with a *Sedimentibacter* sp. *FEMS Microbiology Ecology* **54,** 87–95.

Vilanova, E., Coutinho, C., Maia, G., and Mourao, P. A. S. (2010). Sulfated polysaccharides from marine sponges: Conspicuous distribution among different cell types and involvement on formation of *in vitro* cell aggregates. *Cell and Tissue Research* **340,** 523–531.

Voultsiadou, E. (2007). Sponges: An historical survey of their knowledge in Greek antiquity. *Journal of the Marine Biological Association of the United Kingdom* **87,** 1757–1763.

Wang, G., Li, Q., and Zhu, P. (2008). Phylogenetic diversity of culturable fungi associated with the Hawaiian sponges *Suberites zeteki* and *Gelliodes fibrosa. Antonie van Leeuwenhoek* **93,** 163–174.

Watson, S. W., Novitsky, T. J., Quinby, H. L., and Valois, F. W. (1977). Determination of bacterial number and biomass in the marine environment. *Applied and Environmental Microbiology* **33,** 940–946.

Webster, N., Cobb, R., Soo, R., Anthony, S., Battershill, C., Whalan, S., and Evans-Illidge, E. (2011). Bacterial community dynamics in the marine sponge *Rhopaloeides odorabile* under *in situ* and *ex situ* cultivation. *Marine Biotechnology* **13,** 296–304.

Webster, N. S., and Hill, R. T. (2001). The culturable microbial community of the Great Barrier Reed sponge *Rhodopaloeides odorabile* is dominated by an alpha-proteobacterium. *Marine Biology* **138,** 843–851.

Webster, N. S., Negri, A. P., Webb, R. I., and Hill, R. T. (2002). A spongin-boring alpha-proteobacterium is the etiological agent of disease in the Great Barrier Reef sponge *Rhopaloeides odorabile. Marine Ecology Progress Series* **232,** 305–309.

Webster, N. S., Wilson, K. J., Blackall, L. L., and Hill, R. T. (2001). Phylogenetic diversity of bacteria associated with the marine sponge *Rhopaloeides odorabile*. *Applied and Environmental Microbiology* **67,** 434–444.

Wichels, A., Würtz, S., Döpke, H., Schütt, C., and Gerdts, G. (2006). Bacterial diversity in the breadcrumb sponge *Halichondria panicea* (Pallas). *FEMS Microbiology Ecology* **56,** 102–118.

Wiese, J., Ohlendorf, B., Blümel, M., Schmaljohann, R., and Imhoff, J. F. (2011). Phylogenetic identification of fungi isolated from the marine sponge *Tethya aurantium* and identification of their secondary metabolites. *Marine Drugs* **9,** 561–585.

Wilkinson, C. R., and Garrone, R. (1980). Nutrition of marine sponges. Involvement of symbiotic bacteria in the uptake of dissolved carbon. In "Nutrition in the Lower Metazoa" (D. C. Smith and Y. Tiffon, eds), pp. 157–161. Pergamon, Oxford.

Willoughby, R., and Pomponi, S. A. (2000). Quantitative assessment of marine sponge cells *in vitro*: Development of improved growth medium. *In Vitro Cellular and Developmental Biology—Animal* **36,** 194–200.

Wilson, H. V. (1907). On some phenomena of coalescence and regeneration in sponges. *Journal of Experimental Zoology* **5,** 245–258.

Xue, L., and Zhang, W. (2009). Growth and survival of early juveniles of the marine sponge *Hymeniacidon perlevis* (Demospongiae) under controlled conditions. *Marine Biotechnology* **11,** 640–649.

Yahel, G., Sharp, J. H., Marie, D., Hase, C., and Genin, A. (2003). *In situ* feeding and element removal in the symbiont-bearing sponge *Theonella swinhoei*: Bulk DOC is the major source for carbon. *Limnology and Oceanography* **48,** 141–149.

Zea, S., Castellanos, L., Valderrama, K., Puentes, C. A., Gomez-Leon, J., and Pomponi, S. A. (2010). Potential of antitumoral (+)-discodermalide production by the Caribbean sponge *Discodermia dissoluta*. In "VIII World Sponge Conference, Girona."

Zhang, H.-B., Yang, M.-X., Shi, W., Zheng, Y., Sha, T., and Zhao, Z.-W. (2007). Bacterial diversity in mine tailings compared by cultivation and cultivation-independent methods and their resistance to lead and cadmium. *Microbial Ecology* **54,** 705–712.

Zhang, W., Zhang, X., Cao, X., Xu, J., Zhao, Q., Yu, X., Jin, M., and Deng, M. (2003a). Optimizing the formation of *in vitro* sponge primmorphs from the Chinese sponge *Stylotella agminata* (Ridley). *Journal of Biotechnology* **100,** 161–168.

Zhang, X., Cao, X., Zhang, W., Yu, X., and Jin, M. (2003b). Primmorphs from archaeocytes-dominant cell population of the sponge *Hymeniacidon perleve*: Improved cell proliferation and spiculogenesis. *Biotechnology and Bioengineering* **84,** 583–590.

Zhang, X., Zhang, W., Xue, L., Zhang, B., Jin, M., and Fu, W. (2010). Bioremediation of bacteria pollution using the marine sponge *Hymeniacidon perlevis* in the intensive mariculture water system of turbot *Scophthalmus maximus*. *Biotechnology and Bioengineering* **105,** 59–68.

Zhang, X. Y., Le Pennec, G., Steffen, R., Muller, W. E. G., and Zhang, W. (2004). Application of a MTT assay for screening nutritional factors in growth media of primary sponge cell culture. *Biotechnology Progress* **20,** 151–155.

Zhao, Q. Y., Zhang, W., Jin, M. F., Yu, X. J., and Deng, M. C. (2005). Formulation of a basal medium for primary cell culture of the marine sponge *Hymeniacidon perleve*. *Biotechnology Progress* **21,** 1008–1012.

Zientz, E., Dandekar, T., and Gross, R. (2004). Metabolic interdependence of obligate intracellular bacteria and their insect hosts. *Microbiology and Molecular Biology Reviews* **68,** 745–770.

Subject Index

Note: Page numbers followed by *f* and *t* indicate figures and tables, respectively.

C

D

E

N

T

W

Taxonomic Index

Note: Page numbers followed by *f* and *t* indicate figures and tables, respectively.

FSC
www.fsc.org
MIX
Papier | Fördert
gute Waldnutzung
FSC® C083411

Kolibri 360 GmbH
Ferdinand-Jühlke-Straße 7
99095 Erfurt, Deutschland
produktsicherheit@kolibri360.de